T0322009

MATHEMATICAL METHODS for MECHANICAL SCIENCES

MATHEMATICAL METHODS for MECHANICAL SCIENCES

Michael Howe
Boston University, USA

Imperial College Press

ICP

Published by

Imperial College Press
57 Shelton Street
Covent Garden
London WC2H 9HE

Distributed by

World Scientific Publishing Co. Pte. Ltd.
5 Toh Tuck Link, Singapore 596224
USA office: 27 Warren Street, Suite 401-402, Hackensack, NJ 07601
UK office: 57 Shelton Street, Covent Garden, London WC2H 9HE

Library of Congress Cataloging-in-Publication Data
Howe, Michael (Acoustical engineer)
  Mathematical methods for mechanical sciences / Michael Howe (Boston University, USA).
    pages cm
  Includes bibliographical references.
  ISBN 978-1-78326-664-7 (alk. paper)
  1. Engineering mathematics--Textbooks. 2. Engineering--Mathematical models. 3. Engineering--Study and teaching (Higher). 4. Differential equations--Textbooks. I. Title.
  TA347.D45H69 2015
  515--dc23

                         2015019890

British Library Cataloguing-in-Publication Data
A catalogue record for this book is available from the British Library.

In-house Editors: Thomas Stottor/Chandrima Maitra

Typeset by Stallion Press
Email: enquiries@stallionpress.com

Printed in Singapore

CONTENTS

PREFACE

A mathematical model of a physical system provides the engineer with the insight and intuitive understanding generally required to make efficient system design changes or other modifications. A simple formula is often worth a thousand numerical simulations, and can reveal connections between different control parameters that might otherwise take hours or weeks to deduce from a computational analysis. This book is intended to supply the undergraduate engineer with the basic mathematical tools for developing and understanding such models. A firm grasp of the topics covered will also enable the working engineer (educated to bachelor's degree level) to understand, write and otherwise make sensible use of technical reports and papers.

The book was orginally written for students taking the Boston University senior level, one-semester course in engineering mathematics for mechanical and aerospace engineers. This course marks the final exposure of these students to formal mathematical training prior to graduation, and includes material taken principally from Chapters 1–4. The intention is to consolidate earlier courses in ordinary differential equations, vector calculus, Fourier series and transforms, and linear algebra, and to introduce more advanced topics, including complex variable theory, partial differential equations and elementary generalised functions leading to Green's functions. The book has also formed the basis of a review course for first-year engineering graduate students. It is not possible to cover in a one-semester class all subjects with which an 'educated' engineer might reasonably be expected to be familiar; additional topics are included in the text, mainly for reference, on conformal transformations, special functions and variational methods. However, an overriding objective has been compactness of presentation, and to avoid the currently fashionable trend of attempting to achieve encyclopaedic coverage with a text that typically runs to a thousand or more pages.

M. S. Howe

GREEK ALPHABET

alpha	$\alpha,\ A$		nu	$\nu,\ N$
beta	$\beta,\ B$		xi	$\xi,\ \Xi$
gamma	$\gamma,\ \Gamma$		omicron	$o,\ O$
delta	$\delta,\ \Delta$		pi	$\pi,\ \Pi$
epsilon	$\epsilon,\ E$		rho	$\rho,\ P$
zeta	$\zeta,\ Z$		sigma	$\sigma,\ \Sigma$
eta	$\eta,\ H$		tau	$\tau,\ T$
theta	$\theta,\ \Theta$		upsilon	$\upsilon,\ \Upsilon$
iota	$\iota,\ I$		phi	$\phi,\ \Phi$
kappa	$\kappa,\ K$		chi	$\chi,\ X$
lambda	$\lambda,\ \Lambda$		psi	$\psi,\ \Psi$
mu	$\mu,\ M$		omega	$\omega,\ \Omega$

MATHEMATICAL CONSTANTS

Euler's $\qquad \gamma = 0.5772\ 15665$

Exponential $\quad e = 2.7182\ 81828$

$\pi = 3.1415\ 92654$

1

LINEAR ORDINARY DIFFERENTIAL EQUATIONS

1.1 First-Order Equations

General form:

$$\frac{dy}{dx} + p(x)y = r(x), \quad \text{or} \quad y' + p(x)y = r(x), \quad \text{where } y' = \frac{dy}{dx}.$$

Homogeneous equation $y' + p(x)y = 0.$

Solve by separating the variables:

$$\int \frac{dy}{y} = -\int p(x)dx + C_1, \quad C_1 = \text{constant}$$

$$\therefore \quad \ln y = -\int p(x)dx + C_1$$

$\therefore$ The general solution is $y = Ce^{-\int p(x)dx}$,

$$C = e^{C_1} = \text{arbitrary constant}$$

This solution may also be derived by means of an integrating factor, as described below for the inhomogeneous equation.

Example Find the general solution of $y' + x^2 y = 0.$

$$\int \frac{dy}{y} = -\int x^2 dx + C_1,$$

$$\therefore \quad \ln y = -\frac{1}{3}x^3 + C_1$$

$$\therefore \quad y = Ce^{-\frac{x^3}{3}}.$$

If $y = 2$ when $x = 0$, then $C = 2$ and $y = 2e^{-\frac{x^3}{3}}$.

Inhomogeneous equation $y' + p(x)y = r(x)$.

This is solved by multiplying by the integrating factor $f(x) \equiv e^{\int p(x)dx}$:

$$fy' + fpy \equiv \frac{d}{dx}\left(y(x)e^{\int p(x)dx}\right) = r(x)e^{\int p(x)dx}$$

$$\therefore \quad y(x)e^{\int p(x)dx} = \int r(x)e^{\int p(x)dx} \, dx + C$$

$$\therefore \quad y = e^{-\int p(x)dx}\int r(x)e^{\int p(x)dx} \, dx + Ce^{-\int p(x)dx}$$

$$= \text{particular integral}$$

$$+ \text{ solution of the homogeneous equation}$$

Example   Find the general solution of $y' + x^2y = x^2$.

$$\text{Integrating factor} = e^{\int x^2 dx} = e^{\frac{x^3}{3}}$$

$$\therefore \quad \frac{d}{dx}\left(y(x)e^{\frac{x^3}{3}}\right) = x^2 e^{\frac{x^3}{3}}$$

$$\therefore \quad y(x)e^{\frac{x^3}{3}} = \int x^2 e^{\frac{x^3}{3}} \, dx + C$$

$$\therefore \quad y = 1 + Ce^{-\frac{x^3}{3}}$$

If $y = 2$ when $x = 0$, then $C = 1$ and $y = 1 + e^{-\frac{x^3}{3}}$.

Problems 1A

Find the general solution of:

1. $y' - 4y = 2x - 4x^2$. $[y = x^2 + Ce^{4x}]$
2. $xy' + 2y = 4e^{x^2}$. $[y = (C + 2e^{x^2})/x^2]$
3. $y' + 2y\tan x = \sin^2 x$. $[y = C\cos^2 x + \cos^2 x(\tan x - x)]$
4. $y' + y\cot x = \sin 2x$. $[y = \frac{2}{3}\sin^2 x + C\mathrm{cosec}x]$
5. $\sin xy' - y\cos x = \sin 2x$. $[y = 2\sin x \ln(\sin x) + C\sin x]$
6. $x\ln xy' + y = 2\ln x$. $[y = \ln x + C/\ln x]$
7. $y' + \frac{2y}{x} = e^x$. $[y = C/x^2 + (1 - 2/x + 2/x^2)e^x]$
8. $(x - 1)y' + 3y = x^2$. $[(x - 1)^3y = C + x^5/5 - x^4/2 + x^3/3]$
9. $(x + 1)y' + (2x - 1)y = e^{-2x}$. $[e^{2x}y = C(x + 1)^3 - \frac{1}{3}]$

10. $y' + \frac{y}{x} = \frac{1}{2}\sin\left(\frac{x}{2}\right)$. $\quad \left[y = -\cos\frac{x}{2} + \frac{2}{x}\sin\frac{x}{2} + \frac{C}{x}\right]$

11. $(1 - x^2)y' + x(y - a) = 0$. $\quad \left[y = a + C(1 - x^2)^{\frac{1}{2}}\right]$

12. $y' - (1 + \cot x)y = 0$. $\quad [y = Ce^x \sin x]$

13. $(1 + x^2)y' + xy = 3x + 3x^3$. $\quad \left[y = 1 + x^2 + C(1 + x^2)^{-\frac{1}{2}}\right]$

14. $\sin x \cos x y' + y = \cot x$. $\quad [y = (C + \ln\tan x)/\tan x]$

Solve:

15. $y' + 2xy = 4x$, $y(0) = 3$. $\quad \left[y = 2 + e^{-x^2}\right]$

16. $y' \coth 2x = 2y - 2$, $y(0) = 0$. $\quad [y = 1 - \cosh 2x]$

17. $y' + ky = e^{-kx}$, $y(0) = 1$. $\quad \left[y = (1 + x)e^{-kx}\right]$

18. $y' = a(y - g)$, $y(0) = b$. $\quad \left[y = g + (b - g)e^{ax}\right]$

19. $yy' = 2a$, $y(0) = 0$. $\quad [y^2 = 4ax]$

20. $yy' + x = 0$, $y(0) = a$. $\quad [x^2 + y^2 = a^2]$

21. $yy' + \frac{b^2 x}{a^2} = 0$, $y(0) = b$. $\quad \left[\frac{x^2}{a^2} + \frac{y^2}{b^2} = 1\right]$

22. $(x + 1)y' = y - 3$, $y(0) = 8$. $\quad [y = 5x + 8]$

23. $2xy' + y = 0$, $y(1) = 1$. $\quad [xy^2 = 1]$

24. $(1 + x^2)y' = \sqrt{y}$, $y(0) = 0$. $\quad \left[y = \frac{1}{4}(\tan^{-1} x)^2\right]$

25. $\frac{di}{dt} + 3i = \sin 2t$, $i = 0$ when $t = 0$. $\left[i = \{\sin(2t - \alpha) + e^{-3t}\sin\alpha\}/\sqrt{13}\right.$, where $\tan\alpha = \frac{2}{3}\Big]$

26. Water runs out through a hole in the base of a circular cylindrical tank at speed $\sqrt{2gh}$ ft/s, where $g = 32$ ft/s^2 and h is the water depth. If the tank is 2 ft in height, 1 ft in diameter and is full at time $t = 0$, calculate the time at which half the water has run out when the effective area of the hole is 0.25 in^2. [47 s]

27. The current i in a circuit satisfies $Ldi/dt + Ri = E$, where L, R, E are constants. Show that when t is large the current is approximately equal to E/R.

If, instead, $E = E_o \cos\omega t$, where E_o, ω are constants, show that when t is large

$$i \approx \frac{E_o \cos(\omega t - \epsilon)}{\sqrt{R^2 + \omega^2 L^2}}, \quad \text{where } \tan\epsilon = \frac{\omega L}{R}.$$

1.2 Second-Order Equations with Constant Coefficients

Homogeneous equation $\quad y'' + ay' + by = 0$, $\quad a, \ b = \ $ constants.

Inhomogeneous equation $\quad y'' + ay' + by = r(x)$.

General solution:

$$y = Ay_1(x) + By_2(x) + y_p(x), \quad A, \ B = \text{constant}$$

where y_1, y_2 are any two linearly independent solutions of the homogeneous equation, called *basis functions* or *complementary functions*, and y_p is a *particular integral* that yields $r(x)$ when substituted into the equation.

Solution of the homogeneous equation   Because $d(e^{\lambda x})/dx = \lambda e^{\lambda x}$, $y = e^{\lambda x}$ will be a solution of the homogeneous equation if λ is a solution of the *characteristic equation*

$$\lambda^2 + a\lambda + b = 0, \quad \text{i.e. for } \lambda = \frac{-a \pm \sqrt{a^2 - 4b}}{2} = \lambda_1, \lambda_2. \quad (1.2.1)$$

Case 1 $\lambda_1 \neq \lambda_2$:

$y_1 = e^{\lambda_1 x}$ and $y_2 = e^{\lambda_2 x}$ are linearly independent and the general solution is therefore

$$y = A e^{\lambda_1 x} + B e^{\lambda_2 x}. \quad (1.2.2)$$

The values of the constants A, B are fixed by the *boundary conditions*.

Example   Solve $y'' + 2y' - 8y = 0$, $\quad y(0) = 1$, $y'(0) = 0$.

Characteristic equation : $\lambda^2 + 2\lambda - 8 = 0$

$$\therefore \quad \lambda = -4,\ 2$$

$$\therefore \quad y(x) = A e^{-4x} + B e^{2x}.$$

At $x = 0$: $y = 1$, and $y' = 0$

$$\therefore \quad A + B = 1$$

$$\text{and} \quad -4A + 2B = 0.$$

$$\therefore \quad y = \frac{e^{-4x} + 2e^{2x}}{3}.$$

Case 2 $\lambda_1 = \lambda_2 \equiv \lambda$:

The two solutions in (1.2.2) are not independent. The differential equation can now be written in the factored form

$$y'' + ay' + b \equiv \left(\frac{d}{dx} - \lambda \right) \left(\frac{d}{dx} - \lambda \right) y = 0.$$

If $z = \dfrac{dy}{dx} - \lambda y$, then $z' - \lambda z = 0$, i.e. $z = B e^{\lambda x}$, $B = \text{constant}$,

$$\therefore \quad y' - \lambda y = B e^{\lambda x}.$$

An integrating factor is : $\mathrm{e}^{-\lambda x}$

$$\therefore \quad \frac{d}{dx}\left(y(x)\mathrm{e}^{-\lambda x}\right) = B,$$

i.e. the general solution is

$$y = (A + Bx)\mathrm{e}^{\lambda x}, \quad A, \ B = \text{arbitrary constants.} \tag{1.2.3}$$

Case 3 Complex roots λ of the characteristic equation:

When a and b are real and $a^2 - 4b < 0$ the roots (1.2.1) of the characteristic equation are *complex conjugates* $\lambda = -a/2 \pm i\sqrt{4b - a^2}/2 \equiv -a/2 \pm i\Omega$, say, where $i = \sqrt{-1}$, and the general solution assumes either of the forms

$$y = \mathrm{e}^{-\frac{ax}{2}}\left(A'\mathrm{e}^{i\Omega x} + B'\mathrm{e}^{-i\Omega x}\right), \quad \Omega = \frac{\sqrt{4b - a^2}}{2}, \quad A', \ B' = \text{constants}$$

$$= \mathrm{e}^{-\frac{ax}{2}}\left(A\cos(\Omega x) + B\sin(\Omega x)\right), \quad A, \ B = \text{constants,} \tag{1.2.4}$$

Example The two forms of the solution (1.2.4) are related by Euler's formula

$$\mathrm{e}^{ix} = 1 + ix + \frac{(ix)^2}{2!} + \frac{(ix)^3}{3!} + \frac{(ix)^4}{4!} + \frac{(ix)^5}{5!} + \frac{(ix)^6}{6!} + \frac{(ix)^7}{7!} + \cdots$$

$$= \left(1 - \frac{x^2}{2!} + \frac{x^4}{4!} - \frac{x^6}{6!} + \cdots\right) + i\left(x - \frac{x^3}{3!} + \frac{x^5}{5!} - \frac{x^7}{7!} + \cdots\right)$$

$$\equiv \cos x + i\sin x. \tag{1.2.5}$$

Example Simple harmonic motion at (a real-valued) radian frequency ω is described by the equation

$$\frac{d^2 y}{dt^2} + \omega^2 y = 0, \quad \text{where } t \text{ denotes time.}$$

The roots of the characteristic equation are $\lambda = \pm i\omega$, with the general solution

$$y = A\cos(\omega t) + B\sin(\omega t) \equiv A'\mathrm{e}^{i\omega t} + B'\mathrm{e}^{-i\omega t}.$$

Problems 1B

Find the general solution of:

1. $y'' + 10y' + 25y = 0$. $[y = (A + Bx)\mathrm{e}^{-5x}]$

2. $y'' + 4y' + 9y = 0$. $[y = (A\cos\sqrt{5}x + B\sin\sqrt{5}x)\mathrm{e}^{-2x}]$

3. $y'' - 6y' + 8y = 0$. $[y = A\mathrm{e}^{4x} + B\mathrm{e}^{2x}]$

4. $y'' - 6y' + 25y = 0.$ $[y = (A\cos 4x + B\sin 4x)e^{3x}]$

5. $y'' - 4y = 0.$ $[y = Ae^{2x} + Be^{-2x}]$

6. $y'' + 4y = 0.$ $[y = A\cos 2x + B\sin 2x]$

7. $y'' - y' + y = 0.$ $[y = (A\cos\frac{\sqrt{3}}{2}x + B\sin\frac{\sqrt{3}}{2}x)e^{\frac{1}{2}x}]$

8. $y'' + 3y' = 0.$ $[y = A + Be^{-3x}]$

9. Transform the equation $y'' + x^2 + y + 2 = 0$ by making the substitution $y(x) = z(x) - x^2$, and hence find the general solution. $[y = -x^2 + A\cos x + B\sin x]$

Solve:

10. $4(y'' - y') + y = 0,\ y(0) = 0,\ y(2) = 2.$ $[y = xe^{\frac{1}{2}x-1}]$

11. $y'' - 16y = 0,\ y(0) = 1,\ y'(0) = 20.$ $[y = 3e^{4x} - 2e^{-4x}]$

12. $y'' + 6y' + 9y = 0,\ y(0) = -4,\ y'(0) = 14.$ $[y = (2x - 4)e^{-3x}]$

13. $y'' - 16y = 0,\ y(0) = 5,\ y(\frac{1}{4}) = 5e.$ $[y = 5e^{4x}].$

14. $y'' + 9y = 0,\ y(\pi) = -2,\ y'(\pi) = 3.$ $[y = 2\cos 3x - \sin 3x].$

15. $y'' - 2y' + 2y = 0,\ y(0) = -3,\ y(\pi/2) = 0.$ $[y = -3e^x\cos x].$

1.3   Euler's Homogeneous Equation

The equation

$$x^2 y'' + axy' + by = r(x), \quad a,\ b = \text{constant}, \qquad (1.3.1)$$

is equivalent to

$$x\frac{d}{dx}\left(x\frac{d}{dx}\right)y + (a-1)x\frac{dy}{dx} + by = r(x),$$

which is reduced to a constant coefficient equation by the substitution

$$x = e^z, \quad \text{which implies that } x\frac{d}{dx} = \frac{d}{dz}.$$

Thus,

$$\frac{d^2y}{dz^2} + (a-1)\frac{dy}{dz} + by = r(e^z).$$

The homogeneous form of this equation is solved by the method of §1.2 using the characteristic equation

$$\lambda^2 + (a-1)\lambda + b = 0.$$

Example Find the general solution of $x^2y'' + 9xy' + 16y = 0$.

The substitution $x = e^z$ reduces the equation to

$$\frac{d^2y}{dz^2} + 8\frac{dy}{dz} + 16y = 0, \quad \text{with characteristic equation } \lambda^2 + 8\lambda + 16 = 0,$$

$$\therefore \quad \lambda = -4, \; -4.$$

$$\therefore \quad y = (A + Bz)e^{-4z} = \frac{(A + B\ln x)}{x^4}.$$

Problems 1C

Find the general solution of:

1. $x^2y'' + 6.2xy' + 6.76y = 0.$ $[y = (A + B\ln x)/x^{2.6}]$
2. $x^2y'' + xy' + y = 0.$ $[y = A\cos(\ln x) + B\sin(\ln x)]$
3. $x^2y'' + xy' - 9y = 0.$ $[y = Ax^3 + B/x^3]$
4. $x^2y'' - 2xy' + 2y = 0.$ $[y = Ax + Bx^2]$
5. $(x+1)^2y'' - 2(x+1)y' - 10y = 0.$ $[y = A(x+1)^5 + B(x+1)^{-2}]$
6. $x^2y'' - 3xy' + 4y = 0.$ $[y = x^2(A + B\ln x)]$
7. $x^2y'' + xy' - 4y = 0.$ $[y = Ax^2 + B/x^2]$
8. $x^2y'' - 2xy' - 4y = 0.$ $[y = Ax^4 + B/x]$
9. $x^2y'' - 20y = 0.$ $[y = Ax^5 + B/x^4]$
10. $x^2y'' - xy' + 2y = 0.$ $[y = x\{A\cos(\ln x) + B\sin(\ln x)\}]$
11. $y'' + \frac{2}{x}y' = 0.$ $[y = A + B/x]$
12. $x^2y''' + 3xy'' + y' = 0.$ $[y = A(\ln x)^2 + B\ln x + C]$
13. $x^2y'' + 9xy' + 25y = 0.$ $[y = \{A\cos(3\ln x) + B\sin(3\ln x)\}/x^4]$
14. $(1+2x)^2y'' - 6(1+2x)y' + 16y = 0.$ $[y = (1+2x)^2\{A\ln(1+2x) + B\}]$
15. $(1+x)^2y'' + (1+x)y' + y = 0.$ $[y = A\cos\{\ln(1+x) + \alpha\}]$

Solve:

16. $4x^2y'' + 4xy' - y = 0, \; y(4) = 2, \; y'(4) = -\frac{1}{4}.$ $[y = 4/\sqrt{x}]$
17. $x^2y'' - xy' + 2y = 0, \; y(1) = -1, \; y'(1) = -1.$ $[y = -x\cos(\ln x)]$

1.4 Method of Reduction of Order

Let $y = y_1(x)$ be any solution of the homogeneous equation

$$y'' + p(x)y' + q(x)y = 0. \tag{1.4.1}$$

The general solution $y(x)$ can be found from $y_1(x)$ by the following procedure.

Set
$$y = y_1(x)v(x),$$
$$y' = y_1'v + y_1v',$$
$$y'' = y_1''v + 2y_1'v' + y_1v''.$$

Substitute into (1.4.1) and collect terms:

$$v''y_1 + v'\left(2y_1' + py_1\right) + v\left(y_1'' + py_1' + qy_1\right) = 0.$$

Because y_1 is a solution of (1.4.1) the coefficient of v is zero. Hence,

$$\frac{dU}{dx} = -U\left(\frac{2y_1'}{y_1} + p(x)\right), \quad \text{where } U = v'.$$

Integrating this first-order equation for U:

$$\frac{dv}{dx} \equiv U = \frac{Be^{-\int p(x)dx}}{y_1^2}, \quad B = \text{constant},$$

so that

$$v(x) = A + B\int \frac{e^{-\int p(x)dx}}{y_1^2}\,dx, \quad A = \text{constant}.$$

Hence, the general solution $y = y_1v$ is

$$y(x) = Ay_1(x) + By_1(x)\int \frac{e^{-\int p(x)dx}}{y_1^2}\,dx. \qquad (1.4.2)$$

Example   Find the general solution $y = Ay_1(x) + By_2(x)$ of

$$(x^2 - 1)y'' - 2xy' + 2y = 0, \quad \text{given that } y_1 = x.$$

The equation for $U = v'$ is

$$\frac{1}{U}\frac{dU}{dx} = \frac{-2}{x} + \frac{2x}{x^2 - 1}, \quad \text{i.e. } v' = B\left(1 - \frac{1}{x^2}\right);$$

integrating, and setting the constant of integration equal to zero,

$$v = B\left(x + \frac{1}{x}\right), \quad \text{i.e. } y_2 = xv(x) = B(1 + x^2), \quad \therefore\ y = Ax + B(1 + x^2).$$

Problems 1D

Find the general solution $y = Ay_1(x) + By_2(x)$ given $y_1(x)$:

1. $(x+1)^2 y'' - 2(x+1)y' + 2y = 0$, $\quad y_1 = 1 + x$. $\quad [y = A(1+x) + B(1+x)^2]$

2. $xy'' + 2y' + xy = 0$, $\quad y_1 = \frac{\sin x}{x}$. $\quad \left[y = A\frac{\sin x}{x} + B\frac{\cos x}{x}\right]$

3. $(x+2)y'' - (2x+5)y' + 2y = 0$, $\quad y_1 = e^{2x}$. $\quad [y = Ae^{2x} + B(2x+5)]$

4. $x^2 y'' - (x^2 + 2x)y' + (x+2)y = 0$, $\quad y_1 = x$. $\quad [y = x(A + Be^x)]$

5. $xy'' - 2(x+1)y' + (x+2)y = 0$, $\quad y_1 = e^x$. $\quad [y = (A + Bx^3)e^x]$

6. $x^2 y'' + xy' - 9y = 0$, $\quad y_1 = x^3$. $\quad [y = Ax^3 + B/x^3]$

7. $x(x\cos x - 2\sin x)y'' + (x^2 + 2)y' \sin x - 2y(x\sin x + \cos x) = 0$, $\quad y_1 = x^2$. $[y = Ax^2 + B\sin x]$

8. $(x+1)y'' - 2xy' + (x-1)y = 0$, $\quad y_1 = e^x$. $\quad [y = \{A + B/(x+1)\}e^x]$

9. $x^2 y'' + x^2 y' + (x-2)y = 0$, $\quad y_1 = 1/x$. $\quad [y = A/x + B\{x + 2 + 2/x\}e^{-x}]$

10. $xy'' - (2x+1)y' + (x+1)y = 0$, $\quad y_1 = e^x$. $\quad [y = (A + Bx^2)e^x]$

11. $x(x+1)y'' - 2y' - 2y = 0$, $\quad y_1 = 1/(1+x)$. $\quad [y = (A + Bx^3)/(1+x)]$

12. Solve $4x^2 y'' + 4xy' + (4x^2 - 1)y = 0$ by making the substitution $y = z/\sqrt{x}$. $[z'' + z = 0, y = (A\cos x + B\sin x)/\sqrt{x}]$

13. Set $y = x^n z$ in the equation $x^2 y'' + 2x(x+2)y' + 2(x+1)^2 y = 0$ and choose n so that the equation for z has constant coefficients. Hence solve the given equation.

$$[n = -2, \ z'' + 2z' + 2z = 0, \ y = x^{-2}e^{-x}(A\cos x + B\sin x)]$$

1.5 Particular Integrals of Second-Order Equations

Consider the problem of finding a particular integral $y_p(x)$ in the general solution

$$y = Ay_1(x) + By_2(x) + y_p(x),$$

of

$$y'' + ay' + by = r(x), \quad a, \ b \text{ constant.} \tag{1.5.1}$$

The particular integral can be found in simple form for a certain class of functions $r(x)$. When $r(x)$ is a linear combination of the terms in the first column of Table 1.1, $y_p(x)$ will generally consist of a linear combination of the corresponding terms in the second column (see §1.6 for $r(x)$ of arbitrary functional form).

Table 1.1   Particular integrals (C, a_j, K, M, α, Ω are constants; $m \geq 0$ is an integer)

$r(x)$	$y_p(x)$
$e^{\alpha x}$	$Ce^{\alpha x}$
x^m	$a_0 + a_1 x + a_2 x^2 + \cdots + a_m x^m$
$x^m e^{\alpha x}$	$e^{\alpha x}(a_0 + a_1 x + a_2 x^2 + \cdots + a_m x^m)$
$\cos \Omega x$	$K \cos \Omega x + M \sin \Omega x$
$\sin \Omega x$	$K \cos \Omega x + M \sin \Omega x$
$e^{\alpha x} \cos \Omega x$	$e^{\alpha x}[K \cos \Omega x + M \sin \Omega x]$
$e^{\alpha x} \sin \Omega x$	$e^{\alpha x}[K \cos \Omega x + M \sin \Omega x]$

Example   Find the general solution of $y'' - y = 3e^{2x}$.
The solution of the homogeneous equation is $y = Ae^x + Be^{-x}$. Set $y_p = Ce^{2x}$, where the constant C is to be found. Substituting into the left-hand side of the equation:

$$4Ce^{2x} - Ce^{2x} \equiv 3e^{2x}.$$

$$\therefore \quad C = 1$$

$$\therefore \quad \text{general solution} \quad y = Ae^x + Be^{-x} + e^{2x}.$$

Example   Find a particular integral of $y'' + 5y' + 6y = 9x^4 - x$.

$$y_p = a + bx + cx^2 + dx^3 + ex^4,$$
$$y_p' = b + 2cx + 3dx^2 + 4ex^3,$$
$$y_p'' = 2c + 6dx + 12ex^2.$$

Substitute into the left-hand side of the equation and equate coefficients of x^m, $m = 0, 1, \ldots, 4$ on both sides:

$$6a + 5b + 2c = 0 \quad x^0$$
$$6b + 10c + 6d = -1 \quad x^1$$
$$6c + 15d + 12e = 0 \quad x^2$$
$$6d + 20e = 0 \quad x^3$$
$$6e = 9 \quad x^4.$$

$$\therefore \quad a = 6, \quad b = -11, \quad c = \frac{19}{2}, \quad d = -5, \quad e = \frac{3}{2}$$

$$y_p = 6 - 11x + \frac{19}{2}x^2 - 5x^3 + \frac{3}{2}x^4.$$

Example   Find the general solution of $y'' + 2y' + 5y = \sin 2x$.
Characteristic equation: $\lambda^2 + 2\lambda + 5 = 0$,

$$\therefore \quad \lambda = -1 \pm 2i \quad \text{and} \quad y = e^{-x}\left(A \sin 2x + A \cos 2x\right) + y_p,$$

where

$$y_p = K\cos 2x + M\sin 2x$$
$$y_p' = -2K\sin 2x + 2M\cos 2x$$
$$y_p'' = -4K\cos 2x - 4M\sin 2x.$$

Substitute into the equation and equate coefficients of $\sin 2x$, $\cos 2x$ on both sides:

$$M - 4K = 1, \quad 4M + K = 0, \quad \text{i.e. } K = \frac{-4}{17}, \quad M = \frac{1}{17}$$

$$\therefore \quad y = e^{-x}\left(A\sin 2x + A\cos 2x\right) - \frac{4}{17}\cos 2x + \frac{1}{17}\sin 2x.$$

Resonant forcing occurs when $r(x)$ is proportional to $y_1(x)$ or $y_2(x)$.

When $r(x) = e^{\lambda x}$ (where λ is a root of the characteristic equation) there are two possibilities:

Case 1 $\lambda_1 \neq \lambda_2$:

$$y_p = Axe^{\lambda x}. \tag{1.5.2}$$

Case 2 $\lambda_1 = \lambda_2$:

$$y_p = Ax^2 e^{\lambda x}. \tag{1.5.3}$$

Case 2 arises only for real values of λ when a and b are real.

More generally, if $r(x) = x^m e^{\lambda x}$, $m \geq 0$, we have:

Case 3 $\lambda_1 \neq \lambda_2$:

$$y_p = e^{\lambda x}(a_0 + a_1 x + a_2 x^2 + \cdots + a_{m+1}x^{m+1}). \tag{1.5.4}$$

Case 4 $\lambda_1 = \lambda_2$:

$$y_p = Ax^{m+2}e^{\lambda x}. \tag{1.5.5}$$

For complex $\lambda = \alpha \pm i\Omega$ and $r(x) = e^{\alpha x}\cos\Omega x$ or $e^{\alpha x}\sin\Omega x$, the particular integral is given by:

Case 5 $\lambda = \alpha \pm i\Omega$:

$$y_p = xe^{\alpha x}[K\cos\Omega x + M\sin\Omega x], \tag{1.5.6}$$

and more generally, for $r(x) = x^m e^{\alpha x} \cos \Omega x$ or $x^m e^{\alpha x} \sin \Omega x$ (where $m \geq 0$) by:

Case 6 $\lambda = \alpha \pm i\Omega$:

$$y_p = e^{\alpha x}\Big[(a_0 + a_1 x + a_2 x^2 + \cdots + a_{m+1} x^{m+1}) \cos \Omega x$$
$$+ (b_0 + b_1 x + a_2 x^2 + \cdots + b_{m+1} x^{m+1}) \sin \Omega x\Big]. \qquad (1.5.7)$$

Example Find the general solution of $y'' - 2y' + y = e^x$.
 Characteristic equation: $\lambda^2 - 2\lambda + 1 = 0$, i.e. $\lambda = 1, 1$.

$$\therefore \quad y = (A + Bx)e^x + y_p, \quad y_p = Cx^2 e^x.$$

C is found by substituting into the differential equation:

$$C\Big(2 + 4x + x^2\Big) - 2C\Big(2x + x^2\Big) + C\Big(x^2\Big) \equiv 1, \quad \therefore \quad C = \frac{1}{2}.$$

Example Forced simple harmonic motion: $\frac{d^2 y}{dt^2} + \Omega^2 y = \sin \Omega t$.
 Roots of the characteristic equation $\lambda = \pm i\Omega$

$$\therefore \quad y = A \sin \Omega t + B \cos \Omega t + t\Big(K \cos \Omega t + M \sin \Omega t\Big)$$

K and M are found by substituting into the differential equation: $K = -\frac{1}{2\Omega}, M = 0$,

$$\therefore \quad y = A \sin \Omega t + B \cos \Omega t - \frac{t}{2\Omega} \cos \Omega t.$$

Problems 1E

Find the general solution of:

1. $y'' + y = 3x^2$. $[y = A \sin x + B \cos x + 3x^2 - 6]$
2. $y'' - 4y = e^{2x}$. $[y = A e^{2x} + B e^{-2x} + \frac{x}{2} e^{2x}]$
3. $y'' + 4y' + y = 2 \sin x - 4 \cos x$. $[y = e^{-2x}(A e^{\sqrt{3}x} + B e^{-\sqrt{3}x}) - \sin x - \frac{1}{2} \cos x]$
4. $y'' + 9y = \cos 3x$. $[y = A \sin 3x + B \cos 3x + \frac{x}{6} \sin 3x]$
5. $y'' + 8y' + 16y = 6e^{-4x}$. $[y = e^{-4x}(A + Bx + 3x^2)]$
6. $y'' + 4y = \sin x \sin 3x$. $[y = \frac{1}{24} \cos 4x + A \cos 2x + (B + \frac{1}{8}x) \sin 2x]$
7. $y'' - 3y' + 18y = \sinh 2x$. $[y = e^{\frac{3}{2}x}\{A \cos(\frac{3\sqrt{7}}{2}x) + B \sin(\frac{3\sqrt{7}}{2}x)\} + \frac{1}{32} e^{2x} - \frac{1}{56} e^{-2x}]$
8. $y'' - 6y' + 8y = e^{4x} - \cos 2x$. $[y = A e^{2x} + (B + \frac{1}{2}x)e^{4x} - \frac{1}{40}(\cos 2x - 3 \sin 2x)]$
9. $y'' + 3y' + 2y = e^{-x} \sin x$. $[y = e^{-x}(A - \frac{1}{2} \sin x - \frac{1}{2} \cos x) + B e^{-2x}]$
10. $3y'' - 5y' + 2y = x^2 e^x$. $[y = e^x(A + 18x - 3x^2 + \frac{1}{3}x^3) + B e^{\frac{2}{3}x}]$

11. $y'' + 6y' + 9y = (1+x)e^{-3x}$. $\left[y = e^{-3x}\left(A + Bx + \frac{1}{2}x^2 + \frac{1}{6}x^3\right)\right]$

12. $y'' + y' + y = e^x(x + \cos x)$. $\left[y = e^{-\frac{1}{2}x}\left\{A\cos\left(\frac{\sqrt{3}}{2}x\right) + B\sin\left(\frac{\sqrt{3}}{2}x\right)\right\} + e^x\left\{\frac{1}{3}(x-1) + \frac{1}{13}(2\cos x + 3\sin x)\right\}\right]$

13. $y'' - 4y' + 4y = 8x^2e^{2x}\sin 2x$. $\left[y = e^{2x}\{A + Bx + (3 - 2x^2)\sin 2x - 4x\cos 2x\}\right]$

14. $y'' - 3y' - 4y = 10\cos 2x$. $\left[y = Ae^{4x} + Be^{-x} - \frac{1}{5}(4\cos 2x + 3\sin 2x)\right]$

15. $y'' - 5y' + 6y = 4x^2e^x$. $\left[y = Ae^{2x} + Be^{3x} + e^x(2x^2 + 6x + 7)\right]$

16. $y'' - 10y' + 29y = e^{5x}\sin 2x$. $\left[y = e^{5x}\left\{A\cos 2x + B\sin 2x - \frac{1}{4}x\cos 2x\right\}\right]$

17. $x^2y'' - 20y = 7x^3$. $\left[y = Ax^5 + Bx^{-4} - \frac{1}{2}x^3\right]$

18. $x^2y'' - xy' + 2y = x\ln x$. $\left[y = x\{\ln x + A\cos(\ln x) + B\sin(\ln x)\}\right]$

19. $x^2y'' - 2xy' + 2y = x^3\cos(\ln x)$. $\left[y = Ax + Bx^2 + \frac{1}{10}x^3\{\cos(\ln x) + 3\sin(\ln x)\}\right]$

20. $x^2y''' + 3xy'' + y' = \ln x$. $\left[y = \frac{1}{24}(\ln x)^4 + A(\ln x)^2 + B\ln x + C\right]$

21. $y'' - 5y' + 6y = x^3e^{2x}$. $\left[y = Ae^{3x} + e^{2x}\left(B - 6x - 3x^2 - x^3 - \frac{1}{4}x^4\right)\right]$

22. $y'' - 6y' + 13y = 8e^{3x}\sin 2x$. $\left[y = e^{3x}(A\cos 2x + B\sin 2x - 2x\cos 2x)\right]$

Solve:

23. $y'' - y' - 2y = 10\sin x$, $y(\frac{\pi}{2}) = -3$, $y'(\frac{\pi}{2}) = -1$. $[y = \cos x - 3\sin x]$

24. $y'' - 6y' + 8y = e^{2x} + \sin 2x$, $y(0) = 0$, $y'(0) = 0$. $\left[y = \frac{3}{10}e^{4x} - \left(\frac{3}{8} + \frac{1}{2}x\right)e^{2x} + \frac{1}{40}(3\cos 2x + \sin 2x)\right]$

25. $2y'' - 5y' + 3y = 4e^{2x}$, $y(0) = 0$, $y'(0) = 0$. $[y = 4(e^{2x} + e^x - 2e^{\frac{3}{2}x})]$.

26. $y'' + 16y = \cos 4x$, $y(\frac{\pi}{4}) = 0$. $\left[y = \left(A + \frac{1}{8}x\right)\sin 4x\right]$

1.6   Method of Variation of Parameters

This method enables a particular integral of the equation

$$y'' + p(x)y' + q(x)y = r(x), \qquad (1.6.1)$$

to be found when the basis functions $y_1(x)$, $y_2(x)$ are known.

Set $y_p = u(x)y_1(x) + v(x)y_2(x)$, where the functions $u(x)$ and $v(x)$ are assumed to be chosen such that $y'_p = uy'_1 + vy'_2$. Then

$$u'(x)y_1(x) + v'(x)y_2(x) = 0 \qquad (1.6.2)$$

and

$$y_p = uy_1 + vy_2$$
$$y'_p = uy'_1 + vy'_2$$
$$y''_p = uy''_1 + vy''_2 + u'y'_1 + v'y'_2.$$

Substituting into (1.6.1) and collecting terms

$$u(x)\left[y_1'' + p(x)y_1' + q(x)y_1\right]$$

$$+ v(x)\left[y_2'' + p(x)y_2' + q(x)y_2\right] + u'(x)y_1' + v'(x)y_2' = r(x)$$

$$\therefore \quad u'(x)y_1'(x) + v'(x)y_2'(x) = r(x), \tag{1.6.3}$$

because the terms in square brackets are identically zero.

Solving the simultaneous Equations (1.6.2) and (1.6.3) for u' and v':

$$u'(x) = \frac{-y_2(x)r(x)}{[y_1(x)y_2'(x) - y_1'(x)y_2(x)]}$$

$$v'(x) = \frac{y_1(x)r(x)}{[y_1(x)y_2'(x) - y_1'(x)y_2(x)]}$$

$$\therefore \quad u(x) = \int \frac{-y_2(x)r(x)dx}{[y_1(x)y_2'(x) - y_1'(x)y_2(x)]}$$

$$v(x) = \int \frac{y_1(x)r(x)dx}{[y_1(x)y_2'(x) - y_1'(x)y_2(x)]},$$

so that

$$y_p(x) = y_1(x)\int \frac{-y_2(x)r(x)dx}{[y_1(x)y_2'(x) - y_1'(x)y_2(x)]}$$

$$+ y_2(x)\int \frac{y_1(x)r(x)dx}{[y_1(x)y_2'(x) - y_1'(x)y_2(x)]}. \tag{1.6.4}$$

Wronskian   The determinant

$$\mathcal{W}(y_1, y_2) = y_1(x)y_2'(x) - y_1'(x)y_2(x) \equiv \begin{vmatrix} y_1(x) & y_2(x) \\ y_1'(x) & y_2'(x) \end{vmatrix}, \tag{1.6.5}$$

is called the *Wronskian*. It is non-zero if and only if $y_1(x)$ and $y_2(x)$ form a linearly independent set of basis functions. Thus, the particular integral can also be written

$$y_p(x) = y_1(x)\int \frac{-y_2(x)r(x)dx}{\mathcal{W}(y_1(x), y_2(x))} + y_2(x)\int \frac{y_1(x)r(x)dx}{\mathcal{W}(y_1(x), y_2(x))}.$$

Example Find the general solution of $y'' + y = \sec x$.

Because $y_1 = \cos x$, $y_2 = \sin x$, we set

$$y_p = u \cos x + v \sin x,$$

$$y_p' = -u \sin x + v \cos x,$$

$$y_p'' = -u' \sin x - u \cos x + v' \cos x - v \sin x.$$

Substituting into the differential equation, we find that $u(x)$, $v(x)$ are determined by

$$u'(x) \cos x + v'(x) \sin x = 0, \quad -u'(x) \sin x + v'(x) \cos x = \sec x,$$

$$\therefore \quad u'(x) = -\tan x, \quad v'(x) = 1$$

integrating : $\quad u(x) = \ln|\cos x|, \quad v(x) = x.$

Hence, $\quad y_p(x) = \cos x \ln|\cos x| + x \sin x,$

$$\therefore \quad y = A \sin x + B \cos x + \cos x \ln|\cos x| + x \sin x.$$

Problems 1F

Find the general solutions of:

1. $y'' - 2y' + y = x^{\frac{3}{2}} e^x$. $\quad [y = (A + Bx)e^x + \frac{4}{35} x^{\frac{7}{2}} e^x]$
2. $y'' + 4y' + 4y = e^{-2x}/x^2$. $\quad [y = (A + Bx)e^{-2x} - \ln|x| e^{-2x}]$
3. $x^2 y'' - 2xy' + 2y = 5x^3 \cos x$. $\quad [y = Ax + Bx^2 - 5x \cos x]$
4. $y'' + y = \tan x$. $\quad [y = A \sin x + B \cos x - \cos x \ln|\sec x + \tan x|]$
5. $x^2 y'' - 2xy' + 2y = x^3 \ln x$. $\quad [y = Ax + Bx^2 + x^3 (\frac{1}{2} \ln x - \frac{3}{4})]$
6. $y'' \sin 4x - 4y' \cos^2 2x + 2y = \tan x$. $\quad [y = A + B \cos 2x + \frac{1}{16} (\ln|\tan x| + \cos 2x \ln|(1 - \frac{1}{2} \sec^2 x) \cot 2x|)]$
7. $(1 - x^2)y'' - 2xy' = 2x$, $|x| < 1$. $\quad [y = A + B \ln\{(1 + x)/(1 - x)\} - x]$
8. $(1 + x^2)y'' - 2xy' + 2y = x^3 + 3x$. $\quad [y = Ax + B(x^2 - 1) + \frac{1}{2}x^3]$
9. $(x - 1)y'' - xy' + y = (x - 1)^2$. $\quad [y = Ax + Be^x - (1 + x + x^2)]$
10. $(x + 2)y'' - (2x + 5)y' + 2y = (x + 1)e^x$. $\quad [y = A(2x + 5) + Be^{2x} - e^x]$

1.7 Method of Frobenius

It is possible to derive a series solution

$$u(x, \sigma) = x^\sigma \left\{ a_0 + a_1 x + a_2 x^2 + a_3 x^3 + \cdots \right\} \equiv \sum_{n=0}^{\infty} a_n x^{n+\sigma}, \quad (1.7.1)$$

of the equation

$$y'' + p(x)y' + q(x)y = 0, \quad (1.7.2)$$

whenever $p(x)$ and $q(x)$ can be expanded in the form

$$p(x) = \frac{p_0}{x} + p_1 + p_2 x + p_3 x^2 + \cdots \equiv \sum_{k=0}^{\infty} p_k x^{k-1},$$

$$q(x) = \frac{q_0}{x^2} + \frac{q_1}{x} + q_2 + q_3 x + q_4 x^2 + \cdots \equiv \sum_{k=0}^{\infty} q_k x^{k-2}.$$

(1.7.3)

These functions become infinite at $x = 0$ if $p_0 \neq 0$ and q_0 or $q_1 \neq 0$. They are said to be *singular* at $x = 0$. However, if the series expansions converge to finite values for $0 < |x| < R$, the expansion (1.7.1) will also converge for the same values of x, and possibly also for larger values of $|x|$; it may also remain finite at $x = 0$.

The point $x = 0$ is called a *regular singularity* of the differential equation whenever at least one of p_0, q_0, $q_1 \neq 0$. If they vanish, so that $p(x)$ and $q(x)$ are finite at $x = 0$, $x = 0$ is said to be an *ordinary point* of the equation.

Introduce the shorthand notation

$$\mathcal{L} = \frac{d^2}{dx^2} + p(x)\frac{d}{dx} + q(x),$$

then

$$\mathcal{L}u(x, \sigma) = \sum_{n=0}^{\infty} (n + \sigma)(n + \sigma - 1)a_n x^{n+\sigma-2}$$

$$+ \left(\sum_{k=0}^{\infty} p_k x^{k-1} \right) \left(\sum_{n'=0}^{\infty} (n' + \sigma)a_{n'} x^{n'+\sigma-1} \right)$$

$$+ \left(\sum_{k=0}^{\infty} q_k x^{k-2} \right) \left(\sum_{n'=0}^{\infty} a_{n'} x^{n'+\sigma} \right).$$

To see how the product series in this expression are simplified, consider the final product

$$\left(\sum_{k=0}^{\infty} q_k x^{k-2} \right) \left(\sum_{n'=0}^{\infty} a_{n'} x^{n'+\sigma} \right) = \sum_{k=0}^{\infty} \sum_{n'=0}^{\infty} q_k a_{n'} x^{k+n'+\sigma-2}.$$

The coefficient of $x^{n+\sigma-2}$ on the right-hand side is obtained by setting $k = n - n'$ and summing over all n' for which $q_{n-n'}a_{n'} \neq 0$. Because $q_{n-n'} = 0$ for $n' > n$, the coefficient is just $\sum_{n'=0}^{n} q_{n-n'}a_{n'}$, i.e.

$$\left(\sum_{k=0}^{\infty} q_k x^{k-2} \right) \left(\sum_{n'=0}^{\infty} a_{n'} x^{n'+\sigma} \right) = \sum_{n=0}^{\infty} \sum_{n'=0}^{n} q_{n-n'} a_{n'} x^{n+\sigma-2}$$

$$\equiv \sum_{n=0}^{\infty} \sum_{k=0}^{n} q_{n-k} a_k x^{n+\sigma-2}.$$

The same argument shows that

$$\left(\sum_{k=0}^{\infty} p_k x^{k-1} \right) \left(\sum_{n'=0}^{\infty} (n' + \sigma) a_{n'} x^{n'+\sigma-1} \right) = \sum_{n=0}^{\infty} \sum_{k=0}^{n} p_{n-k} a_k (k+\sigma) x^{n+\sigma-2}.$$

Hence

$$\mathcal{L}u(x, \sigma) = \sum_{n=0}^{\infty} \left\{ (n+\sigma)(n+\sigma-1) a_n \right.$$

$$\left. + \sum_{k=0}^{n} a_k \Big((k+\sigma) p_{n-k} + q_{n-k} \Big) \right\} x^{n+\sigma-2}. \quad (1.7.4)$$

It can now be seen that $u(x, \sigma)$ will be a solution of equation (1.7.2), i.e. of $\mathcal{L}y = 0$, provided the coefficient of each power of x in this expansion is zero. For $n = 0$ we must have

$$\mathcal{F}(\sigma) a_0 = 0, \quad (1.7.5)$$

where

$$\mathcal{F}(\sigma) = \sigma^2 + (p_0 - 1)\sigma + q_0. \quad (1.7.6)$$

For $n \geq 1$

$$\mathcal{F}(n+\sigma) a_n = \sum_{k=0}^{n-1} -\Big((k+\sigma) p_{n-k} + q_{n-k} \Big) a_k. \quad (1.7.7)$$

This *recurrence relation* determines a_n ($n \geq 1$) in terms of all preceding values of a_n. It is therefore essential to impose the condition $a_0 \neq 0$, in which case equation (1.7.5) can be satisfied only if σ is a root of the quadratic *indicial equation*

$$\mathcal{F}(\sigma) \equiv \sigma^2 + (p_0 - 1)\sigma + q_0 = 0. \tag{1.7.8}$$

Let the roots be $\sigma = \alpha, \beta$, and suppose that $\alpha - \beta = \nu > 0$ (if the roots are complex, let the real part of α be greater than the real part of β). Now

$$\mathcal{F}(\sigma) = (\sigma - \alpha)(\sigma - \beta),$$

so that

$$\mathcal{F}(n + \beta) \equiv n\{n - (\alpha - \beta)\} = n(n - \nu). \tag{1.7.9}$$

This shows that the coefficient of a_ν in the recurrence relation (1.7.7) will vanish when $\sigma = \beta$ and ν is a positive integer. In general, three different cases must be considered:

1. ν is not an integer,
2. $\nu = 0$,
3. ν is a positive integer.

Case 3 includes the important special case when $x = 0$ is an ordinary point, when the indicial equation reduces to

$$\mathcal{F}(\sigma) \equiv \sigma(\sigma - 1) = 0, \quad \text{i.e.} \quad \sigma = 0, \ 1. \tag{1.7.10}$$

Case 1 $\nu = \alpha - \beta \neq$ an integer:

The recurrence relation (1.7.7) can be solved for $\sigma = \alpha$ or β and the expansion (1.7.1) yields two independent solutions of the differential equation.

Example   Obtain series expansions about $x = 0$ of the general solution of the equation

$$(2x + x^3)\frac{d^2y}{dx^2} - \frac{dy}{dx} - 6xy = 0.$$

This has a regular singularity at $x = 0$.

It is usually more convenient to substitute the Frobenius expansion (1.7.1) directly into the equation without first determining the series expansions of $p(x)$ and $q(x)$. This yields

$$\sum_{n=0}^{\infty}(n+\sigma)(2n+2\sigma-3)a_n x^{n+\sigma-1} + \sum_{n=0}^{\infty}[(n+\sigma)(n+\sigma-1)-6]a_n x^{n+\sigma+1} = 0.$$

By replacing the summation variable n in the second sum by $n = n'-2$ the general term in each series has the same power of x, i.e.

$$\sum_{n=0}^{\infty}(n+\sigma)(2n+2\sigma-3)a_n x^{n+\sigma-1} + \sum_{n=2}^{\infty}(n+\sigma)(n+\sigma-5)a_{n-2} x^{n+\sigma-1} = 0.$$

The coefficients of successive powers of x are now equated to zero, the coefficient for $n = 0$ giving the indicial equation:

$n = 0:$ $\qquad \sigma(2\sigma-3)a_0 = 0$ $\qquad\qquad \therefore \quad \sigma = 0, \frac{3}{2}$

$n = 1:$ $\qquad (1+\sigma)(2\sigma-1)a_1 = 0$ $\qquad\qquad \therefore \quad a_1 = 0$

$n \geq 2:$ $\qquad\qquad a_n = -\dfrac{a_{n-2}(n+\sigma-5)}{(2n+2\sigma-3)}$ $\qquad$ general recurrence relation.

For $\sigma = 0$: $\quad u(x,\sigma) = a_0 y_1(x)$, where $\quad y_1 = 1 + 3x^2 + \frac{3}{5}x^4 - \frac{1}{15}x^6 + \frac{1}{64}x^8 + \cdots$.

For $\sigma = \frac{3}{2}$: $\quad u(x,\sigma) = a_0' y_2(x)$, where $\quad y_2 = x^{\frac{3}{2}}\left(1 + \frac{3}{8}x^2 + \frac{1.3}{8.16}x^4 + \frac{1.3.5}{8.16.24}x^6 + \cdots\right)$.

Range of convergence To determine the range of values of x for which the series expansion converges we can make use of the *ratio test*. The infinite series $S = \sum_{n=0}^{\infty} u_n$ converges provided

$$\left|\frac{u_n}{u_{n-1}}\right| < 1, \quad \text{as } n \to \infty.$$

In the previous example the ratio of successive terms is

$$\left|\frac{a_{2n}x^{2n}}{a_{2n-2}x^{2n-2}}\right| = \frac{(\sigma+2n-5)x^2}{(2\sigma+4n-3)} \to \frac{x^2}{2}, \quad \text{as } n \to \infty,$$

so that the series converge for

$$|x| < \sqrt{2}.$$

This is the same range as for the expansions

$$p(x) = \frac{-1}{x(2+x^2)} = \frac{-1}{2x}\left(1 - \frac{x^2}{2} + \frac{x^4}{2^2} - \frac{x^6}{2^3} + \cdots\right),$$

$$q(x) = \frac{-6}{(2+x^2)} = \frac{-3}{x}\left(1 - \frac{x^2}{2} + \frac{x^4}{2^2} - \frac{x^6}{2^3} + \cdots\right).$$

Case 2 $\nu = \alpha - \beta = 0$, equal roots:

When $\alpha = \beta$, $\mathcal{F}(n+\sigma) \equiv n^2$ and the recurrence relation (1.7.7) permits a_n $(n \geq 1)$ to be calculated in terms of a_0, but only one solution $u(x, \beta)$ of the equation is determined by the resulting expansion. However, when the coefficients are assumed to be determined in terms of a_0 by (1.7.7) for *arbitrary* values of σ, equation (1.7.4) becomes

$$\mathcal{L}u(x, \sigma) = \mathcal{F}(\sigma)a_0 \equiv (\sigma - \beta)^2 a_0. \qquad (1.7.11)$$

This not only confirms that $u(x, \sigma)$ is a solution of the differential equation for arbitrary values of a_0 when $\sigma = \beta$, but also that

$$\left(\frac{\partial u}{\partial \sigma}(x, \sigma)\right)_{\sigma=\beta} \quad \text{is a solution, because} \quad \frac{\partial}{\partial \sigma}\mathcal{L}u \equiv \mathcal{L}\left(\frac{\partial u}{\partial \sigma}\right).$$

Thus,

$$\left(\frac{\partial u}{\partial \sigma}(x, \sigma)\right)_{\sigma=\beta} = u(x, \beta)\ln x + \sum_{n=1}^{\infty}\left(\frac{\partial a_n}{\partial \sigma}(\sigma)\right)_{\sigma=\beta} x^{n+\beta} \qquad (1.7.12)$$

is a second independent solution of the differential equation.

Example   Obtain series expansions about $x = 0$ of the general solution of the equation

$$(x - x^2)\frac{d^2y}{dx^2} + (1 - 5x)\frac{dy}{dx} - 4y = 0.$$

This has a regular singularity at $x = 0$.

After substituting the Frobenius expansion (1.7.1) into the equation we find

$$\sum_{n=0}^{\infty}(n + \sigma)^2 a_n x^{n+\sigma-1} - \sum_{n=1}^{\infty}(n + \sigma + 1)^2 a_{n-1}x^{n+\sigma-1} = 0.$$

Hence,

$$n = 0: \qquad \sigma^2 a_0 = 0 \qquad\qquad\qquad \therefore \qquad \sigma = 0,\ 0$$
$$n \geq 1: \qquad a_n = a_{n-1}(n + \sigma + 1)^2/(n + \sigma)^2 \qquad \text{general recurrence relation.}$$

Thus,

$$u(x, \sigma) = a_0 x^\sigma \left[1 + \left(\frac{\sigma + 2}{\sigma + 1} \right)^2 x + \left(\frac{\sigma + 3}{\sigma + 1} \right)^2 x^2 + \left(\frac{\sigma + 4}{\sigma + 1} \right)^2 x^3 + \cdots \right]$$

$$= a_0 x^\sigma \sum_{n=0}^{\infty} \left(1 + \frac{n}{\sigma + 1} \right)^2 x^n,$$

and

$$\frac{\partial u}{\partial \sigma}(x, \sigma) = u(x, \sigma) \ln x + \sum_{n=1}^{\infty} \frac{-2n(n + \sigma + 1)}{(\sigma + 1)^3} x^{n+\sigma}.$$

Setting $\sigma = 0$ we find the two independent solutions

$$y_1 = 1 + 2^2 x + 3^2 x^2 + 4^2 x^3 + \cdots + (n + 1)^2 x^n + \cdots,$$

$$y_2 = y_1(x) \ln x - 2 \left[1.2x + 2.3x^2 + 3.4x^3 + \cdots + n(n + 1)x^n + \cdots \right].$$

The expansions converge for $|x| < 1$.

Case 3 $\nu = \alpha - \beta =$ an integer > 0:

There are two subcases: (i) $a_\nu = \infty$ for $\sigma = \beta$, which can happen because the coefficient $\mathcal{F}(n + \beta) = n(n - \nu)$ is zero at $n = \nu$; (ii) a_ν becomes indeterminate for $\sigma = \beta$, because the right-hand side of the recurrence formula (1.7.7) is also zero when $n = \nu$.

For (i): Set

$$a_0 = \bar{a}_0 (\sigma - \beta), \quad \text{where } \bar{a}_0 \neq 0, \tag{1.7.13}$$

then

$$u(x, \sigma) = x^\sigma \left[\bar{a}_0 (\sigma - \beta) + \sum_{n=1}^{\infty} a_n x^n \right].$$

When $\sigma = \beta$ the recurrence relation (1.7.7) now shows that $a_n = 0$ for $1 < n < \nu$. In place of (1.7.11) we find

$$\mathcal{L}u(x, \sigma) = \mathcal{F}(\sigma)(\sigma - \beta) \bar{a}_0 \equiv (\sigma - \beta)^2 (\sigma - \alpha) \bar{a}_0.$$

Two independent solutions are therefore:

$$y_1(x) = u(x, \beta) = x^\beta \sum_{n=\nu}^{\infty} a_n x^n \equiv x^\alpha \sum_{n=0}^{\infty} a_{n+\nu} x^n,$$

$$y_2(x) = \left(\frac{\partial u}{\partial \sigma}(x, \sigma)\right)_{\sigma=\beta} = u(x, \beta) \ln x + x^\beta \left[\bar{a}_0 + \sum_{n=1}^{\infty} \frac{\partial a_n}{\partial \sigma}(\sigma) x^n\right]_{\sigma=\beta}.$$

$$(1.7.14)$$

The solution $y_3(x) = u(x, \alpha)$ obtained by using the larger root of the indicial equation may be shown to be a multiple of $y_1(x)$ (see the examples).

For (ii): When $\sigma = \beta$ in the indicial equation (1.7.7) the coefficient of a_n on the left vanishes when $n = \nu$ and the right-hand side is also zero. In this case a_ν is *indeterminate*, and may be taken to be an arbitrary constant. The recurrence relation for $\sigma = \beta$ then determines *two* independent series solutions, the first starting with $a_0 x^\beta$ and the second with $a_\nu x^\alpha$. The latter is a multiple of that obtained by setting $\sigma = \alpha$ in the recurrence relation.

Example   Obtain series expansions about $x = 0$ for Bessel's equation of order 1:

$$x^2 \frac{d^2y}{dx^2} + x\frac{dy}{dx} + (x^2 - 1)y = 0.$$

This has a regular singularity at $x = 0$.
   Substitute the Frobenius expansion (1.7.1) into the equation to obtain

$$\sum_{n=0}^{\infty}(n + \sigma - 1)(n + \sigma + 1)a_n x^{n+\sigma} + \sum_{n=2}^{\infty} a_{n-2} x^{n+\sigma} = 0.$$

Then,

$$\begin{array}{llll} n = 0: & (\sigma + 1)(\sigma - 1)a_0 = 0 & \therefore & \sigma = 1, -1 \\ n = 1: & \sigma(\sigma + 2)a_1 = 0 & \therefore & a_1 = 0 \end{array}$$

$$n \geq 2: \qquad a_n = -\frac{a_{n-2}}{[(n + \sigma - 1)(n + \sigma + 1)]}$$

general recurrence relation.

Hence, $\nu = 2$ and for general σ

$$u(x, \sigma) = a_0 x^\sigma \left[1 - \frac{x^2}{(\sigma + 1)(\sigma + 3)} + \frac{x^4}{(\sigma + 1)(\sigma + 3)^2(\sigma + 5)}\right.$$

$$\left. - \frac{x^6}{(\sigma + 1)(\sigma + 3)^2(\sigma + 5)^2(\sigma + 7)} + \cdots \right].$$

The coefficient $a_2 = \infty$ when $\sigma = \beta \equiv -1$. Therefore, set $a_0 = \bar{a}_0(\sigma+1)$, whereupon

$$u(x,\sigma)$$
$$= \bar{a}_0 x^\sigma \left[(\sigma+1) - \frac{x^2}{(\sigma+3)} + \frac{x^4}{(\sigma+3)^2(\sigma+5)} - \frac{x^6}{(\sigma+3)^2(\sigma+5)^2(\sigma+7)} + \cdots \right].$$

The root $\sigma = \beta = -1$ therefore supplies the two independent solutions (taking $\bar{u}_0 = 1$)

$$y_1 = -\frac{x}{2}\left[1 - \frac{x^2}{2.2^2} + \frac{x^4}{3.2^2 \cdot 4^2} + \cdots \right],$$

$$y_2 = \left(\frac{\partial u}{\partial \sigma} \right)_{\sigma=-1}$$

$$= y_1(x)\ln x + \frac{1}{x}\left[1 + \frac{x^2}{2^2} + \left(\frac{2}{2} + \frac{1}{4} \right)\frac{x^4}{2^2.4} + \left(\frac{2}{2} + \frac{2}{4} + \frac{1}{6} \right)\frac{x^6}{2^2.4^2.6} + \cdots \right].$$

The expansions converge for $|x| < \infty$.

When $\sigma = \alpha = +1$ we find

$$y_3 = 2x\left[1 - \frac{x^2}{2.2^2} + \frac{x^4}{3.2^2.4^2} + \cdots \right] = -4y_1(x).$$

Example   Obtain series expansions about $x = 0$ for the equation:

$$(1 - x^2)\frac{d^2y}{dx^2} + 2x\frac{dy}{dx} + y = 0.$$

$x = 0$ is an *ordinary point* of the equation.

Substitute the Frobenius expansion (1.7.1) into the equation to obtain

$$\sum_{n=0}^{\infty}(n+\sigma)(n+\sigma-1)a_n x^{n+\sigma-2} - \sum_{n=2}^{\infty}a_{n-2}[(n+\sigma-2)(n+\sigma-5)-1]x^{n+\sigma-2} = 0.$$

Hence,

$$n = 0: \qquad \sigma(\sigma-1)a_0 = 0 \qquad \therefore \quad \sigma = 0,\ 1$$
$$n = 1: \qquad \sigma(\sigma+1)a_1 = 0 \qquad \therefore \quad a_1 = \text{arbitrary for } \sigma = 0$$
$$n \geq 2: \ (n+\sigma)(n+\sigma-1)a_n = -\frac{a_{n-2}}{[(n+\sigma)(n+\sigma-1)]}$$
$$\text{general recurrence relation.}$$

For $\sigma = \beta = 0$ we obtain the two independent solutions

$$y_1 = 1 - \frac{x^2}{2} + \frac{x^4}{8} + \frac{x^6}{80} + \cdots,$$

$$y_2 = x - \frac{x^3}{2} + \frac{x^5}{40} + \frac{3x^7}{560} + \cdots.$$

The expansions converge for $|x| < 1$.

When $\sigma = \alpha = +1$ we find $a_1 = 0$ and

$$y_3 = x\left[1 - \frac{x^2}{2} + \frac{x^4}{40} + \frac{3x^6}{560} + \cdots\right] = y_2(x).$$

Summary of the Frobenius procedure for $y'' + p(x)y' + q(x)y = 0$, where

$$p(x) = \frac{p_0}{x} + p_1 + p_2 x + p_3 x^2 + \cdots, \qquad q(x) = \frac{q_0}{x^2} + \frac{q_1}{x} + q_2 + q_3 x + q_4 x^2 + \cdots.$$

Case 1 $\alpha - \beta = \nu = $ non-integer:

$$y_1(x) = x^\beta \sum_{n=0}^\infty a_n x^n, \quad y_2(x) = x^\alpha \sum_{n=0}^\infty a'_n x^n, \quad a_0 = a'_0 = 1.$$

Case 2 $\alpha = \beta$:

$$u(x, \sigma) = x^\sigma \sum_{n=0}^\infty a_n(\sigma)x^n, \quad a_0 = 1,$$

$$y_1(x) = u(x, \beta),$$

$$y_2(x) = \left(\frac{\partial u}{\partial \sigma}\right)_{\sigma=\beta} = y_1(x)\ln x + \sum_{n=1}^\infty \left(\frac{\partial a_n}{\partial \sigma}\right)_{\sigma=\beta} x^{n+\beta}.$$

Case 3(i) $\alpha - \beta = \nu = $ integer > 0, $a_\nu = \infty$:

$$u(x, \sigma) = x^\sigma \left[(\sigma - \beta) + \sum_{n=1}^\infty a_n(\sigma)x^n\right], \quad a_n(\beta) = 0 \text{ for } n < \nu,$$

$$y_1(x) = u(x, \beta) = \sum_{n=0}^\infty a_{n+\nu}x^{n+\alpha},$$

$$y_2(x) = \left(\frac{\partial u}{\partial \sigma}\right)_{\sigma=\beta} = y_1(x)\ln x + x^\beta\left[1 + \sum_{n=1}^\infty \left(\frac{\partial a_n}{\partial \sigma}\right)_{\sigma=\beta} x^n\right].$$

Case 3(ii) $\alpha - \beta = \nu = $ integer > 0, a_ν indeterminate:

The expansion for $\sigma = \beta$ contains the two arbitrary constants a_0 and a_ν and gives two independent solutions

$$y_1(x) = x^\beta \sum_{n=0}^\infty a_n x^n, \quad y_2(x) = x^\alpha \sum_{n=0}^\infty a'_n x^n, \quad a_0 = a'_0 = 1.$$

This includes the case in which $x = 0$ is an *ordinary point*, for which $\alpha = 1,\ \beta = 0,\ \nu = 1$.

Problems 1G

Find the general solution near $x = 0$ by the method of Frobenius; state the range of convergence

1. $y'' + y = 0$. $[\sigma = 0,\ 1,\ y_1 = \cos x,\ y_2 = \sin x,\ \text{all } x]$
2. $4xy'' + 2y' + y = 0$. $[\sigma = 0,\ \frac{1}{2},\ y_1 = 1 - \frac{x}{2!} + \frac{x^2}{4!} - \cdots = \cos\sqrt{x}, y_2 = x^{\frac{1}{2}}\left[1 - \frac{x}{3!} + \frac{x^2}{5!} - \cdots\right] = \sin\sqrt{x},\ \text{all } x]$
3. $xy'' + y' + xy = 0$ (Bessel's equation of order 0). $[\sigma = 0,\ 0,\ y_1 = 1 - \frac{x^2}{2^2} + \frac{x^4}{2^2.4^2} - \frac{x^6}{2^2.4^2.6^2} \cdots = J_0(x),\ y_2 = y_1(x)\ln x + \frac{x^2}{2^2} - \frac{x^4}{2^2.4^2}\left(1 + \frac{1}{2}\right) + \frac{x^6}{2^2.4^2.6^2}\left(1 + \frac{1}{2} + \frac{1}{3}\right) - \cdots,\ \text{all } x]$
4. $x(1-x)y'' - 3xy' - y = 0$. $[\sigma = 0,\ 1,\ y_1 = x + 2x^2 + 3x^3 + \cdots = x/(1-x)^2,\ y_2 = y_1(x)\ln x + 1 + x + x^2 + x^3 + \cdots = y_1(x)\ln x + 1/(1-x),\ |x| < 1]$
5. $x^2y'' + xy' + (x^2 - \frac{1}{4})y = 0$. $[\sigma = \frac{1}{2},\ -\frac{1}{2},\ y_1 = x^{\frac{1}{2}}(1 - \frac{x^2}{3!}x + \frac{x^4}{5!} + \cdots) = x^{-\frac{1}{2}}\sin x,\ y_2 = x^{-\frac{1}{2}}\cos x,\ \text{all } x]$
6. $4x(1-x)y'' + 2(1-2x)y' + y = 0$. $[\sigma = 0,\ \frac{1}{2},\ y_1 = x^{\frac{1}{2}},\ y_2 = (1-x)^{\frac{1}{2}},\ |x| < 1]$
7. $(2x + 4x^3)y'' - y' - 24xy = 0$. $[\sigma = 0,\ \frac{3}{2},\ y_1 = 1 + 12x^2 + \frac{48}{5}x^4 - \frac{64}{15}x^6 + \cdots,\ y_2 = x^{\frac{3}{2}}(1 + \frac{3}{2}x^2 - \frac{1.3}{2.4}x^4 + \frac{1.3.5}{2.4.6}x^6 + \cdots),\ |x| < \frac{1}{2}]$
8. $x^2(1+x)y'' - x(1+2x)y' + y = 0$. $[\sigma = 1, 2, y_1 = x, y_2 = x^2 + x\ln x,\ \text{all } x]$
9. $2(2-x)x^2y'' - (4-x)xy' + (3-x)y = 0$. $[\sigma = \frac{1}{2}, \frac{1}{2}, y_1 = x^{\frac{1}{2}}, y_2 = x^{\frac{1}{2}}(1 - \frac{1}{2}x)^{\frac{1}{2}}, |x| < 2]$
10. $xy'' + (1 + 4x^2)y' + 4x(1+x^2)y = 0$. $[\sigma = 0, 0, y_1 = e^{-x^2}, y_2 = e^{-x^2}\ln x,\ \text{all } x]$

1.8   Bessel Functions of Integer Order

Bessel's equation of order n

$$x^2y'' + xy' + (x^2 - n^2)y = 0, \tag{1.8.1}$$

has a regular singularity at $x = 0$. The coefficients of the Frobenius solution

$$u(x, \sigma) = \sum_{k=0}^{\infty} a_k x^{k+\sigma},$$

satisfy

$$\sum_{k=0}^{\infty} a_k\left[(k+\sigma)^2 - n^2\right]x^{k+\sigma} + \sum_{k=2}^{\infty} a_{k-2}x^{k+\sigma} = 0.$$

The roots $\alpha = n$, $\beta = -n$ of the indicial equation $\sigma^2 = n^2$ differ by an integer or zero ($\nu = \alpha - \beta = 2n$) when n is an integer. For $k = 1$

$$a_1 \left[(1 + \sigma)^2 - n^2 \right] = 0, \quad \therefore \quad a_1 = 0,$$

and for $k \geq 2$

$$a_k = \frac{-a_{k-2}}{(k + \sigma - n)(k + \sigma + n)},$$

which shows that $a_\nu \equiv a_{2n} = \infty$ when $\sigma = -n$, corresponding to Case 3(i) of the Frobenius method.

The solution $y_1(x)$ that is finite at $x = 0$ starts with the term of order x^n. When the coefficient of this term is $1/2^n n!$ the series defines the Bessel function *of the first kind*

$$J_n(x) = \left(\frac{x}{2} \right)^n \sum_{k=0}^{\infty} \frac{\left(-\frac{1}{4} x^2 \right)^k}{k! (k + n)!}, \quad n = 0, 1, 2, \ldots . \tag{1.8.2}$$

The series converges for all real (or complex) values of x. If n is replaced by $-n$ in this expansion, the first n terms of the series are zero, because $(k - n)! = \infty$ for $k < n$ (see §5.1). By introducing a new summation parameter $k' = k - n$, we find

$$J_{-n}(x) = (-1)^n J_n(x).$$

If (1.8.2) is multiplied by x^n and differentiated we obtain

$$\frac{d}{dx} \left(x^n J_n(x) \right) = x^n J_{n-1}(x). \tag{1.8.3}$$

The second solution given by the Frobenius method in Case 3(i) defines the Bessel function of the second kind

$$Y_n(x) = \frac{2}{\pi} J_n(x) \left(\ln \frac{x}{2} + \gamma \right) - \frac{\left(\frac{1}{2} x \right)^{-n}}{\pi} \sum_{k=0}^{n-1} \frac{(n - k - 1)!}{k!} \left(\frac{1}{4} x^2 \right)^k$$

$$- \frac{\left(\frac{1}{2} x \right)^n}{\pi} \sum_{k=0}^{\infty} \left\{ \bar{\psi}(k) + \bar{\psi}(k+1) \right\} \frac{\left(-\frac{1}{4} x^2 \right)^k}{k! (n + k)!}, \tag{1.8.4}$$

where $\bar{\psi}(0) = 0, \bar{\psi}(k) = 1 + \frac{1}{2} + \frac{1}{3} + \cdots + \frac{1}{k}, k \geq 1$ and

$$\gamma = \lim_{k \to \infty} \left[1 + \frac{1}{2} + \frac{1}{3} + \frac{1}{4} + \cdots + \frac{1}{k} - \ln k \right] \approx 0.57721$$

is *Euler's constant.*

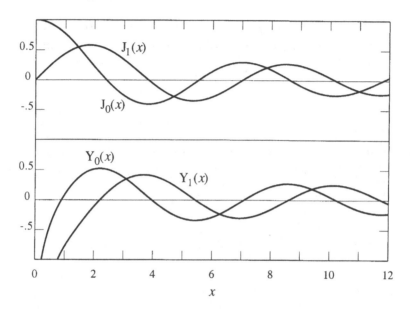

As $x \to \infty$ Bessel's equation approximates to $y'' + y = 0$, and the Bessel functions are found to resemble trigonometric functions of slowly varying amplitudes:

$$J_n(x) \approx \sqrt{\frac{2}{\pi x}} \cos \left(x - \frac{(2n + 1)\pi}{4} \right),$$

$$Y_n(x) \approx \sqrt{\frac{2}{\pi x}} \sin \left(x - \frac{(2n + 1)\pi}{4} \right), \quad x \to \infty.$$

1.9 The Sturm–Liouville Equation

The Sturm–Liouville equation

$$\frac{d}{dx} \left(p(x) \frac{dy}{dx} \right) + \left(q(x) + \lambda r(x) \right) y = 0, \quad a \leq x \leq b, \qquad (1.9.1)$$

arises when the method of *separation of variables* is used to solve linear, second-order partial differential equations (§4.3). $p(x)$, $q(x)$ and $r(x)$ are given functions and λ is a parameter. The solution $y(x)$ typically satisfies two-point boundary conditions of the general form

$$a_1 y(a) + a_2 y'(a) = 0, \quad b_1 y(b) + b_2 y'(b) = 0, \quad (1.9.2)$$

where a_1, a_2, b_1, b_2 are real constants. These conditions can determine $y(x)$ to within an arbitrary multiplicative constant, and can normally be satisfied only for certain values

$$\lambda = \lambda_n, \quad n = 1, 2, 3, \ldots,$$

called *eigenvalues*.

Let $y_n(x)$ and $y_m(x)$ be solutions corresponding to different eigenvalues λ_n and λ_m. These are called *eigenfunctions*, and they satisfy the following *orthogonality* relation

$$\int_a^b r(x) y_n(x) y_m(x) dx = 0, \quad \lambda_n \neq \lambda_m. \quad (1.9.3)$$

To prove this, write

$$\Big(p(x) y_n'(x) \Big)' + \Big(q(x) + \lambda_n r(x) \Big) y_n = 0,$$

$$\Big(p(x) y_m'(x) \Big)' + \Big(q(x) + \lambda_m r(x) \Big) y_m = 0.$$

Multiply the first equation by $y_m(x)$, the second by $y_n(x)$, subtract, and integrate over $a < x < b$. The result can be rearranged in the form

$$(\lambda_m - \lambda_n) \int_a^b r(x) y_n(x) y_m(x) dx$$

$$= \int_a^b \left\{ y_m \frac{d}{dx} \left(p \frac{dy_n}{dx} \right) - y_n \frac{d}{dx} \left(p \frac{dy_m}{dx} \right) \right\} dx$$

$$= \int_a^b \frac{d}{dx} \left(p y_m \frac{dy_n}{dx} - p y_n \frac{dy_m}{dx} \right) dx = \Big[p \Big(y_m y_n' - y_n y_m' \Big) \Big]_a^b,$$

which vanishes because of the boundary conditions (1.9.2). It also vanishes if the functions y_n, y'_n are merely *bounded* at $x = a$, b provided that $p(a) = p(b) = 0$, or if p vanishes at either $x = a$ or $x = b$ and one of the conditions (1.9.2) is satisfied at the other end (respectively at $x = b$ or $x = a$).

Example   Waves on a stretched string Consider small amplitude vibrations of a stretched string fixed at its ends $x = 0$ and $x = 1$. The lateral displacement $\zeta(x, t)$ of the string is a function of both x and the time t, and is governed by the *wave equation*

$$\frac{\partial^2 \zeta}{\partial x^2} - \frac{\partial^2 \zeta}{\partial t^2} = 0, \quad 0 < x < 1; \quad \text{where } \zeta = 0 \text{ at } x = 0, 1.$$

It is assumed in this equation that the length and time scales have been adjusted to make the 'wave speed' equal to 1.

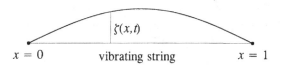

$$x = 0 \qquad \text{vibrating string} \qquad x = 1$$

A plucked string can vibrate at certain discrete frequencies ω_n. They can be determined by looking for solutions of the form

$$\zeta = y(x) \cos \omega t.$$

Substituting this into the wave equation and dividing through by $\cos \omega t$, we find that $y(x)$ must satisfy the Sturm–Liouville problem (1.9.1) (with $p(x) = 1, q(x) = 0$, $r(x) = 1$)

$$y'' + \omega^2 y = 0, \quad y(0) = 0, \quad y(1) = 0,$$

where ω^2 is equivalent to the eigenvalue parameter λ.

The general solution $y = A \cos \omega x + B \sin \omega x$ involves *three* unknown constants A, B and ω. The *two* boundary conditions $y(0) = y(1) = 0$ correspond to the Sturm–Liouville conditions (1.9.2), and give

$$A = 0, \quad B \sin \omega = 0.$$

But, if the string vibrates we cannot have $B = 0$, so that the admissible frequencies ω are solutions of the eigenvalue equation

$$\sin \omega = 0, \quad \text{i.e.} \quad \omega = \pi, 2\pi, 3\pi, \dots.$$

Stretched string eigenfunctions (determined only up to a multiplicative constant) are therefore

$$y_n(x) = \sin(n\pi x), \quad \text{and} \quad \lambda_n = \omega_n^2 = n^2\pi^2, \quad n = 1, 2, \ldots.$$

The most general motion of the string ($\propto \cos\omega t$) is described by a linear combination of terms of the type $\sin(n\pi x)\cos(\omega_n t) \equiv \sin(n\pi x)\cos(n\pi t)$, i.e.

$$\zeta(x,t) = \sum_{n=1}^{\infty} B_n \sin(n\pi x)\cos(n\pi t), \tag{1.9.4}$$

where the coefficients B_n are constants.

The orthogonality condition (1.9.3) can be used to determine the values of the B_n if the shape of the string is known at some value of the time t. In particular, (1.9.4) provides the solution to the following initial value problem:

$$\zeta(x,0) = f(x), \quad \left(\frac{\partial\zeta}{\partial t}(x,t)\right)_{t=0} = 0, \quad \text{for } 0 < x < 1.$$

This corresponds to a string which is released from rest when initially displaced into the shape of the curve $y = f(x)$.

The time derivative of (1.9.4) is zero at $t = 0$, so that the series automatically satisfies the second of our initial conditions. For the first, we set $t = 0$ in (1.9.4) to obtain

$$\sum_{n=1}^{\infty} B_n \sin(n\pi x) \equiv \sum_{n=1}^{\infty} B_n y_n(x) = f(x), \quad 0 < x < 1.$$

The mth coefficient B_m is calculated by multiplying this equation by $r(x)y_m(x) \equiv \sin(m\pi x)$ and integrating over the length $(0,1)$ of the string. Because of the orthogonality condition (1.9.3) all members of the series integrate to zero except the one for which $n = m$.

$$\therefore \quad B_m \int_0^1 \sin^2(m\pi x)dx = \int_0^1 f(x)\sin(m\pi x)dx.$$

Now $\int_0^1 \sin^2(m\pi x)dx = \frac{1}{2}$, therefore the solution of the initial value problem can be written

$$\zeta(x,t) = \sum_{n=1}^{\infty} B_n \sin(n\pi x)\cos(n\pi t), \quad B_n = 2\int_0^1 f(x)\sin(n\pi x)dx, \quad t \geq 0. \tag{1.9.5}$$

The infinite series can be summed explicitly, but further discussion is postponed to §4.3.

Eigenfunction expansions The procedure described above for the string can be used to express any function $f(x)$ defined over the interval $a < x < b$ as an *eigenfunction expansion* in terms of the eigenfunctions and eigenvalues of a Sturm–Liouville problem defined over the same interval. Set

$$f(x) = \sum_n A_n y_n(x), \quad a < x < b,$$

where the summation is over all of the eigenfunctions. By multiplying both sides by $r(x)y_m(x)$ and integrating over (a, b) we find (from [1.9.3])

$$A_m \int_a^b r(x)y_m^2(x)dx = \int_a^b f(x)r(x)y_m(x)dx.$$

Hence,

$$f(x) = \sum_n A_n y_n(x), \quad a < x < b,$$

where

$$A_n = \frac{\int\limits_a^b f(x)r(x)y_n(x)dx}{\int\limits_a^b r(x)y_n^2(x)dx}. \tag{1.9.6}$$

When the eigenfunctions $y_n(x)$ are normalised to make $\int_a^b r(x)y_n^2(x)dx = 1$, the eigenfunctions are said to be *orthonormal*.

Problems 1H

Find the eigenvalues and eigenfunctions of:

1. $y'' + \lambda y = 0$, $y(0) = 0$, $y'(1) = 0$. $[\lambda_n = \left(\frac{(2n+1)\pi}{2}\right)^2$, $n = 0, 1, 2, \ldots, y_n = \sin(\sqrt{\lambda_n}x)]$

2. $y'' + \lambda y = 0$, $y'(0) = 0$, $y'(\pi) = 0$. $[\lambda_n = n^2$, $n = 0, 1, 2, \ldots, y_n = \cos(nx)]$

3. Transform the equation $y'' + 2y' + (1 - \lambda)y = 0$, $(y(0) = 0, y(1) = 0)$ into Sturm–Liouville form by multiplying by e^{2x}. Calculate the eigenvalues and eigenfunctions and show that $\int_0^1 e^{2x}y_n(x)y_m(x)dx = 0, m \neq n$.
$[\lambda_n = -n^2\pi^2, n = 1, 2, 3, \ldots, y_n = e^{-x}\sin(n\pi x)]$

4. Transform the equation $4y'' - 4y' + (1 + \lambda)y = 0$, $(y(-1) = 0, y(1) = 0)$ into Sturm–Liouville form. Calculate the eigenvalues and eigenfunctions and show that $\int_{-1}^1 e^{-x}y_n(x)y_m(x)dx = 0, m \neq n$.
$[\lambda_n = n^2\pi^2, y_n = e^{\frac{1}{2}x}\sin(\frac{1}{2}n\pi x), n = 2, 4, 6, \ldots, y_n = e^{\frac{1}{2}x}\cos(\frac{1}{2}n\pi x), n = 1, 3, 5, \ldots]$

5. Transform the equation $y'' + 2y' + (1 - \lambda)y = 0$, $(y'(0) = 0,\ y'(\pi) = 0)$ into Sturm–Liouville form. Calculate the eigenvalues and eigenfunctions and show that $\int_0^\pi e^{2x} y_n(x) y_m(x)\, dx = 0$, $m \neq n$.
   $[\lambda_n = -n^2,\ y_n = e^{-x}(n \cos nx + \sin nx),\ n = 1,\ 2,\ 3,\ldots,\ \lambda_0 = 1,\ y_0 = 1]$

6. Transform the equation $x^2 y'' + xy' + \lambda y = 0$, $(y(1) = 0,\ y(a) = 0)$ into Sturm–Liouville form. Calculate the eigenvalues and eigenfunctions and show that $\int_1^a y_n(x) y_m(x) \frac{dx}{x} = 0$, $m \neq n$.
   $[\lambda_n = (n\pi/\ln a)^2,\ y_n = \sin(n\pi \ln x / \ln a),\ n = 1,\ 2,\ 3,\ldots]$

1.10   Fourier Series

Consider the Sturm–Liouville problem with *periodic* boundary conditions

$$y'' + k^2 y = 0, \qquad \begin{cases} y(x + 2\pi) = y(x), \\ y'(x + 2\pi) = y'(x). \end{cases} \tag{1.10.1}$$

The general solution satisfying these conditions is $y = A \cos nx + B \sin nx$, for integer values of n. The eigenvalues are therefore $\lambda = k^2 = n^2$ where $n = 0, 1, 2, 3, \ldots$, and each value of n is associated with the eigenfunction pair $\{\cos nx,\ \sin nx\}$. The eigenfunctions satisfy the orthogonality relations over *any interval* of length 2π:

$$\int_a^{a+2\pi} \cos nx \sin mx\, dx = 0, \quad \text{for all } n,\ m,$$

$$\int_a^{a+2\pi} \cos nx \cos mx\, dx = 0, \quad \text{for all } n \neq m,$$

$$\int_a^{a+2\pi} \sin nx \sin mx\, dx = 0, \quad \text{for all } n \neq m,$$

$$\int_a^{a+2\pi} \sin^2 nx\, dx = \pi, \quad \text{for all } n \geq 1,$$

$$\int_a^{a+2\pi} \cos^2 nx\, dx = \begin{cases} \pi, & \text{for all } n \geq 1 \\ 2\pi, & \text{for } n = 0. \end{cases}$$

The series obtained when a function $f(x)$ of period 2π is expanded in terms of these eigenfunctions is called a *Fourier series*, and is usually written

$$f(x) = a_0 + \sum_{n=1}^{\infty}(a_n \cos nx + b_n \sin nx), \qquad (1.10.2)$$

where the orthogonality conditions supply

$$a_0 = \frac{1}{2\pi}\int_{-\pi}^{\pi} f(x)dx;$$

$$a_n = \frac{1}{\pi}\int_{-\pi}^{\pi} f(x)\cos nx\, dx, \qquad (1.10.3)$$

$$b_n = \frac{1}{\pi}\int_{-\pi}^{\pi} f(x)\sin nx\, dx, \quad n \geq 1.$$

Fourier series of arbitrary period The corresponding Fourier series expansion for a function of period 2ℓ is obtained by replacing x in the above formulae by $\pi x/\ell$, so that:

$$f(x) = a_0 + \sum_{n=1}^{\infty}\left(a_n \cos \frac{n\pi x}{\ell} + b_n \sin \frac{n\pi x}{\ell}\right), \qquad (1.10.4)$$

where

$$a_0 = \frac{1}{2\ell}\int_{-\ell}^{\ell} f(x)dx,$$

$$a_n = \frac{1}{\ell}\int_{-\ell}^{\ell} f(x)\cos \frac{n\pi x}{\ell}\, dx,\ n \geq 1; \qquad (1.10.5)$$

$$b_n = \frac{1}{\ell}\int_{-\ell}^{\ell} f(x)\sin \frac{n\pi x}{\ell}\, dx.$$

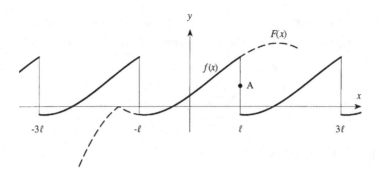

It is frequently required to represent a continuous, *non-periodic* function $F(x)$ by a Fourier series over a finite interval $-\ell < x < \ell$, say, of the x-axis. In this case the series defines a periodic function $f(x)$ which agrees with $F(x)$ for $-\ell < x < \ell$, but gives a periodic extension of $F(x)$ outside this interval, as illustrated in the figure. This periodic extension is often discontinuous at $x = -\ell + 2n\ell, n = 0, \pm 1, \pm 2, \ldots$. At such a point the Fourier series converges to the *mean* of the limiting values of $f(x)$ as x approaches the discontinuity from either side. Thus, at the point labelled A in the figure (at $x = \ell$), the Fourier series converges to

$$\lim_{\epsilon \to 0} \frac{1}{2} [f(\ell - \epsilon) + f(\ell + \epsilon)] = \frac{1}{2} [F(-\ell) + F(\ell)].$$

Example   Derive the Fourier series expansion of $f(x) = x$ over the interval $-\ell < x < \ell$.

Because $f(x) = x$ is an odd function,

$$a_0 = \frac{1}{2\ell} \int_{-\ell}^{\ell} x\,dx = 0, \quad a_n = \frac{1}{\ell} \int_{-\ell}^{\ell} x \cos\left(\frac{n\pi x}{\ell}\right) dx = 0, \quad n > 0;$$

but, integrating by parts,

$$b_n = \frac{1}{\ell} \int_{-\ell}^{\ell} x \sin\left(\frac{n\pi x}{\ell}\right) dx = \frac{1}{\ell} \left(\left[\frac{x \cos(n\pi x/\ell)}{(-n\pi/\ell)} \right]_{-\ell}^{\ell} + \frac{\ell}{n\pi} \int_{-\ell}^{\ell} \cos\left(\frac{n\pi x}{\ell}\right) dx \right)$$

$$= \frac{-2\ell \cos(n\pi)}{n\pi} + 0 \equiv \frac{2\ell}{n\pi} (-1)^{n+1}.$$

$$\therefore \quad x = \frac{2\ell}{\pi} \sum_{n=1}^{\infty} \frac{(-1)^{n+1}}{n} \sin\left(\frac{n\pi x}{\ell}\right)$$

$$= \frac{2\ell}{\pi} \left\{ \sin\left(\frac{\pi x}{\ell}\right) - \frac{1}{2}\sin\left(\frac{2\pi x}{\ell}\right) + \frac{1}{3}\sin\left(\frac{3\pi x}{\ell}\right) - \cdots \right\}.$$

Half-range series   An arbitrary function $F(x)$ defined over the interval $0 < x < \ell$ can be expanded in a Fourier series of the type (1.10.4) by replacing ℓ in (1.10.4) and (1.10.5) by $\ell/2$ and taking the limits of integration to be $x = 0$ and $x = \ell$ in the formulae (1.10.5) for the Fourier coefficients a_n and b_n. The series then defines a periodic function $f(x)$ of period ℓ. However, expansions of $F(x)$ in $0 < x < \ell$ involving, respectively, only cosines or only sines can also be derived by *defining* $F(x) = \pm F(-x)$ in $-\ell < x < 0$.

In the first of these alternatives a periodic function $f(x)$ is defined over one period $(-\ell < x < \ell)$ as the *even* function

$$f(x) = \begin{cases} F(-x), & -\ell < x < 0, \\ F(x), & 0 < x < \ell, \end{cases}$$

and we obtain the *cosine series*

$$f(x) = a_0 + \sum_{n=1}^{\infty} a_n \cos\frac{n\pi x}{\ell},$$

where

$$a_0 = \frac{1}{\ell} \int_0^{\ell} f(x)dx, \quad a_n = \frac{2}{\ell} \int_0^{\ell} f(x) \cos\frac{n\pi x}{\ell} \, dx, \ n \geq 1. \qquad (1.10.6)$$

For the second alternative $f(x)$ is defined over $(-\ell < x < \ell)$ as the *odd* function

$$f(x) = \begin{cases} -F(-x), & -\ell < x < 0, \\ F(x), & 0 < x < \ell, \end{cases}$$

leading to the *sine series*

$$f(x) = \sum_{n=1}^{\infty} b_n \sin\frac{n\pi x}{\ell}, \quad \text{where } b_n = \frac{2}{\ell} \int_0^{\ell} f(x) \sin\frac{n\pi x}{\ell} \, dx. \qquad (1.10.7)$$

Example Let $F(x) = 1$ for $0 < x < \ell$.

(i) When this is extended to all values of x as a periodic function of period ℓ the coefficients of the Fourier series (1.10.4) (in which ℓ is replaced by $\ell/2$ and the limits of integration by $0 < x < \ell$) are

$$a_0 = \frac{1}{\ell} \int_0^\ell dx = 1,$$

$$a_n = \frac{2}{\ell} \int_0^\ell \cos\left(\frac{2n\pi x}{\ell}\right) dx = 0,$$

$$b_n = \frac{2}{\ell} \int_0^\ell \sin\left(\frac{2n\pi x}{\ell}\right) dx = 0 \ \ n > 0;$$

and the series reduces to one term:

$$f(x) = 1;$$

(ii) When extended to all values of x as an *even* periodic function of period 2ℓ the coefficients of the Fourier cosine series (1.10.6) are

$$a_0 = \frac{1}{\ell} \int_0^\ell dx = 1, \quad a_n = \frac{2}{\ell} \int_0^\ell \cos\left(\frac{n\pi x}{\ell}\right) dx = 0, \quad n > 0;$$

and again the series becomes

$$f(x) = 1;$$

(iii) When extended to $-\ell < x < 0$ as an *odd* function of period 2ℓ the coefficients b_n in the Fourier sine series (1.10.7) are

$$b_n = \frac{2}{\ell} \int_0^\ell \sin\left(\frac{n\pi x}{\ell}\right) dx = \frac{2}{n\pi}\left(1 - \cos(n\pi)\right) = \begin{cases} 0 & n \text{ even} \\ \frac{4}{n\pi} & n \text{ odd.} \end{cases}$$

Hence, setting $n = 2N + 1$, we obtain

$$f(x) = \frac{4}{\pi} \sum_{N=0}^{\infty} \frac{\sin[(2N+1)\pi x/\ell]}{2N+1}$$

$$= \frac{4}{\pi}\left[\frac{\sin(\pi x/\ell)}{1} + \frac{\sin(3\pi x/\ell)}{3} + \frac{\sin(5\pi x/\ell)}{5} + \cdots\right].$$

All of these representations define the same function in $0 < x < \ell$.

Problems 1I

Verify the following Fourier series expansions for:

1. $f(x) = \begin{cases} 1, & |x| < \frac{\pi}{2}, \\ 0, & |x| > \frac{\pi}{2}, \end{cases} = \frac{1}{2} + \frac{2}{\pi} \left[\frac{\cos x}{1} - \frac{\cos 3x}{3} + \frac{\cos 5x}{5} + \cdots \right], \quad -\pi < x < \pi.$

2. $f(x) = x = 2 \left[\frac{\sin x}{1} - \frac{\sin 2x}{2} + \frac{\sin 3x}{3} - \frac{\sin 4x}{4} + \cdots \right], \quad |x| < \pi.$

3. $f(x) = x^2 = \frac{\pi^2}{3} + 4 \sum_{n=1}^{\infty} \frac{(-1)^n \cos nx}{n^2}, \quad |x| < \pi.$

4. $f(x) = |x| = \frac{\pi}{2} - \frac{4}{\pi} \sum_{n=0}^{\infty} \frac{\cos(2n+1)x}{(2n+1)^2}, \quad |x| < \pi.$

5. $f(x) = \frac{\pi - x}{2} = \sum_{n=1}^{\infty} \frac{\sin nx}{n}, \quad 0 < x < 2\pi.$

6. $f(x) = \ln\left(2\cos\frac{x}{2}\right) = \sum_{n=1}^{\infty} (-1)^{n-1} \frac{\cos nx}{n}, \quad |x| < \pi.$

7. $f(x) = \frac{\pi}{4} = \sum_{n=1}^{\infty} \frac{\sin(2n-1)x}{(2n-1)}, \quad 0 < x < \pi.$

8. $f(x) = \frac{1}{2}\ln\cot\frac{x}{2} = \sum_{n=1}^{\infty} \frac{\cos(2n-1)x}{(2n-1)}, \quad 0 < x < \pi.$

9. $f(x) = \frac{1}{2}\ln\left(\frac{1}{2(1-\cos x)}\right) = \sum_{n=1}^{\infty} \frac{\cos nx}{n}, \quad 0 < x < 2\pi.$

10. $f(x) = \frac{1}{2}\ln\tan\left(\frac{\pi}{4} + \frac{x}{2}\right) = \sum_{n=1}^{\infty} (-1)^{n-1} \frac{\sin(2n-1)x}{(2n-1)}, \quad |x| < \frac{\pi}{2}.$

11. $f(x) = e^x = \frac{2\sinh\pi}{\pi} \left[\frac{1}{2} + \sum_{n=1}^{\infty} \frac{(-1)^n}{1+n^2} (\cos nx - n\sin nx) \right], \quad |x| < \pi.$

12. $f(x) = x = \frac{\pi}{2} - \frac{4}{\pi} \sum_{n=1}^{\infty} \frac{\cos(2n-1)x}{(2n-1)^2}, \quad 0 \le x < \pi.$

13. $f(x) = \sin x = \frac{2}{\pi} - \frac{4}{\pi} \sum_{n=1}^{\infty} \frac{\cos 2nx}{4n^2 - 1}, \quad 0 \le x < \pi.$

14. $f(x) = \dfrac{\pi^2 x}{6} - \dfrac{\pi x^2}{4} + \dfrac{x^3}{12} = \displaystyle\sum_{n=1}^{\infty} \dfrac{\sin nx}{n^3}, \quad 0 \le x < 2\pi.$

15. $f(x) = \dfrac{\pi^2}{6} - \dfrac{\pi x}{2} + \dfrac{x^2}{4} = \displaystyle\sum_{n=1}^{\infty} \dfrac{\cos nx}{n^2}, \quad 0 \le x < 2\pi.$

1.11   Generalised Functions and Green's Function

The *delta function* $\delta(x - y)$ is defined by the relations

$$\int_a^b \delta(x - y)\, dy = \begin{cases} 1, & \text{when } a < x < b, \\ 0, & \text{otherwise,} \end{cases} \tag{1.11.1}$$

$$\int_a^b F(y)\delta(x - y)\, dy = F(x), \quad \text{when } a < x < b. \tag{1.11.2}$$

No 'ordinary' function exhibits properties of this kind, but we can regard $\delta(x - y)$ as the limit as $\epsilon \to 0$ of the 'epsilon sequence'

$$\delta_\epsilon(x - y) = \frac{\epsilon}{\pi[(x - y)^2 + \epsilon^2]}, \quad \epsilon > 0. \tag{1.11.3}$$

When ϵ is very small this function has a large peak $\delta_\epsilon(x - y) = 1/\pi\epsilon$ at $x = y$, and tends to zero as $\epsilon \to 0$ when $x \ne y$.

For any positive value of ϵ

$$\int_a^b \delta_\epsilon(x - y)\, dy = \frac{1}{\pi}\left[\tan^{-1}\left(\frac{y - x}{\epsilon} \right) \right]_a^b$$

$$= \frac{1}{\pi}\left[\tan^{-1}\left(\frac{b - x}{\epsilon} \right) - \tan^{-1}\left(\frac{a - x}{\epsilon} \right) \right],$$

which establishes (1.11.1), because

$$\lim_{\epsilon \to 0} \tan^{-1}\left(\frac{X}{\epsilon} \right) = \pm\frac{\pi}{2} \quad \text{for } X \gtrless 0.$$

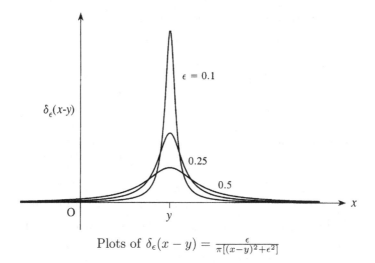

Plots of $\delta_\epsilon(x - y) = \frac{\epsilon}{\pi[(x-y)^2+\epsilon^2]}$

Similarly, because $\delta_\epsilon(x - y) \to 0$ as $\epsilon \to 0$ when $x \neq y$, we see that, for any 'smooth' function $F(x)$,

$$\lim_{\epsilon \to 0} \int_a^b F(y)\delta_\epsilon(x - y)dy$$

$$= F(x)\lim_{\epsilon \to 0} \int_a^b \delta_\epsilon(x - y)dy = F(x) \quad \text{when } a < x < b.$$

In applications we are not actually required to calculate the 'value' of $\lim_{\epsilon \to 0} \delta_\epsilon(x)$ at any particular value of x, but only its contribution to an integral, and the limit is taken after the integral is evaluated. A mathematical entity defined by this procedure is called a *generalised function*. The δ-function is the particular generalised function defined by the ϵ-sequence (1.11.3) with the properties (1.11.1), (1.11.2). Note, however, that the defining sequence is by no means unique: for example, the following ϵ-sequences also define the δ-function:

$$\delta_\epsilon(x - y) = \frac{1}{\epsilon\sqrt{\pi}}\exp\left[-\frac{(x-y)^2}{\epsilon^2}\right], \quad \frac{\sin\{(x-y)/\epsilon\}}{\pi(x-y)}, \quad \frac{\text{H}(\epsilon - |x-y|)}{2\epsilon},$$

where H(x) is the *Heaviside step function*

$$H(x) = \begin{cases} 0, & \text{for } x < 0, \\ 1, & \text{for } x > 0. \end{cases} \tag{1.11.4}$$

An alternative procedure is to define the generalised function $f(x)$ by reference to the value of the integral $\int f(x)F(x)dx$, where $F(x)$ is any member of a set of 'ordinary' functions (which may, however, be required to vary in a special way, depending on the application), and the integration is over a region relevant to the problem at hand. When $f(x)$ is defined by an ϵ-sequence $f_\epsilon(x)$, this 'value' is defined by

$$\int f(x)F(x)dx = \lim_{\epsilon \to 0} \int f_\epsilon(x)F(x)dx, \tag{1.11.5}$$

and $F(x)$ is said to be a *test* function.

Example $\delta(x-y)$ is defined over $-\infty < x < \infty$ by the ϵ-sequence

$$\delta_\epsilon(x-y) = \frac{\epsilon}{\pi[(x-y)^2 + \epsilon^2]},$$

because, for any bounded 'smooth' test function $F(x)$ we find by making the change of variable $y - x = \epsilon Y$

$$\int_{-\infty}^{\infty} F(y)\delta(x-y)dy = \lim_{\epsilon \to 0} \int_{-\infty}^{\infty} \frac{\epsilon F(y)dy}{\pi[(x-y)^2 + \epsilon^2]}$$

$$= \lim_{\epsilon \to 0} \int_{-\infty}^{\infty} \frac{F(x+\epsilon Y)dY}{\pi[Y^2 + 1]} = F(x)\int_{-\infty}^{\infty} \frac{dY}{\pi[Y^2 + 1]} = F(x).$$

Example Verify that $\delta(x-y)$ can be defined by the ϵ-sequence

$$\delta_\epsilon(x-y) = \frac{\sin\{(x-y)/\epsilon\}}{\pi(x-y)}.$$

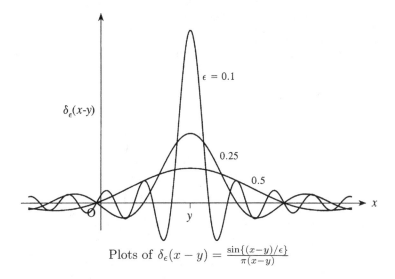

Plots of $\delta_\epsilon(x-y) = \dfrac{\sin\{(x-y)/\epsilon\}}{\pi(x-y)}$

Example The Heaviside function is the generalised function defined by the ϵ-sequence

$$H_\epsilon(x) = \int_{-\infty}^{x} \frac{\epsilon dy}{\pi(y^2 + \epsilon^2)} = \frac{1}{2} + \frac{1}{\pi}\tan^{-1}\left(\frac{x}{\epsilon}\right).$$

Example The Heaviside step function can be defined by

$$H(x) = \int_{-\infty}^{x} \delta(y)dy = \begin{cases} 0, & \text{for } x < 0, \\ 1, & \text{for } x > 0. \end{cases} \quad \text{so that} \quad \delta(x) = \frac{dH(x)}{dx}. \qquad (1.11.6)$$

Example The following argument can also be used to show that $\delta(x) = dH(x)/dx$, i.e. that when $a < 0 < b$, $\int_a^b F(x)\frac{dH}{dx}\,dx = F(0)$:

$$\int_a^b F(x)\frac{dH}{dx}\,dx = \int_a^b \left(\frac{d}{dx}(HF) - H(x)\frac{dF}{dx}\right)dx$$

$$= [H(x)F(x)]_a^b - \int_0^b \frac{dF}{dx}(x)dx = F(b) - [F(x)]_0^b = F(0).$$

Example $\delta(-x) = \delta(x)$, because

$$\delta(-x) = \lim_{\epsilon \to 0} \frac{\epsilon}{\pi[(-x)^2 + \epsilon^2]} = \lim_{\epsilon \to 0} \frac{\epsilon}{\pi[(x)^2 + \epsilon^2]} \equiv \delta(x).$$

Example The function $\text{sgn}(x) = 2H(x) - 1$ ($= -1$ for $x < 0$, $+1$ for $x > 0$). Then

$$\frac{d}{dx}|x| = \text{sgn}(x) \quad \text{and} \quad \frac{d^2}{dx^2}|x| \equiv \frac{d}{dx}\text{sgn}(x) = \frac{d}{dx}\Big[2H(x) - 1\Big] = 2\delta(x).$$

Example $\delta'(x - y)$ is defined by the ϵ-sequence

$$\delta'_\epsilon(x - y) = \frac{-2\epsilon(x - y)}{\pi[(x - y)^2 + \epsilon^2]^2}.$$

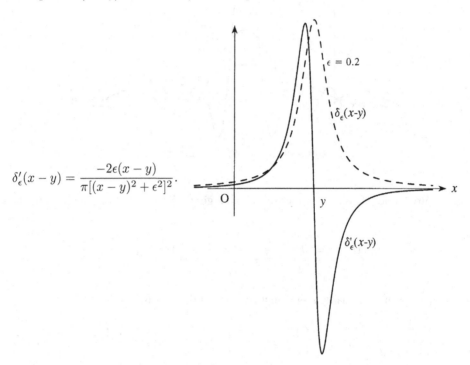

Example Evaluate $\int_{-\infty}^{\infty} F(x)\delta'(x)dx$, where $\delta'(x) = d\delta(x)/dx$.

$$\int\limits_{-\infty}^{\infty} F(x)\delta'(x)dx = \int\limits_{-\infty}^{\infty} \left[\frac{d}{dx}\Big(F(x)\delta(x)\Big) - F'(x)\delta(x)\right]dx$$

$$= \Big[F(x)\delta(x)\Big]_{-\infty}^{\infty} - F'(0) = -F'(0),$$

because $\delta(x) \equiv 0$ at $x = \pm\infty$.

Example $\delta(\alpha x) = \frac{1}{|\alpha|}\delta(x)$ for real α. Because for any test function $F(x)$ we have (setting $y = \alpha x$)

$$\int\limits_{-\infty}^{\infty} F(x)\delta(\alpha x)dx = \frac{1}{|\alpha|}\int\limits_{-\infty}^{\infty} F\left(\frac{y}{\alpha}\right)\delta(y)dy = \frac{1}{|\alpha|}F(0).$$

Example Evaluate $\int_{-\infty}^{\infty} F(x)\delta(x^2 - a^2)dx$ for real a.

For infinitesimal $\Delta > 0$

$$\int\limits_{-\infty}^{\infty} F(x)\delta(x^2 - a^2)dx$$

$$\equiv \int\limits_{-a-\Delta}^{-a+\Delta} F(x)\delta(x^2 - a^2)dx + \int\limits_{a-\Delta}^{a+\Delta} F(x)\delta(x^2 - a^2)dx$$

$$= \int\limits_{-a-\Delta}^{-a+\Delta} F(x)\delta(2a[x + a])dx + \int\limits_{a-\Delta}^{a+\Delta} F(x)\delta(2a[x - a])dx$$

$$= \frac{1}{2|a|} \int\limits_{-a-\Delta}^{-a+\Delta} F(x)\delta(x + a)dx + \frac{1}{2|a|} \int\limits_{a-\Delta}^{a+\Delta} F(x)\delta(x - a)dx$$

$$= \frac{1}{2|a|}\Big(F(-a) + F(a)\Big).$$

Example Evaluate $\int_0^5 x^2\delta\Big((x - 3)(x + 2)\Big)dx$.

$$\int\limits_0^5 x^2\delta\Big((x - 3)(x + 2)\Big)dx = \int\limits_0^5 x^2\delta\Big(5(x - 3)\Big)dx = \frac{1}{5}\int\limits_0^5 x^2\delta(x - 3)dx = \frac{9}{5}.$$

Example The generalised function equal to the *principal value* of $\frac{1}{x}$ is defined by the epsilon sequence

$$P\left(\frac{1}{x}\right) = \lim_{\epsilon \to 0} \frac{x}{x^2 + \epsilon^2}, \tag{1.11.7}$$

because, for $\epsilon > 0$,

$$\int\limits_{-\infty}^{\infty} P\left(\frac{1}{x}\right) f(x)dx = \lim_{\epsilon \to 0} \int\limits_{-\infty}^{\infty} \frac{xf(x)dx}{x^2 + \epsilon^2}$$

$$\equiv \lim_{\epsilon \to 0}\left(\int\limits_{-\infty}^{-\epsilon} \frac{f(x)dx}{x} + \int\limits_{\epsilon}^{\infty} \frac{f(x)dx}{x}\right) = \int\limits_{-\infty}^{\infty} \frac{f(x)dx}{x}.$$

Example A constant C is the generalised function defined by

$$C = \lim_{\epsilon \to +0} Ce^{-\epsilon|x|}.$$

This can be used to give meaning to otherwise formally divergent integrals. The following *Fourier* integral (in which $C = 1$) is particularly important:

$$\frac{1}{2\pi} \int_{-\infty}^{\infty} e^{ikx}\, dk = \lim_{\epsilon \to 0} \frac{1}{2\pi} \int_{-\infty}^{\infty} e^{-\epsilon|k|+ikx}\, dk$$

$$= \lim_{\epsilon \to 0} \frac{1}{2\pi} \left(\int_{-\infty}^{0} e^{(\epsilon+ix)k}\, dk + \int_{0}^{\infty} e^{-(\epsilon-ix)k}\, dk \right)$$

$$= \lim_{\epsilon \to 0} \frac{1}{2\pi} \left(\frac{1}{\epsilon + ix} + \frac{1}{\epsilon - ix} \right) = \lim_{\epsilon \to 0} \frac{\epsilon}{\pi(x^2 + \epsilon^2)}$$

$$\therefore \quad \frac{1}{2\pi} \int_{-\infty}^{\infty} e^{ikx}\, dk = \delta(x). \tag{1.11.8}$$

Green's function Consider the equation

$$\frac{d^2 G}{dt^2} + \omega^2 G = \delta(t), \quad -\infty < t < \infty. \tag{1.11.9}$$

The particular solution $G(t)$ that vanishes for $t < 0$ (the *causal* solution when t denotes time) is found by first recalling that the solution of the homogeneous equation is $A \cos \omega t + B \sin \omega t$, where A, B are arbitrary constants. Thus

$$G(t) = \mathrm{H}(t)\Big\{ A \cos \omega t + B \sin \omega t \Big\},$$

vanishes for $t < 0$ and satisfies the equation identically for $t > 0$. The constants A and B are found by substitution:

$$\frac{dG}{dt} = \delta(t)\{A \cos \omega t + B \sin \omega t\} + \mathrm{H}(t)\omega\Big\{ -A \sin \omega t + B \cos \omega t \Big\}$$

$$\equiv A\delta(t) + \omega \mathrm{H}(t)\Big\{ -A \sin \omega t + B \cos \omega t \Big\},$$

$$\therefore \quad \frac{d^2 G}{dt^2} = A\delta'(t) + \omega\delta(t)\Big\{ -A \sin \omega t + B \cos \omega t \Big\}$$

$$- \omega^2 \mathrm{H}(t)\Big\{ A \cos \omega t + B \sin \omega t \Big\}$$

$$\equiv A\delta'(t) + \omega B\delta(t) - \omega^2 G.$$

The insertion of this expression for d^2G/dt^2 into equation (1.11.9) yields the identity

$$A\delta'(t) + \omega B\delta(t) \equiv \delta(t)$$

$$\therefore \quad A = 0, \quad B = \frac{1}{\omega}$$

$$\text{i.e.} \quad G(t) = \frac{1}{\omega}H(t)\sin \omega t.$$

It is now obvious that the solution of

$$\frac{d^2G}{dt^2} + \omega^2 G = \delta(t - \tau), \quad \text{where} \ \ G = 0 \ \text{for} \ t < \tau, \qquad (1.11.10)$$

is

$$G(t - \tau) = \frac{1}{\omega}H(t - \tau)\sin \omega(t - \tau).$$

This is called the *causal Green's function*.

$G(t - \tau)$ can be used to obtain the causal solution of

$$\frac{d^2\varphi}{dt^2} + \omega^2\varphi = f(t),$$

where the 'force' $f(t)$ is assumed to vanish as $t \to -\infty$. To do this multiply both sides of the Green's function equation (1.11.10) by $f(\tau)$ and integrate over $-\infty < \tau < \infty$:

$$\left(\frac{d^2}{dt^2} + \omega^2\right) \int_{-\infty}^{\infty} G(t - \tau)f(\tau)d\tau = \int_{-\infty}^{\infty} \delta(t - \tau)f(\tau)d\tau \equiv f(t)$$

$$\therefore \quad \varphi(t) = \int_{-\infty}^{\infty} G(t - \tau)f(\tau)d\tau$$

$$= \frac{1}{\omega} \int_{-\infty}^{\infty} H(t - \tau)\sin \omega(t - \tau)f(\tau)d\tau$$

$$\text{i.e.} \quad \varphi(t) = \frac{1}{\omega} \int_{-\infty}^{t} f(\tau)\sin \omega(t - \tau)d\tau.$$

$$(1.11.11)$$

The last line shows that the solution $\varphi(t)$ at time t depends only on the behaviour of the forcing function $f(t)$ at *earlier* times.

Example Find the causal solution of

$$\frac{d^2\varphi}{dt^2} + \omega^2\varphi = \mathrm{H}(t)\mathrm{e}^{-t}.$$

From (1.11.11)

$$\varphi(t) = \frac{1}{\omega}\int_{-\infty}^{t} \mathrm{H}(\tau)\mathrm{e}^{-\tau}\sin\omega(t-\tau)d\tau = \frac{\mathrm{H}(t)}{\omega}\mathrm{Im}\left(\int_{0}^{t}\mathrm{e}^{-\tau+i\omega(t-\tau)}d\tau\right)$$

$$= \frac{\mathrm{H}(t)}{\omega}\mathrm{Im}\left(\frac{\mathrm{e}^{i\omega t}}{1+i\omega} - \frac{\mathrm{e}^{-t}}{1+i\omega}\right) = \mathrm{H}(t)\left(\frac{\sin\omega t - \omega\cos\omega t}{\omega(1+\omega^2)} + \frac{\mathrm{e}^{-t}}{(1+\omega^2)}\right).$$

Two-point boundary value problems A similar procedure can be used to solve

$$\frac{d^2 G}{dx^2} + \omega^2 G = \delta(x-y), \quad 0 < x,\ y < a, \quad \text{where } G = 0 \text{ at } x = 0,\ a.$$
(1.11.12)

The respective solutions of the homogeneous equation that vanish at $x = 0 < y$ and $x = a > y$ are

$$G = A\sin\omega x \quad \text{and} \quad G = B\sin\omega(x-a).$$

We therefore consider the trial solution

$$G = A\mathrm{H}(y-x)\sin\omega x + B\mathrm{H}(x-y)\sin\omega(x-a)$$

$$\therefore \quad \frac{dG}{dx} = -A\delta(x-y)\sin\omega y + \omega A\mathrm{H}(y-x)\cos\omega x$$

$$+ B\delta(x-y)\sin\omega(y-a) + \omega B\mathrm{H}(x-y)\cos\omega(x-a),$$

$$\frac{d^2 G}{dx^2} = -A\delta'(x-y)\sin\omega y - \omega A\delta(x-y)\cos\omega y$$

$$- \omega^2 A\mathrm{H}(y-x)\sin\omega x + B\delta'(x-y)\sin\omega(y-a)$$

$$+ \omega B\delta(x-y)\cos\omega(y-a) - \omega^2 B\mathrm{H}(x-y)\sin\omega(x-a)$$

$$= \delta'(x-y)\Big\{ -A\sin\omega y + B\sin\omega(y-a)\Big\}$$

$$+ \omega\delta(x-y)\Big\{ -A\cos\omega y + \omega B\cos\omega(y-a)\Big\} - \omega^2 G.$$

Substituting into the differential equation of (1.11.12)

$$\delta'(x-y)\Big\{-A\sin\omega y + B\sin\omega(y-a)\Big\}$$

$$+\omega\delta(x-y)\Big\{-A\cos\omega y + B\cos\omega(y-a)\Big\} \equiv \delta(x-y)$$

$$\therefore \quad -A\sin\omega y + B\sin\omega(y-a) = 0 \quad \text{and}$$

$$-A\cos\omega y + B\cos\omega(y-a) = \frac{1}{\omega}$$

$$\therefore \quad A = \frac{\sin\omega(y-a)}{\omega\sin\omega a} \quad \text{and} \quad B = \frac{\sin\omega y}{\omega\sin\omega a}.$$

Hence,

$$G(x,y) = \frac{H(y-x)\sin\omega(y-a)\sin\omega x}{\omega\sin\omega a} + \frac{H(x-y)\sin\omega(x-a)\sin\omega y}{\omega\sin\omega a}.$$

$$(1.11.13)$$

This satisfies the *reciprocity principle* $G(x,y) = G(y,x)$.

Example Solve the boundary value problem

$$\frac{d^2\varphi}{dx^2} + \omega^2\varphi = f(x), \quad 0 < x < a, \text{ where } \varphi(0) = \varphi(a) = 0.$$

Use Green's function (1.11.13):

$$\varphi(x) = \int_0^a G(x,y)f(y)dy$$

$$= \frac{\sin\omega x}{\omega\sin\omega a}\int_x^a f(y)\sin\omega(y-a)dy + \frac{\sin\omega(x-a)}{\omega\sin\omega a}\int_0^x f(y)\sin\omega y\, dy.$$

By using the expansion $\sin(A-B) = \sin A\cos B - \cos A\sin B$ this can be written

$$\varphi(x) = \frac{\cos\omega a\sin\omega x}{\omega\sin\omega a}\int_0^a f(y)\sin\omega y\, dy$$

$$-\frac{\sin\omega x}{\omega}\int_x^a f(y)\cos\omega y\, dy - \frac{\cos\omega x}{\omega}\int_0^x f(y)\sin\omega y\, dy. \quad (1.11.14)$$

This shows that if $\sin\omega a = 0$ (so that ω and $\sin\omega x$ are, respectively, an *eigenvalue* and *eigenfunction* of the homogeneous problem $d^2\varphi/dx^2 + \omega^2\varphi = 0$,

$\varphi(0) = \varphi(a) = 0)$ a solution exists if and only if $\int_0^a f(y) \sin \omega y \, dy = 0$. But the solution is now *not unique* because any multiple of $\sin \omega x$ can be added.

Example Solve

$$\frac{d^2\varphi}{dx^2} + \omega^2 \varphi = x, \quad 0 < x < a, \quad \text{where } \varphi(0) = \varphi(a) = 0.$$

$$\varphi(x) = \int_0^a G(x,y) y \, dy$$

$$= \frac{\sin \omega x}{\omega \sin \omega a} \int_x^a y \sin \omega(y-a) dy + \frac{\sin \omega(x-a)}{\omega \sin \omega a} \int_0^x y \sin \omega y \, dy$$

$$= \frac{1}{\omega^2} \left(x - a \frac{\sin \omega x}{\sin \omega a} \right).$$

Example Solve

$$\frac{d^2\varphi}{dx^2} + \pi^2 \varphi = \sin 2\pi x, \quad 0 < x < 1, \quad \text{where } \varphi(0) = \varphi(1) = 0.$$

$\sin \pi x$ is a solution of the homogeneous problem (an eigenfunction), but $\int_0^1 \sin 2\pi x \sin \pi x \, dx = 0$. Therefore, (1.11.14) gives

$$\varphi(x) = A \sin \pi x - \frac{\sin \pi x}{\pi} \int_x^1 \sin 2\pi y \cos \pi y \, dy - \frac{\cos \pi x}{\pi} \int_0^x \sin 2\pi y \sin \pi y \, dy$$

$$= B \sin \pi x - \frac{\sin 2\pi x}{3\pi^2},$$

where A and B are arbitrary constants.

Non-homogeneous problems The Green's function (1.11.13) for the boundary value problem (1.11.12) can also be used to solve the following problem with non-homogeneous boundary conditions:

$$\frac{d^2\varphi}{dx^2} + \omega^2 \varphi = f(x), \quad 0 < x < a, \quad \text{where } \varphi(0) = \varphi_0 \text{ and } \varphi(a) = \varphi_a.$$

To do this recall that $G(x,y) \equiv G(y,x)$, so that G also satisfies equation (1.11.12) with x and y interchanged. We can therefore rewrite our equations with y as the independent variable for a fixed value of x

in $0 < x < a$:

$$\frac{d^2\varphi}{dy^2} + \omega^2\varphi = f(y), \quad 0 < y < a,$$

$$\frac{d^2 G}{dy^2} + \omega^2 G = \delta(x - y), \quad 0 < x, \ y < a.$$

Now multiply the first equation by $G(x, y) \equiv G(y, x)$, the second equation by $\varphi(y)$, subtract the first from the second, and integrate over $0 < y < a$:

$$\int_0^a \left(\varphi(y)\frac{d^2 G}{dy^2}(x, y) - G(x, y)\frac{d^2\varphi}{dy^2}(y) \right) dy$$

$$= \int_0^a \varphi(y)\delta(x - y)dy - \int_0^a G(x, y)f(y)dy.$$

The integrand on the left is an exact differential, because

$$\varphi(y)\frac{d^2 G}{dy^2}(x, y) - G(x, y)\frac{d^2\varphi}{dy^2}(y) = \frac{d}{dy}\left(\varphi(y)\frac{dG}{dy}(x, y) - G(x, y)\frac{d\varphi}{dy}(y) \right).$$

Therefore, because $\int_0^a \varphi(y)\delta(x - y)dy = \varphi(x)$, we find

$$\varphi(x) = \int_0^a G(x, y)f(y)dy$$

$$+ \left[\varphi(y)\frac{dG}{dy}(x, y) - G(x, y)\frac{d\varphi}{dy}(y) \right]_0^a \quad 0 < x < a. \quad (1.11.15)$$

The final term is known, because $G(x, 0) = G(x, a) = 0$ and $\varphi(0) = \varphi_0$, $\varphi(a) = \varphi_a$, i.e.

$$\varphi(x) = \int_0^a G(x, y)f(y)dy + \varphi_a \frac{dG}{dy}(x, a) - \varphi_0 \frac{dG}{dy}(x, 0)$$

$$= \int_0^a G(x, y)f(y)dy + \frac{\varphi_a \sin\omega x - \varphi_0 \sin\omega(x - a)}{\sin\omega a}.$$

This solution is seen to satisfy the boundary conditions because the integrated term vanishes at $x = 0$, a, and the remaining term reduces, respectively, to φ_0 and φ_a at $x = 0$, a.

In deriving the relation (1.11.15) the condition $G(x, y) = G(y, x)$ is used, but no use is made of the conditions that $G(x, y) = 0$ at $x = 0$, a. It is easy to show that the *reciprocal formula* $G(x, y) = G(y, x)$ is satisfied for any 'self adjoint' boundary value problem of the form

$$\frac{d}{dx}\left(p(x)\frac{dG}{dx}\right) + q(x)G = \delta(x - y), \quad 0 < x,\ y < a,$$

$$a_1 G(0) + a_2\frac{dG}{dx}(0) = 0, \quad b_1 G(a) + b_2\frac{dG}{dx}(a) = 0,$$

where a_1, a_2, b_1, b_2 are constants. Therefore, (1.11.15) also represents the solution for more general conditions at $x = 0$, a. For example, if the values of $d\varphi/dx$ instead of φ are prescribed at $x = 0, a$, the solution is given by (1.11.15) in terms of the Green's function that satisfies $dG/dy = 0$ at $x = 0$, a, namely,

$$\varphi(x) = \int\limits_0^a G(x, y)f(y)dy - \varphi_a' G(x, a) + \varphi_0' G(x, 0)$$

where

$$\frac{d\varphi}{dx}(0) = \varphi_0', \quad \frac{d\varphi}{dx}(a) = \varphi_a'.$$

Problems 1J

Show that:

1. $\delta(x - y) = \delta(y - x)$.

2. $x\delta(x) = 0$.

3. If $xf(x) = 0$ for all values of x, then $f(x) = A\delta(x)$, where A is an arbitrary constant. $\left[\int_{-\infty}^{\infty} g(x)f(x)dx = \int_{-\infty}^{\infty} f(x)\left\{x\left(\frac{g(x)-g(0)}{x}\right) + g(0)\right\}dx = 0 + g(0)\int_{-\infty}^{\infty} f(x)dx = \text{constant} \times g(0), \ \therefore \ f(x) = \text{constant} \times \delta(x)\right]$

4. $\delta(ax) = \frac{1}{|a|}\delta(x)$.

5. $F(x)\delta(x - a) = F(a)\delta(x - a)$.

6. $\delta(x^2 - a^2) = \frac{1}{2|a|}[\delta(x-a) + \delta(x+a)]$.

7. $\frac{d}{dx}H(f(x)) = \frac{df(x)}{dx}\delta(f(x))$.

8. $\nabla H(f(\mathbf{x})) = \nabla f(\mathbf{x})\delta(f(\mathbf{x}))$.

9. $\frac{d^2|x|}{dx^2} = 2\delta(x)$.

10. $\int_{-\infty}^{\infty} x\delta'(x-a)dx = -1$.

11. $\int_{-\infty}^{\infty} \delta^{(n)}(x-a)f(x)dx = (-1)^n f^{(n)}(a)$.

12. $\lim_{n\to\infty} \frac{\sin nx}{\pi x} = \delta(x)$.

13. $\lim_{\epsilon\to+0} \frac{-2\epsilon x}{\pi(x^2+\epsilon^2)^2} = \delta'(x)$.

14. $P\left(\frac{1}{x}\right) + \pi i\delta(x) = \lim_{\epsilon\to+0} \frac{1}{x-i\epsilon}$.

15. $\int_0^{10} x^3\delta((x-5)(x^2+2))dx = \left(\frac{5}{3}\right)^3$.

16. $\int_0^3 x^2\delta(x^2-4)dx = 1$.

17. Let $\ln x$ be defined by its principal value when $x < 0$, i.e. $\ln x = \ln|x| + i\pi$. Deduce that
$$\frac{d}{dx}\ln x = P\left(\frac{1}{x}\right) - i\pi\delta(x).$$

18. The three-dimensional delta function $\delta(\mathbf{x}-\mathbf{y}) = \delta(x_1-y_1)\delta(x_2-y_2)\delta(x_3-y_3)$, where $\mathbf{x} = (x_1, x_2, x_3)$, $\mathbf{y} = (y_1, y_2, y_3)$.

19. A volume V is bounded by the closed surface S: $f(x, y, z) \equiv f(\mathbf{x}) = 0$, where $f(\mathbf{x}) \lessgtr 0$ according $\mathbf{x}$ is within or outside V. The unit *outward* normal to S is $\mathbf{n} = \nabla f(\mathbf{x})/|\nabla f(\mathbf{x})|$ evaluated on S. Show that
$$\oint_S \mathbf{F}(\mathbf{x}) \cdot \mathbf{n}\, dS = \int_{-\infty}^{\infty} \mathbf{F} \cdot \nabla H(f(\mathbf{x}))\, dx\, dy\, dz.$$

20. If $f(x)$ has simple zeros at $x = a_1$, a_2, a_3, etc., show that
$$\delta(f(x)) = \sum_n \frac{\delta(x-a_n)}{|f'(a_n)|}. \qquad (1.11.16)$$

21. Find the causal solution of
$$\frac{d^2\varphi}{dt^2} = \delta(t) + \delta(t-\tau). \quad \left[tH(t) + (t-\tau)H(t-\tau)\right]$$

22. Find the causal solution of
$$\frac{d\varphi}{dt} + 2t\varphi = \delta(t-\tau). \quad \left[H(t-\tau)e^{\tau^2-t^2}\right]$$

23. Use the Green's function $G(t-\tau) = \frac{1}{\omega}H(t-\tau)\sin\omega(t-\tau)$ to show that the causal solution of
$$\frac{d^2\varphi}{dt^2} + \omega^2\varphi = e^{-|t|}, \quad -\infty < t < \infty, \quad \text{is } \varphi = \frac{2H(t)\sin\omega t}{\omega(1+\omega^2)} + \frac{e^{-|t|}}{1+\omega^2}.$$

24. Solve the boundary value problem

$$\frac{d^2G}{dx^2} + \omega^2 G = \delta(x - y), \quad 0 < x, \ y < a, \quad \text{where} \quad \frac{dG}{dx} = 0 \ \text{at} \ x = 0, \ a.$$

$$\left[\frac{H(y-x)\cos\omega(y-a)\cos\omega x}{\omega \sin \omega a} + \frac{H(x-y)\cos\omega(x-a)\cos\omega y}{\omega \sin \omega a}\right]$$

25. Use the solution of Problem 24 to solve

$$\frac{d^2\varphi}{dx^2} + \omega^2\varphi = x, \quad 0 < x < a, \quad \text{where} \quad \frac{d\varphi}{dx} = 0 \ \text{at} \ x = 0, \ a.$$

$$\left[\frac{x}{\omega^2} - \frac{\sin \omega x}{\omega^3} + \frac{(1-\cos \omega a)\cos \omega x}{\omega^3 \sin \omega a}\right]$$

26. Solve the boundary value problem

$$\frac{d^2G}{dx^2} - \omega^2 G = \delta(x - y), \quad 0 < x, \ y < a, \quad \text{where} \quad G = 0 \ \text{at} \ x = 0, \ a.$$

$$\left[\frac{H(y-x)\sinh\omega(y-a)\sinh\omega x}{\omega \sinh \omega a} + \frac{H(x-y)\sinh\omega(x-a)\sinh\omega y}{\omega \sinh \omega a}\right]$$

27. Solve the boundary value problem

$$\frac{d^2G}{dx^2} = \delta(x - y), \quad 0 < x, \ y < a, \quad \text{where} \quad G = 0 \ \text{at} \ x = 0, \ a.$$

$$\left[\frac{H(y-x)x(y-a)}{a} + \frac{H(x-y)y(x-a)}{a}\right]$$

28. Solve the boundary value problem

$$\frac{d^2G}{dx^2} = \delta(x - y), \quad 0 < x, \ y < a, \quad \text{where} \quad G(0) = 0, \ \frac{dG}{dx}(a) = 0.$$

$$\left[-xH(y-x) - yH(x-y)\right]$$

29. Use (1.11.15) to solve

$$\frac{d^2\varphi}{dx^2} + \omega^2\varphi = x, \quad 0 < x < a, \quad \text{where} \quad \varphi(0) = \varphi_0 \ \text{and} \ \varphi(a) = \varphi_a.$$

$$\left[\frac{1}{\omega^2}\left(x - a\frac{\sin \omega x}{\sin \omega a}\right) + \frac{\varphi_a \sin \omega x - \varphi_0 \sin \omega(x-a)}{\sin \omega a}\right]$$

30. Use the solution of Problem 24 to solve

$$\frac{d^2\varphi}{dx^2} + \omega^2\varphi = x, \quad 0 < x < a, \quad \text{where} \quad \frac{d\varphi}{dx}(0) = \varphi_0' \ \text{and} \ \frac{d\varphi}{dx}(a) = \varphi_a'.$$

$$\left[\frac{x}{\omega^2} - \frac{\sin \omega x}{\omega^3} + \frac{(1-\cos \omega a)\cos \omega x}{\omega^3 \sin \omega a} + \frac{\varphi_0' \cos \omega(x-a) - \varphi_a' \cos \omega x}{\omega \sin \omega a}\right]$$

2

VECTOR CALCULUS

2.1 Elementary Operations with Vectors

A vector $\mathbf{a} = (a_1, a_2, a_3)$ will be referred to a right-handed coordinate system (x, y, z). Unit vectors along the x-, y- and z-directions are respectively denoted by $\mathbf{i}$, $\mathbf{j}$ and $\mathbf{k}$, so that

$$\mathbf{a} = (a_1, a_2, a_3) \equiv a_1 \mathbf{i} + a_2 \mathbf{j} + a_3 \mathbf{k}.$$

$\mathbf{a}$ has magnitude $a = |\mathbf{a}| = \sqrt{a_1^2 + a_2^2 + a_3^2}$, and $\hat{\mathbf{a}} = \mathbf{a}/|\mathbf{a}| = \mathbf{a}/\sqrt{a_1^2 + a_2^2 + a_3^2}$ is a unit vector in the direction of $\mathbf{a}$.

The *scalar product* $\mathbf{a} \cdot \mathbf{b}$ of two vectors $\mathbf{a}$ and $\mathbf{b}$ is defined by

$$\mathbf{a} \cdot \mathbf{b} = ab\cos\theta = a_1 b_1 + a_2 b_2 + a_3 b_3,$$

where θ is the angle between the directions of $\mathbf{a}$ and $\mathbf{b}$. Hence, $\mathbf{i} \cdot \mathbf{j} = \mathbf{j} \cdot \mathbf{k} = \mathbf{k} \cdot \mathbf{i} = 0$, and $|\mathbf{a}| = \sqrt{\mathbf{a} \cdot \mathbf{a}}$.

Two non-parallel vectors $\mathbf{a}$ and $\mathbf{b}$ define a plane with *unit normal* $\mathbf{n}$. Let θ be measured from $\mathbf{a}$ to $\mathbf{b}$ in the *positive* (or *right-handed*) sense about the direction $\mathbf{n}$ (i.e. clockwise when looking along the direction of $\mathbf{n}$). Then the *vector* (or *cross*) *product* $\mathbf{a} \times \mathbf{b}$ is defined by

$$\mathbf{a} \times \mathbf{b} = ab\sin\theta\, \mathbf{n}.$$

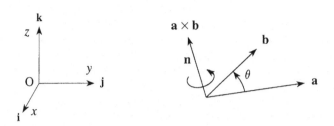

The definition does not depend on the choice of the direction **n**. If the direction of **n** is reversed the angle of rotation is replaced by $2\pi - \theta$ and $\sin\theta$ is replaced by $\sin(2\pi - \theta) = -\sin\theta$, so that magnitude and direction of $\mathbf{a} \times \mathbf{b}$ are unchanged.

Note also that

$$\mathbf{a}\times\mathbf{b} = \begin{vmatrix} \mathbf{i} & \mathbf{j} & \mathbf{k} \\ a_1 & a_2 & a_3 \\ b_1 & b_2 & b_3 \end{vmatrix}, \quad \text{in particular} \quad \begin{cases} \mathbf{a} \times \mathbf{a} = 0, \\ \mathbf{a} \times \mathbf{b} = -\mathbf{b} \times \mathbf{a}, \\ \mathbf{i} \times \mathbf{j} = \mathbf{k}, \ \mathbf{j} \times \mathbf{k} = \mathbf{i}, \ \mathbf{k} \times \mathbf{i} = \mathbf{j}. \end{cases}$$

The *triple scalar product* of three vectors **a**, **b**, **c** is

$$\mathbf{a} \cdot \mathbf{b} \times \mathbf{c} = \mathbf{a} \times \mathbf{b} \cdot \mathbf{c} = \begin{vmatrix} a_1 & a_2 & a_3 \\ b_1 & b_2 & b_3 \\ c_1 & c_2 & c_3 \end{vmatrix}$$

$$= \begin{cases} \text{volume} \\ \text{of the} \\ \text{parallelopiped:} \end{cases}$$

This is because $\mathbf{a}\times\mathbf{b} = ab\sin\theta\,\mathbf{n}$, where **n** is a unit vector perpendicular to **a** and **b**, and therefore

$$\mathbf{a} \cdot \mathbf{b} \times \mathbf{c} = ab\sin\theta\,\mathbf{n}\cdot\mathbf{c} = ab\sin\theta \times c\cos\phi$$

$$= \text{area of base} \times \text{perpendicular height}.$$

The following vector identities are frequently encountered:

$$\mathbf{a} \times (\mathbf{b} \times \mathbf{c}) = (\mathbf{a}\cdot\mathbf{c})\mathbf{b} - (\mathbf{a}\cdot\mathbf{b})\mathbf{c} \quad \text{(triple vector product)}$$

$$(\mathbf{a} \times \mathbf{b}) \times (\mathbf{c} \times \mathbf{d}) = (\mathbf{a}\cdot\mathbf{b} \times \mathbf{d})\mathbf{c} - (\mathbf{a}\cdot\mathbf{b} \times \mathbf{c})\mathbf{d}.$$

Example   Solve the vector equation : $\mathbf{x} + \mathbf{x} \times \mathbf{a} = \mathbf{b}.$ (2.1.1)

Take the cross product with **a**: $\mathbf{x} \times \mathbf{a} + (\mathbf{x} \times \mathbf{a}) \times \mathbf{a} = \mathbf{b} \times \mathbf{a}$
but $(\mathbf{x} \times \mathbf{a}) \times \mathbf{a} = \mathbf{a}(\mathbf{a} \cdot \mathbf{x}) - a^2 \mathbf{x}$
$$\therefore \quad \mathbf{x} \times \mathbf{a} - a^2 \mathbf{x} = -\mathbf{a}(\mathbf{a} \cdot \mathbf{x}) + \mathbf{b} \times \mathbf{a}$$

To find $\mathbf{a} \cdot \mathbf{x}$ take the scalar
product of (2.1.1) with **a**: $\mathbf{a} \cdot \mathbf{x} = \mathbf{a} \cdot \mathbf{b}$
$$\therefore \quad \mathbf{x} \times \mathbf{a} - a^2 \mathbf{x} = -\mathbf{a}(\mathbf{a} \cdot \mathbf{b}) + \mathbf{b} \times \mathbf{a}$$
Subtract this equation from (2.1.1): $(1 + a^2)\mathbf{x} = \mathbf{b} + (\mathbf{a} \cdot \mathbf{b})\mathbf{a} - \mathbf{b} \times \mathbf{a}$

$$\therefore \quad \mathbf{x} = \frac{\mathbf{b} + (\mathbf{a} \cdot \mathbf{b})\mathbf{a} - \mathbf{b} \times \mathbf{a}}{(1 + a^2)}.$$

Problems 2A

1. Find the value of x given that $\mathbf{a} = 3\mathbf{i} - 2\mathbf{j}$ and $\mathbf{b} = 4\mathbf{i} + x\mathbf{j}$ are perpendicular. [6]

2. Find the lengths of $\mathbf{a} = 2\mathbf{i} - \mathbf{j} + 2\mathbf{k}$, $\mathbf{b} = 5\mathbf{i} + 3\mathbf{j} - \mathbf{k}$. What is the angle between the directions of **a** and **b**? [3, $\sqrt{35}$, $\cos^{-1}(5/3\sqrt{35}]$

3. Solve for **x** the vector equation $\mathbf{x} + \mathbf{a}(\mathbf{b} \cdot \mathbf{x}) = \mathbf{c}$, where **a**, **b**, **c** are constant vectors. What happens when $\mathbf{a} \cdot \mathbf{b} = -1$? $[\mathbf{x} = \mathbf{c} - \mathbf{a}(\mathbf{b} \cdot \mathbf{c})/(1 + \mathbf{a} \cdot \mathbf{b})]$

4. Solve $A\mathbf{x} + \mathbf{a} \times \mathbf{x} = \mathbf{b}$, where $A \neq 0$ is a constant. $[\mathbf{x} = \{A^2\mathbf{b} + \mathbf{a}(\mathbf{a} \cdot \mathbf{b}) + A\mathbf{b} \times \mathbf{a}\}/A(A^2 + a^2)]$

5. Solve the simultaneous equations $\mathbf{x} + \mathbf{y} \times \mathbf{p} = \mathbf{a}$, $\mathbf{y} + \mathbf{x} \times \mathbf{p} = \mathbf{b}$.

   $[\mathbf{x} = \{(\mathbf{p} \cdot \mathbf{a})\mathbf{p} + \mathbf{a} - \mathbf{b} \times \mathbf{p}\}/(1 + p^2), \quad \mathbf{y} = \{(\mathbf{p} \cdot \mathbf{b})\mathbf{p} + \mathbf{b} - \mathbf{a} \times \mathbf{p}\}/(1 + p^2)]$

6. Calculate $\dot{\mathbf{r}}$ and $\ddot{\mathbf{r}}$, where the dot denotes differentiation with respect to the time t, when $\mathbf{r} = (t + \sin t)\mathbf{i} + (t - \sin t)\mathbf{j} + \sqrt{2}(1 - \cos t)\mathbf{k}$. Show that $\dot{\mathbf{r}}$ and $\ddot{\mathbf{r}}$ are perpendicular and have constant magnitudes. $[\dot{\mathbf{r}} = (1 + \cos t)\mathbf{i} + (1 - \cos t)\mathbf{j} + \sqrt{2}\sin t\mathbf{k}; \ddot{\mathbf{r}} = -\sin t\mathbf{i} + \sin t\mathbf{j} + \sqrt{2}\cos t\mathbf{k}; \dot{\mathbf{r}} \cdot \ddot{\mathbf{r}} = 0; |\dot{\mathbf{r}}| = 2; |\ddot{\mathbf{r}}| = \sqrt{2}]$

7. Show that $(\mathbf{a} \times \mathbf{b}) \cdot (\mathbf{a} \times \mathbf{c}) \times \mathbf{d} = (\mathbf{a} \cdot \mathbf{d})(\mathbf{a} \cdot \mathbf{b} \times \mathbf{c})$.

8. If $\mathbf{r} = (x, y, z)$, show that the equation of the straight line through the point $\mathbf{r}_o$ in the direction of the unit vector **t** is $\mathbf{r} = \mathbf{r}_o + \lambda \mathbf{t}$, $-\infty < \lambda < \infty$.

9. If $\mathbf{r} = (x, y, z)$ lies on the straight line through $\mathbf{r}_o$ in the direction of the unit vector **t**, show that $(\mathbf{r} - \mathbf{r}_o) \times \mathbf{t} = \mathbf{0}$.

10. Show that the equation of the plane whose *unit normal* is **n** (so that $|\mathbf{n}| = 1$) and which passes through the point $\mathbf{r}_o$ is $(\mathbf{r} - \mathbf{r}_o) \cdot \mathbf{n} = 0$. Show that the perpendicular from the point $\mathbf{r}_1$ intersects the plane at $\mathbf{r} = \mathbf{r}_1 - \mathbf{n}[(\mathbf{r}_1 - \mathbf{r}_o) \cdot \mathbf{n}]$.

11. Show that the equation to the perpendicular line from the point $\mathbf{b}$ to the straight line $\mathbf{r} = \mathbf{a} + \lambda\mathbf{t}$ is

$$\mathbf{r} = \mathbf{b} + \mu\mathbf{t} \times \{(\mathbf{a} - \mathbf{b}) \times \mathbf{t}\}, \quad -\infty < \mu < \infty.$$

12. Show that the straight lines $\mathbf{r} = \mathbf{a} + \lambda\mathbf{u}$ and $\mathbf{r} = \mathbf{b} + \mu\mathbf{v}$ will intersect if $\mathbf{v} \cdot \mathbf{b} \times \mathbf{u} = \mathbf{v} \cdot \mathbf{a} \times \mathbf{u}$, and that the point of intersection can be written in either of the forms

$$\mathbf{a} + \frac{\mathbf{a} \cdot \mathbf{b} \times \mathbf{v}}{\mathbf{v} \cdot \mathbf{a} \times \mathbf{u}}\mathbf{u} = \mathbf{b} + \frac{\mathbf{a} \cdot \mathbf{b} \times \mathbf{u}}{\mathbf{v} \cdot \mathbf{b} \times \mathbf{u}}\mathbf{v}.$$

13. If the straight lines $\mathbf{r} = \mathbf{a} + \lambda\mathbf{u}$ and $\mathbf{r} = \mathbf{b} + \mu\mathbf{v}$ do *not* intersect show that the length of the common perpendicular joining them is $|(\mathbf{b} - \mathbf{a}) \cdot \mathbf{n}|$, where $\mathbf{n} = \mathbf{u} \times \mathbf{v}/|\mathbf{u} \times \mathbf{v}|$.

14. Show that the equation of the plane through the points $\mathbf{r}_1$, $\mathbf{r}_2$, $\mathbf{r}_3$ can be written

$$\mathbf{r} \cdot \mathbf{r}_2 \times \mathbf{r}_3 + \mathbf{r} \cdot \mathbf{r}_3 \times \mathbf{r}_1 + \mathbf{r} \cdot \mathbf{r}_1 \times \mathbf{r}_2 = \mathbf{r}_1 \cdot \mathbf{r}_2 \times \mathbf{r}_3.$$

15. Show that the equation of the sphere of radius a and centre $\mathbf{r}_o$ is

$$(\mathbf{r} - \mathbf{r}_o) \cdot (\mathbf{r} - \mathbf{r}_o) = a^2.$$

16. The foci of an ellipse of major axis $2a$ are at the points $\pm\mathbf{b}$. Show that the point $\mathbf{r}$ lies on the ellipse if

$$a^4 - a^2(r^2 + b^2) + (\mathbf{b} \cdot \mathbf{r})^2 = 0.$$

2.2 Scalar and Vector Fields

Single-valued, scalar and vector functions defined over a region of space are referred to respectively as scalar and vector *fields*. The mass density, temperature, gravitational and electrical potentials, etc., are examples of scalar fields. The velocity of a moving fluid at any instant, and the gravitational and electrical force strengths are examples of vector fields. In applications scalar and vector fields also depend on time, which may be regarded as an independent parameter.

Let $\varphi(x, y, z) \equiv \varphi(\mathbf{x})$ be a scalar field. The equation

$$\varphi(x, y, z) = C = \text{constant},$$

defines a 'level-surface' S. Because $\varphi(x, y, z)$ is a single-valued function of position, two level-surfaces corresponding to different values of C cannot intersect.

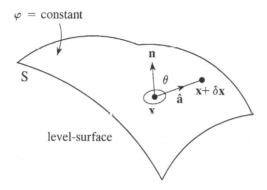

Suppose $\mathbf{x}$ and $\mathbf{x} + \delta\mathbf{x}$ are two neighbouring points on different level-surfaces. The change $\delta\varphi$ in the value of φ in moving from $\mathbf{x}$ to $\mathbf{x} + \delta\mathbf{x}$ is calculated by writing

$$\varphi(x + \delta x, y + \delta y, z + \delta z) = \varphi(x, y, z) + \delta x \frac{\partial\varphi}{\partial x} + \delta y \frac{\partial\varphi}{\partial y} + \delta z \frac{\partial\varphi}{\partial z} + \cdots$$

$$= \varphi(x, y, z)$$
$$+ (\delta x, \delta y, \delta z) \cdot \left(\frac{\partial\varphi}{\partial x}, \frac{\partial\varphi}{\partial y}, \frac{\partial\varphi}{\partial z} \right) + \cdots ,$$

$$\therefore \quad \delta\varphi \approx \delta\mathbf{x} \cdot \nabla\varphi + \cdots , \tag{2.2.1}$$

where the vector

$$\nabla\varphi = \left(\frac{\partial\varphi}{\partial x}, \frac{\partial\varphi}{\partial y}, \frac{\partial\varphi}{\partial z} \right) = \mathbf{i}\frac{\partial\varphi}{\partial x} + \mathbf{j}\frac{\partial\varphi}{\partial y} + \mathbf{k}\frac{\partial\varphi}{\partial z} \tag{2.2.2}$$

is the *gradient* of φ at $\mathbf{x}$ (also denoted by grad φ). The operator ∇ is pronounced 'del' or 'nabla', and is written (with respect to rectangular coordinates)

$$\nabla = \left(\frac{\partial}{\partial x}, \frac{\partial}{\partial y}, \frac{\partial}{\partial z} \right) = \mathbf{i}\frac{\partial}{\partial x} + \mathbf{j}\frac{\partial}{\partial y} + \mathbf{k}\frac{\partial}{\partial z}. \tag{2.2.3}$$

Let $\delta\mathbf{x}$ be in the direction of the unit vector $\hat{\mathbf{a}}$, and put $\delta s = |\delta\mathbf{x}|$, so that $\hat{\mathbf{a}} = \delta\mathbf{x}/\delta s$. Then the rate of change of φ at the point $\mathbf{x}$ in the direction $\hat{\mathbf{a}}$, called the *directional derivative* of φ, is

$$\left(\frac{d\varphi}{ds} \right)_{\hat{\mathbf{a}}} = \lim_{\delta s \to 0} \frac{\delta\varphi}{\delta s} = \frac{\delta\mathbf{x}}{\delta s} \cdot \nabla\varphi \equiv \hat{\mathbf{a}} \cdot \nabla\varphi \tag{2.2.4}$$

When two neighbouring points $\mathbf{x}$ and $\mathbf{x} + \delta\mathbf{x}$ lie on the *same* level-surface, equation (2.2.1) shows that

$$\frac{\delta\mathbf{x}}{\delta s} \cdot \nabla\varphi \to 0 \quad \text{as } \delta\mathbf{x} \to \mathbf{0}$$

for all orientations of the vector $\delta\mathbf{x}$ on S. But in this limit, the vector displacement $\delta\mathbf{x}$ lies in the *tangent plane* to the level-surface through $\mathbf{x}$, so that $\nabla\varphi$ is *normal* to the surfaces of constant φ, and the unit normal is

$$\mathbf{n} = \frac{\nabla\varphi}{|\nabla\varphi|}, \quad \text{in the direction of increasing } \varphi. \qquad (2.2.5)$$

This is the direction in which φ changes most rapidly, because according to (2.2.4)

$$\left(\frac{d\varphi}{ds}\right)_{\hat{\mathbf{a}}} = \hat{\mathbf{a}} \cdot \nabla\varphi = |\nabla\varphi|\cos\theta,$$

which assumes its maximum value when $\theta = 0$ (where θ is the angle between $\mathbf{n} = \nabla\varphi/|\nabla\varphi|$ and the unit vector $\hat{\mathbf{a}}$).

Example Evaluate ∇r, where $r = \sqrt{x^2 + y^2 + z^2}$.

$$\nabla r = \left(\frac{\partial r}{\partial x}, \frac{\partial r}{\partial y}, \frac{\partial r}{\partial z}\right) = \left(\frac{x}{\sqrt{x^2+y^2+z^2}}, \frac{y}{\sqrt{x^2+y^2+z^2}}, \frac{z}{\sqrt{x^2+y^2+z^2}}\right) \equiv \frac{\mathbf{r}}{r}.$$

Example Calculate the unit *outward* normal at (x, y, z) to the ellipsoid

$$\frac{x^2}{a^2} + \frac{y^2}{b^2} + \frac{z^2}{c^2} = 1.$$

$\varphi \equiv \frac{x^2}{a^2} + \frac{y^2}{b^2} + \frac{z^2}{c^2} - 1$ is an increasing function of x, y and z, so that $\nabla\varphi$ is directed *away* from the origin, and the required outward unit normal is therefore

$$\mathbf{n} = \frac{\nabla\varphi}{|\nabla\varphi|} = \frac{\left(\frac{2x}{a^2}, \frac{2y}{b^2}, \frac{2z}{c^2}\right)}{\sqrt{\frac{4x^2}{a^4} + \frac{4y^2}{b^4} + \frac{4z^2}{c^4}}} = \frac{\left(\frac{x}{a^2}, \frac{y}{b^2}, \frac{z}{c^2}\right)}{\sqrt{\frac{x^2}{a^4} + \frac{y^2}{b^4} + \frac{z^2}{c^4}}}.$$

Problems 2B

Calculate the gradients of:

1. $\varphi = x$. $[\mathbf{i}]$

2. $\varphi = x^3 + y^3 + z^3$. $[3(x^2\mathbf{i} + y^2\mathbf{j} + z^2\mathbf{k})]$

3. $\varphi = r^n$, $\mathbf{r} = x\mathbf{i} + y\mathbf{j} + z\mathbf{k}$. $[nr^{n-2}\mathbf{r}]$

4. $\varphi = \mathbf{a} \cdot \mathbf{r}$, $\mathbf{r} = x\mathbf{i} + y\mathbf{j} + z\mathbf{k}$. $[\mathbf{a}]$

5. $\varphi = \mathbf{r} \cdot \nabla(x + y + z)$, $\mathbf{r} = x\mathbf{i} + y\mathbf{j} + z\mathbf{k}$. $[\mathbf{i} + \mathbf{j} + \mathbf{k}]$

6. $\varphi = f(r)$, $r = |x\mathbf{i} + y\mathbf{j} + z\mathbf{k}|$. $[\frac{\mathbf{r}}{r}\frac{df}{dr}]$

Find the directional derivatives in the direction of $\mathbf{a}$ of.

7. $\varphi = e^x \cos y$, $\mathbf{a} = (2, 3, 0)$, $\mathbf{x} = (2, \pi, 0)$. $[-2e^2/\sqrt{13}]$

8. $\varphi = xyz$, $\mathbf{a} = (1, -2, 2)$, $\mathbf{x} = (-1, 1, 3)$. $[7/3]$

Find the unit normal to the surfaces:

9. $z = \sqrt{x^2 + y^2}$ at $\mathbf{x} = (3, 4, 5)$. $[(3, 4, -5)/5\sqrt{2}]$

10. $ax + by + cz + d = 0$ at any $\mathbf{x}$. $[(a, b, c)/\sqrt{a^2 + b^2 + c^2}]$

11. Show that, for the product of scalar functions φ_1, φ_2

$$\nabla(\varphi_1\varphi_2) = \varphi_1\nabla\varphi_2 + \varphi_2\nabla\varphi_1.$$

12. Show that the directional derivative of the vector $\mathbf{F}$ in the direction of the unit vector $\hat{\mathbf{a}}$ is given by

$$\left(\frac{d\mathbf{F}}{ds}\right)_{\hat{\mathbf{a}}} = (\hat{\mathbf{a}} \cdot \nabla)\mathbf{F}, \quad \text{where } \hat{\mathbf{a}} \cdot \nabla = \hat{a}_1\frac{\partial}{\partial x} + \hat{a}_2\frac{\partial}{\partial y} + \hat{a}_3\frac{\partial}{\partial z}.$$

2.3   The Divergence and the Divergence Theorem

Let dS denote the area element on a surface S, with unit normal $\mathbf{n} \equiv \mathbf{n}(\mathbf{x})$. Because the surface has two 'sides', the vector $\mathbf{n}$ can be chosen arbitrarily to be directed away from either side. For a closed surface we usually take $\mathbf{n}$ to be the *outward* normal.

The *flux* of a vector field $\mathbf{F}(\mathbf{x})$ through the surface S is defined by the *surface integral*

$$\int_S \mathbf{n} \cdot \mathbf{F}dS \equiv \int_S \mathbf{F} \cdot d\mathbf{S},$$

where the $d\mathbf{S}$ denotes the vector surface element

$$d\mathbf{S} = \mathbf{n}dS = (n_1, n_2, n_3)dS,$$

orientated in the direction of the surface normal $\mathbf{n}$.

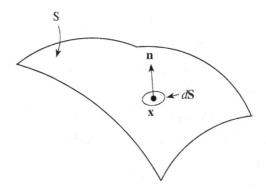

The net flux from a *closed* surface, with outward normal **n**, is obtained by integrating over the whole of the surface, and this operation is signified by the notation $\oint_S \mathbf{F} \cdot d\mathbf{S}$. In the case of a fluid with flow velocity $\mathbf{v}(\mathbf{x}, t)$, $\oint_S \mathbf{v} \cdot d\mathbf{S}$ is equal to the rate of increase of the volume $V(t)$ occupied by the fluid in S. Referring to the following figure, in time δt a surface element of area dS undergoes a vector displacement $\mathbf{v}\delta t$ and sweeps out a volume $dS\,\mathbf{n} \cdot \mathbf{v}\delta t$. The net change δV in the volume of the fluid initially in S is therefore $\delta V = \sum d\mathbf{S} \cdot \mathbf{v}\delta t$, and

$$\frac{\delta V}{\delta t} = \sum dS\,\mathbf{n} \cdot \mathbf{v} \rightarrow \oint_S \mathbf{v} \cdot d\mathbf{S}.$$

Therefore, $\frac{1}{V} \oint_S \mathbf{v} \cdot d\mathbf{S}$ is just equal to the rate of expansion of V per unit volume, $\frac{1}{V}\frac{dV}{dt}$.

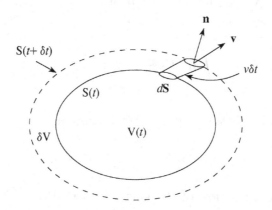

With this interpretation in mind, we define the *divergence* of a vector field **F** at **x** by

$$\operatorname{div} \mathbf{F} = \text{'outflux' per unit volume}$$

$$= \lim_{V \to 0} \frac{1}{V} \oint_S \mathbf{n} \cdot \mathbf{F} dS \equiv \lim_{V \to 0} \frac{1}{V} \oint_S \mathbf{F} \cdot d\mathbf{S}, \qquad (2.3.1)$$

where the integration is over the surface S of an infinitesimal volume element V containing **x**. This is a scalar function of position.

The explicit representation of div **F** in rectangular coordinates is derived by taking a volume element in the shape of a rectangular parallelopiped of sides dx, dy, dz aligned with the coordinate axes, and volume $V = dxdydz$. In the following figure the contribution of the faces S_- and S_+ to the integral (2.3.1) is

$$F_1(x + dx, y, z)dydz - F_1(x, y, z)dydz = \frac{\partial F_1}{\partial x}dxdydz,$$

with analogous expressions involving $\partial F_2/\partial y$, $\partial F_3/\partial z$ for the other faces. Hence

$$\operatorname{div} \mathbf{F} = \frac{\partial F_1}{\partial x} + \frac{\partial F_2}{\partial y} + \frac{\partial F_3}{\partial z} \equiv \nabla \cdot \mathbf{F}. \qquad (2.3.2)$$

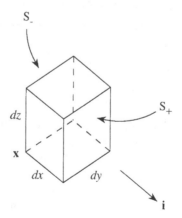

The divergence theorem Let $\mathbf{F}(\mathbf{x})$ be defined on and within the interior V of a closed surface S. Then

$$\int_V \operatorname{div} \mathbf{F} \, dV = \oint_S \mathbf{n} \cdot \mathbf{F} dS \equiv \oint_S \mathbf{F} \cdot d\mathbf{S}. \qquad (2.3.3)$$

The proof is an immediate consequence of the definition (2.3.1).

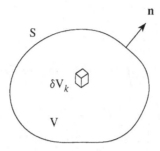

The volume integral on the left of (2.3.3) is evaluated by dividing V into volume elements δV_k with surfaces S_k. For each element

$$\delta V_k (\text{div } \mathbf{F})_k \approx \oint_{S_k} \mathbf{n} \cdot \mathbf{F} dS,$$

where the divergence is evaluated at some point within δV_k. Then as $\delta V_k \to 0$

$$\int_V \text{div } \mathbf{F} \, dV = \sum_k \delta V_k (\text{div } \mathbf{F})_k = \sum_k \oint_{S_k} \mathbf{n} \cdot \mathbf{F} dS.$$

The surface integrals over a shared rectangular surface of adjacent volume elements δV_k are equal and opposite, and the only terms in the final summation on the right-hand side that make a non-zero contribution correspond to the surface elements that approximate to the boundary S as $\delta V_k \to 0$, which yield $\oint_S \mathbf{F} \cdot d\mathbf{S}$ as $\delta V_k \to 0$, thereby establishing the theorem.

Example Show that $V = \frac{1}{3} \oint_S \mathbf{r} \cdot d\mathbf{S}$, where V is the volume bounded by the closed surface S and $\mathbf{r} = (x, y, z)$.

By the divergence theorem: $\frac{1}{3} \oint_S \mathbf{r} \cdot d\mathbf{S} = \frac{1}{3} \int_V \text{div } \mathbf{r} dV \equiv \frac{1}{3} \int_V 3 dV = V$.

Example A vector field whose divergence vanishes is called a *solenoidal vector*. The velocity $\mathbf{v}(\mathbf{x}, t)$ of an incompressible fluid is solenoidal, because the volume occupied by a moving fluid element is invariant, i.e. its *rate of volumetric expansion* is zero, although the mean fluid *density* may vary with position in the fluid.

In an ideal fluid in *irrotational* motion the velocity $\mathbf{v} = \nabla \varphi$, where $\varphi(\mathbf{x}, t)$ is a velocity potential. When the motion is incompressible $\text{div}(\nabla \varphi) = 0$. Now

$$\text{div}(\nabla) = \text{div grad} = \nabla \cdot \nabla \equiv \nabla^2 = \frac{\partial^2}{\partial x^2} + \frac{\partial^2}{\partial y^2} + \frac{\partial^2}{\partial z^2}. \tag{2.3.4}$$

Hence, for incompressible flow φ satisfies *Laplace's equation*

$$\nabla^2 \varphi \equiv \frac{\partial^2 \varphi}{\partial x^2} + \frac{\partial^2 \varphi}{\partial y^2} + \frac{\partial^2 \varphi}{\partial z^2} = 0. \qquad (2.3.5)$$

Problems 2C

Find the divergence of:

1. $\mathbf{F} = y\mathbf{i} + z\mathbf{j} + x\mathbf{k}$.   [0]

2. $\mathbf{F} = \mathbf{x}$.   [3]

3. $\mathbf{F} = (x, y^2, z^3)$. $[1 + 2y + 3z^2]$

4. $\mathbf{F} = 4x^2\mathbf{i} + 4y^2\mathbf{j} - z^2\mathbf{k}$. $[8x + 8y - 2z]$

5. $\mathbf{F} = xyz(\mathbf{i} + \mathbf{j} + \mathbf{k})$. $[yz + xz + xy]$

6. $\mathbf{F} = \mathbf{r}/r^3$, $\mathbf{r} = x\mathbf{i} + y\mathbf{j} + z\mathbf{k}$. $[0,\ r > 0]$

7. $\mathbf{F} = \mathbf{r}(\mathbf{r} \cdot \mathbf{a})$, $\mathbf{r} = x\mathbf{i} + y\mathbf{j} + z\mathbf{k}$, $\mathbf{a} = $ constant. $[4\mathbf{r} \cdot \mathbf{a}]$

Prove that:

8. $\nabla^2(1/r) = 0$, $r > 0$, $\mathbf{r} = x\mathbf{i} + y\mathbf{j} + z\mathbf{k}$.

9. $\nabla^2(r^n) = n(n+1)r^{n-2}$, $\mathbf{r} = x\mathbf{i} + y\mathbf{j} + z\mathbf{k}$.

10. For scalar and vector fields $\varphi(\mathbf{x})$ and $\mathbf{a}(\mathbf{x})$, $\mathrm{div}(\varphi\mathbf{a}) = \varphi\mathrm{div}\,\mathbf{a} + \nabla\varphi \cdot \mathbf{a}$.

11. $\mathrm{div}\,(f\nabla g) - \mathrm{div}\,(g\nabla f) = f\nabla^2 g - g\nabla^2 f$.

12. $\oint_S \varphi\mathbf{n} \cdot \nabla\varphi dS = \int_V (\nabla\varphi)^2 \, dV$, provided $\nabla^2\varphi = 0$.

13. $V = \frac{1}{6}\oint_S \mathbf{n}\cdot\nabla(r^2)dS$, where $\mathbf{r} = x\mathbf{i} + y\mathbf{j} + z\mathbf{k}$ and V is the volume enclosed by S.

14. $\oint_S (x^3\mathbf{i} + y^3\mathbf{j} + z^3\mathbf{k})\cdot d\mathbf{S} = \frac{12}{5}\pi R^5$, where S is a sphere of radius R with centre at the origin.

Evaluate $\oint_S \mathbf{n} \cdot \mathbf{F}dS$ when:

15. $\mathbf{F} = (x, x^2y, -x^2z)$ and S is the surface of the tetrahedron with vertices $(0,0,0)$, $(1,0,0)$, $(0,1,0)$ $(0,0,1)$. $[\frac{1}{6}]$

16. $\mathbf{F} = \frac{1}{3}(x^3, y^3, z^3)$ and S is the surface of the sphere $|\mathbf{x}| = 2$. $[128\pi/5]$

17. $\mathbf{F} = ax\mathbf{i} + by\mathbf{j} + cz\mathbf{k}$ where a, b, c are constants, and S is the unit sphere $|\mathbf{x}| = 1$. $[\frac{4}{3}\pi(a + b + c)]$

Divergence theorem in two dimensions   For a two-dimensional vector field $\mathbf{F}(x, y)$ in the xy-plane, the divergence theorem involves the *line integral* $\oint_C \mathbf{n} \cdot \mathbf{F}ds$ taken around the closed boundary C of an area A of the plane, where $ds = \sqrt{dx^2 + dy^2}$ and $\mathbf{n}$ is the outward normal to the contour:

$$\oint_C \mathbf{n} \cdot \mathbf{F} ds = \int_A \operatorname{div} \mathbf{F} dS = \int_A \left(\frac{\partial F_1}{\partial x} + \frac{\partial F_2}{\partial y} \right) dx dy \qquad (2.3.6)$$

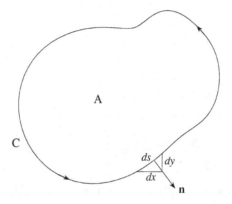

The reader can easily show that $\mathbf{n} ds = (dy - dx)$ when the line integral is taken in the positive sense around C (such that the interior of the curve is to the left when travelling along C), and thereby deduce *Green's theorem*

$$\oint_C F_1 dy - F_2 dx = \int_A \left(\frac{\partial F_1}{\partial x} + \frac{\partial F_2}{\partial y} \right) dx dy.$$

In the usual statement of this formula (F_1, F_2) is replaced by $(-G, F)$, giving the two-dimensional form of *Stokes' theorem* (§2.4):

$$\oint_C F dx + G dy = \int_A \left(\frac{\partial G}{\partial x} - \frac{\partial F}{\partial y} \right) dx dy. \qquad (2.3.7)$$

2.4 Stokes' Theorem and Curl

Circulation of a vector field The line integral

$$\oint_C \mathbf{F} \cdot d\mathbf{r}$$

taken around a closed circuit C (Figure (a)) is called the circulation of $\mathbf{F}$ around C.

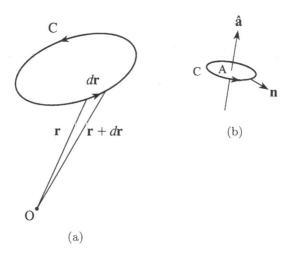

(a)

(b)

For a given vector field $\mathbf{F}(\mathbf{x})$ a new vector field denoted by **curl F** is defined whose value at the point $\mathbf{x}$ is obtained by the following limiting procedure. Choose a direction defined by the unit vector $\hat{\mathbf{a}}$, then

$$\hat{\mathbf{a}} \cdot \mathbf{curl\,F} = \text{circulation around } \hat{\mathbf{a}} \text{ per unit area} = \lim_{A \to 0} \frac{1}{A} \oint_C \mathbf{F} \cdot d\mathbf{r}$$

$$(2.4.1)$$

where C is any closed contour around $\mathbf{x}$ lying in the plane containing $\mathbf{x}$ whose normal is $\hat{\mathbf{a}}$. C encloses an area A of the plane, and is traversed in the positive sense with respect to the direction $\hat{\mathbf{a}}$ (Figure (b)).

To evaluate **curl F** explicitly the vector line element $d\mathbf{r}$ is written in the form

$$d\mathbf{r} = (\hat{\mathbf{a}} \times \mathbf{n})ds,$$

where $ds = |d\mathbf{r}| > 0$ is the arc length on C, and $\mathbf{n}$ is the unit outward normal to C in the plane. Then

$$\hat{\mathbf{a}} \cdot \mathbf{curl\,F} = \lim_{A \to 0} \frac{1}{A} \oint_C \mathbf{F} \cdot (\hat{\mathbf{a}} \times \mathbf{n})ds = \lim_{A \to 0} \frac{1}{A} \oint_C \mathbf{n} \cdot (\mathbf{F} \times \hat{\mathbf{a}})ds.$$

The vector $\mathbf{F} \times \hat{\mathbf{a}}$ lies in the plane of the curve C. We may therefore apply the two-dimensional divergence theorem (2.3.6) to the final integral, to obtain

$$\hat{\mathbf{a}} \cdot \mathbf{curl}\,\mathbf{F} = \lim_{A \to 0} \frac{1}{A} \int_A \mathrm{div}(\mathbf{F} \times \hat{\mathbf{a}})dS = \mathrm{div}(\mathbf{F} \times \hat{\mathbf{a}})$$

$$\equiv \nabla \cdot (\mathbf{F} \times \hat{\mathbf{a}}) = \hat{\mathbf{a}} \cdot (\nabla \times \mathbf{F}).$$

In rectangular coordinates (x, y, z) we find (using (2.2.3)), taking in turn $\hat{\mathbf{a}} = \mathbf{i},\ \mathbf{j},\ \mathbf{k}$:

$$\mathbf{curl}\,\mathbf{F} \equiv \nabla \times \mathbf{F} = \begin{vmatrix} \mathbf{i} & \mathbf{j} & \mathbf{k} \\ \frac{\partial}{\partial x} & \frac{\partial}{\partial y} & \frac{\partial}{\partial z} \\ F_1 & F_2 & F_3 \end{vmatrix}$$

$$= \left(\frac{\partial F_3}{\partial y} - \frac{\partial F_2}{\partial z} \right)\mathbf{i} + \left(\frac{\partial F_1}{\partial z} - \frac{\partial F_3}{\partial x} \right)\mathbf{j}$$

$$+ \left(\frac{\partial F_2}{\partial x} - \frac{\partial F_1}{\partial y} \right)\mathbf{k}. \qquad (2.4.2)$$

Stokes' theorem Let C be a closed contour and S an open two-sided surface bounded by C. The unit normal $\mathbf{n}$ on S is assumed to be orientated in the positive sense relative to that in which the contour C is described (see Figure (a) below). Then

$$\oint_C \mathbf{F}\cdot d\mathbf{r} = \int_S \mathbf{curl}\,\mathbf{F} \cdot d\mathbf{S} \equiv \int_S \mathbf{n} \cdot \mathbf{curl}\,\mathbf{F}\,dS. \qquad (2.4.3)$$

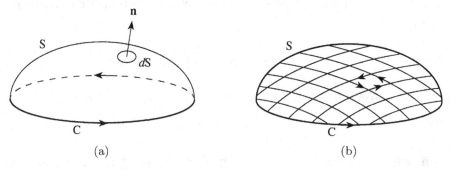

(a)   (b)

The theorem is proved by ruling a mesh over S in the manner illustrated in Figure (b), and summing the contour integrals taken over all of the mesh boundaries in the positive sense. The sum is just equal to $\oint_C \mathbf{F}\cdot d\mathbf{r}$, because the contributions from a mesh side within C is equal and opposite to that of an adjacent mesh. Then, if C_k, δS_k and $\mathbf{n}_k$ denote the boundary, surface area and unit normal for the kth mesh, the definition (2.4.1), with $\hat{\mathbf{a}} = \mathbf{n}_k$, gives as $\delta S_k \to 0$

$$\oint_C \mathbf{F} \cdot d\mathbf{r} = \sum_{l_0} \oint_{C_k} \mathbf{F} \cdot d\mathbf{r} = \sum_k \mathbf{n}_k \cdot \mathbf{curl}\, \mathbf{F} \delta S_k = \int_S \mathbf{curl}\, \mathbf{F} \cdot d\mathbf{S}.$$

Problems 2D

1. Calculate $\mathbf{curl}\, \mathbf{F}$ for $\mathbf{F} = x\mathbf{i}$, $x\mathbf{i} + y\mathbf{j} + z\mathbf{k}$, $(x\mathbf{i} - y\mathbf{j})/(x+y)$. $[\mathbf{0}, \mathbf{0}, \mathbf{k}/(x+y)]$

2. Show that $\mathbf{curl}\,\{\mathbf{r}(\mathbf{a} \cdot \mathbf{r})\} = \mathbf{a} \times \mathbf{r}$, and $\mathbf{curl}\,\mathbf{curl}\,\{\mathbf{r}(\mathbf{a} \cdot \mathbf{r})\} = \mathbf{curl}\,(\mathbf{a} \times \mathbf{r}) = 2\mathbf{a}$, where $\mathbf{r} = x\mathbf{i} + y\mathbf{j} + z\mathbf{k}$ and $\mathbf{a}$ is constant.

3. If $\mathbf{a}$ is a constant *unit vector*, show that $\mathbf{a} \cdot \{\nabla(\mathbf{v} \cdot \mathbf{a}) - \mathbf{curl}\,(\mathbf{v} \times \mathbf{a})\} = \mathrm{div}\,\mathbf{v}$.

4. Prove that $(\mathbf{v} \cdot \nabla)\mathbf{v} = \nabla(\frac{1}{2}v^2) + \mathbf{curl}\,\mathbf{v} \times \mathbf{v}$ where $(\mathbf{v} \cdot \nabla) = v_1 \partial/\partial x + v_2 \partial/\partial y + v_3 \partial/\partial z$ and $\mathbf{v} = (v_1,\ v_2,\ v_3)$.

5. Use the relation $\oint_C \nabla \varphi \cdot d\mathbf{r} = 0$, where C is any simple closed curve (i.e. one that can be 'shrunk' to a point without crossing any boundaries) to deduce that $\mathbf{curl}\,\nabla\varphi \equiv \mathbf{0}$.

6. Show that if $\mathbf{curl}\,\mathbf{v} = \mathbf{0}$ where $\mathbf{v} = (xyz)^b(x^a\mathbf{i} + y^a\mathbf{j} + z^a\mathbf{k})$, then either $b = 0$ or $a = -1$.

7. Establish the identities:
$$\mathbf{curl}\,(\varphi\mathbf{F}) = \nabla \times (\varphi\mathbf{F}) = \varphi\mathbf{curl}\,\mathbf{F} + \nabla\varphi \times \mathbf{F};$$
$$\mathbf{curl}\,(\mathbf{curl}\,\mathbf{F}) = \nabla \times (\nabla \times \mathbf{F}) = \nabla(\mathrm{div}\,\mathbf{F}) - \nabla^2\mathbf{F};$$
$$\mathrm{div}(\mathbf{F} \times \mathbf{G}) = \nabla \cdot (\mathbf{F} \times \mathbf{G}) = \mathbf{curl}\,\mathbf{F} \cdot \mathbf{G} - \mathbf{F} \cdot \mathbf{curl}\,\mathbf{G};$$
$$\mathbf{curl}\,(\mathbf{F} \times \mathbf{G}) = \nabla \times (\mathbf{F} \times \mathbf{G}) = (\mathbf{G} \cdot \nabla)\mathbf{F} - (\mathbf{F} \cdot \nabla)\mathbf{G} + \mathbf{F}\,\mathrm{div}\,\mathbf{G} - \mathbf{G}\,\mathrm{div}\,\mathbf{F};$$
$$\mathrm{grad}(\mathbf{F} \cdot \mathbf{G}) = \nabla(\mathbf{F} \cdot \mathbf{G}) = (\mathbf{G} \cdot \nabla)\mathbf{F} + (\mathbf{F} \cdot \nabla)\mathbf{G} + \mathbf{G} \times \mathbf{curl}\,\mathbf{F}$$
$$+ \mathbf{F} \times \mathbf{curl}\,\mathbf{G}.$$
$$(2.4.4)$$

8. Show that $\mathbf{curl}\,(r^\alpha \mathbf{r}) = \mathbf{0}$, for any α, but $\mathrm{div}(r^\alpha \mathbf{r}) = 0$ only for $\alpha = -3$.

9. Show that $\mathrm{div}(\mathbf{curl}\,\mathbf{F}) \equiv 0$ by applying the divergence theorem to $\oint_S \mathbf{n} \cdot \mathbf{curl}\,\mathbf{F} dS$ for a closed surface S. Divide S into two parts S_1, S_2 by any closed curve C drawn once around S. Stokes' theorem shows that the surface integrals over S_1 and S_2 are equal and opposite, because they are equal to the circulations of $\mathbf{F}$ around C in opposite directions.

10. Prove that for a scalar field φ
$$\int_V \nabla\varphi\, dV = \oint_S \varphi\, \mathbf{n} dS,$$
where V is the interior of the closed surface S with outward normal $\mathbf{n}$.

11. By considering the surface integral $\oint_S \hat{\mathbf{a}} \cdot \mathbf{n} \times \mathbf{F} dS$ for an arbitrary unit vector $\hat{\mathbf{a}}$, prove that

$$\int_V \mathbf{curl\,F}\,dV = \oint_S \mathbf{n} \times \mathbf{F} dS,$$

where V is the interior of the closed surface S with outward normal $\mathbf{n}$. Hence, deduce that $\mathbf{curl\,F}$ can be defined by

$$\mathbf{curl\,F} = \lim_{V \to 0} \frac{1}{V} \oint_S \mathbf{n} \times \mathbf{F} dS.$$

Questions 10 and 11 of Problems 2D and the divergence theorem are all variations of one integral transformation, which is clear from the symbolic representations

$$\left.\begin{array}{c} \displaystyle\int_V \nabla\varphi\,dV = \oint_S \mathbf{n}\,\varphi\,dS, \\[2ex] \displaystyle\int_V \nabla \cdot \mathbf{F}\,dV = \oint_S \mathbf{n} \cdot \mathbf{F} dS, \\[2ex] \displaystyle\int_V \nabla \times \mathbf{F}\,dV = \oint_S \mathbf{n} \times \mathbf{F} dS. \end{array}\right\} \qquad (2.4.5)$$

2.5　Green's Identities

Let φ_1 and φ_2 be scalar fields. Then

$$\mathrm{div}(\varphi_1 \nabla \varphi_2) = \nabla\varphi_1 \cdot \nabla\varphi_2 + \varphi_1 \nabla^2 \varphi_2,$$
$$\mathrm{div}(\varphi_2 \nabla \varphi_1) = \nabla\varphi_1 \cdot \nabla\varphi_2 + \varphi_2 \nabla^2 \varphi_1.$$

The divergence theorem yields for a region V bounded by a closed surface S

$$\int_V \nabla\varphi_1 \cdot \nabla\varphi_2\,dV = \oint_S \varphi_1 \nabla\varphi_2 \cdot d\mathbf{S} - \int_V \varphi_1 \nabla^2 \varphi_2\,dV$$

$$= \oint_S \varphi_2 \nabla\varphi_1 \cdot d\mathbf{S} - \int_V \varphi_2 \nabla^2 \varphi_1\,dV. \quad (2.5.1)$$

either of which is known as *Green's first identity*.

The right-hand sides of (2.5.1) supply *Green's second identity*

$$\oint_S \left\{ \varphi_1 \nabla \varphi_2 - \varphi_2 \nabla \varphi_1 \right\} \cdot d\mathbf{S} = \int_V \left\{ \varphi_1 \nabla^2 \varphi_2 - \varphi_2 \nabla^2 \varphi_1 \right\} dV. \quad (2.5.2)$$

When φ_1 and φ_2 both satisfy the same one of the following equations

Laplace:
$$\nabla^2 \varphi = 0,$$

Helmholtz:
$$\nabla^2 \varphi + \kappa^2 \varphi = 0,$$

in V, Green's second identity reduces to the 'reciprocity' relation

$$\oint_S \left\{ \varphi_1 \nabla \varphi_2 - \varphi_2 \nabla \varphi_1 \right\} \cdot d\mathbf{S} = 0.$$

Example An infinite region of incompressible fluid is bounded *internally* by the surface S of a *rigid* body. The body is in motion and causes the fluid to move irrotationally with velocity $\mathbf{v} = \nabla \varphi$, where φ satisfies Laplace's equation $\nabla^2 \varphi = 0$ in the fluid. This implies that $|\nabla \varphi| \sim 1/|\mathbf{x}|^3$ as $|\mathbf{x}| \to \infty$ in the fluid, and the fluid may be assumed to be at rest at infinity.

Green's first identity, with $\varphi_1 = \varphi_2 = \varphi$ shows that the kinetic energy of the fluid is proportional to

$$\int_V \frac{1}{2}(\nabla \varphi)^2 dV = \frac{1}{2} \oint_S \varphi \frac{\partial \varphi}{\partial n} dS,$$

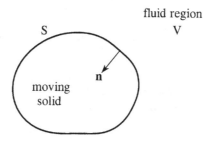

fluid region
V

S

n

moving
solid

where $\partial \varphi / \partial n = \mathbf{n} \cdot \nabla \varphi$ is the normal component of the velocity of S directed *into* the rigid body. If the body is suddenly brought to rest, $\partial \varphi / \partial n$ becomes instantaneously zero, and the equation supplies the unphysical prediction that motion also ceases immediately everywhere in the fluid!

Example The velocity distribution of the irrotational motion of the previous example is uniquely determined by the boundary motion, i.e. if φ and φ' are two velocity potentials found by different means that have the same values of $\partial \varphi / \partial n$ on S, then $\nabla \varphi(\mathbf{x}) \equiv \nabla \varphi'(\mathbf{x})$ in V.

To prove this we set $\Phi = \varphi - \varphi'$, and note that this implies that $\partial\Phi/\partial n = 0$ on S and $\nabla^2\Phi = 0$ in V. Green's first identity, with $\varphi_1 = \varphi_2 = \Phi$, then gives

$$\int_V (\nabla\Phi)^2 \, dV = \oint_S \Phi \frac{\partial\Phi}{\partial n} \, dS - \int_V \Phi\nabla^2\Phi \, dV \equiv 0.$$

The integrand on the left is non-negative, and we must therefore have $\nabla\Phi = \mathbf{0}$ in V, i.e. $\nabla\varphi = \nabla\varphi'$.

Example The kinetic energy of the irrotational motion of the previous two examples is *smaller* than the kinetic energy of any other motion consistent with the boundary conditions.

If ρ is the fluid density and φ the velocity potential for irrotational motion, the kinetic energy is

$$T_0 = \frac{1}{2}\rho \int_V (\nabla\varphi)^2 dV.$$

Let $\mathbf{v}_1(\mathbf{x})$ be any other (rotational) velocity distribution that satisfies $\mathbf{n} \cdot \mathbf{v}_1 = \partial\phi/\partial n$ on S, with kinetic energy $T_1 = \frac{1}{2}\rho \int_V v_1^2 dV$. Then

$$T_1 - T_0 = \frac{1}{2}\rho \int_V \left(v_1^2 - (\nabla\varphi)^2\right) dV$$

$$= \frac{1}{2}\rho \int_V \left(2\nabla\varphi \cdot (\mathbf{v}_1 - \nabla\varphi) + (\mathbf{v}_1 - \nabla\varphi)^2\right) dV$$

$$= \rho \int_V \left[\mathrm{div}\left(\varphi(\mathbf{v}_1 - \nabla\varphi)\right) - \varphi\,\mathrm{div}(\mathbf{v}_1 - \nabla\varphi)\right] dV + \frac{1}{2}\rho \int_V \left(\mathbf{v}_1 - \nabla\varphi\right)^2 dV$$

$$= \rho \oint_S \varphi(\mathbf{v}_1 - \nabla\varphi) \cdot d\mathbf{S} - \rho \int_V \varphi\,\mathrm{div}(\mathbf{v}_1 - \nabla\varphi) dV + \frac{1}{2}\rho \int_V \left(\mathbf{v}_1 - \nabla\varphi\right)^2 dV.$$

On the final line, the first integral is zero because $\mathbf{n} \cdot \mathbf{v}_1 = \partial\phi/\partial n$ on S; the second integral is zero because the fluid is incompressible ($\mathrm{div}(\mathbf{v}_1 - \nabla\varphi) = 0$). The final integral is non-negative, thereby proving that $T_1 \geq T_0$.

2.6 Orthogonal Curvilinear Coordinates

Consider three scalar fields $q_1(\mathbf{x})$, $q_2(\mathbf{x})$, $q_3(\mathbf{x})$. Each has a level-surface that passes through a given point P (see figure below). It is assumed that the functions are such that the three level-surfaces are not coincident or meet in a common curve. The values of q_1, q_2, q_3 on these surfaces accordingly determine the point P, and constitute *curvilinear coordinates* of P. The three surfaces are the *coordinate surfaces* through P, their lines of intersection are the *coordinate lines*, and the tangents of the coordinate lines at P are the *coordinate axes*, whose directions vary for different points P.

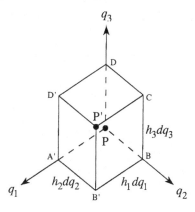

We consider here orthogonal curvilinear coordinates for which, at every point P, the coordinate axes are mutually perpendicular. Choose the positive directions of the orthogonal coordinate axes of (q_1, q_2, q_3) at P to form a right-handed system, with respective unit vectors $\hat{\mathbf{a}}_1$, $\hat{\mathbf{a}}_2, \hat{\mathbf{a}}_3$ in these directions. Let the coordinate surfaces $q_1 + dq_1$, $q_2 + dq_2$, $q_3 + dq_3$ define the coordinates of a neighbouring point P'. Suppose that the infinitesimal rectangular parallelopiped, shown in the figure, formed by these surfaces and the surfaces through P has sides of length $h_1 dq_1$, $h_2 dq_2, h_3 dq_3$ $(h_1, h_2, h_3 > 0)$ respectively parallel to $\hat{\mathbf{a}}_1, \hat{\mathbf{a}}_2, \hat{\mathbf{a}}_3$. Then the *volume element* dV and *line element* ds (the distance PP') are given by

$$dV = h_1 h_2 h_3 dq_1 dq_2 dq_3, \quad ds^2 = h_1^2 dq_1^2 + h_2^2 dq_2^2 + h_3^2 dq_3^2, \quad (2.6.1)$$

and the gradient $\nabla\varphi$ of a scalar field φ by

$$\nabla\varphi = \frac{\hat{\mathbf{a}}_1}{h_1}\frac{\partial\varphi}{\partial q_1} + \frac{\hat{\mathbf{a}}_2}{h_2}\frac{\partial\varphi}{\partial q_2} + \frac{\hat{\mathbf{a}}_3}{h_3}\frac{\partial\varphi}{\partial q_3}. \quad (2.6.2)$$

The corresponding expressions for div $\mathbf{F}$ and **curl F** in orthogonal curvilinear coordinates for a vector field

$$\mathbf{F} = F_1\hat{\mathbf{a}}_1 + F_2\hat{\mathbf{a}}_2 + F_3\hat{\mathbf{a}}_3$$

may now be derived from their definitions (2.3.1) and (2.4.1).

The volume of the parallelopiped is $dV = h_1 h_2 h_3 dq_1 dq_2 dq_3$. The surface integral $\int_{\text{PBCD}} \mathbf{n} \cdot \mathbf{F} dS$ over the face PBCD is

$$-\mathbf{F} \cdot \hat{\mathbf{a}}_1 h_2 h_3 dq_2 dq_3 = -F_1 h_2 h_3 dq_2 dq_3,$$

and over the face A'B'P'D'

$$F_1 h_2 h_3 dq_2 dq_3 + \frac{\partial}{\partial q_1} \left(F_1 h_2 h_3 dq_2 dq_3 \right) dq_1,$$

so that the net contribution from the two faces is

$$\frac{\partial}{\partial q_1} (F_1 h_2 h_3) dq_1 dq_2 dq_3.$$

A similar calculation shows that the contributions from the remaining pairs of opposite faces are

$$\frac{\partial}{\partial q_2} (F_2 h_3 h_1) dq_1 dq_2 dq_3 \quad \text{and} \quad \frac{\partial}{\partial q_3} (F_3 h_1 h_2) dq_1 dq_2 dq_3.$$

Hence, by equating the sum of the surface integrals to div $\mathbf{F}\, h_1 h_2 h_3 dq_1 dq_2 dq_3$ we find

$$\text{div}\, \mathbf{F} = \frac{1}{h_1 h_2 h_3} \left[\frac{\partial}{\partial q_1} (F_1 h_2 h_3) + \frac{\partial}{\partial q_2} (F_2 h_3 h_1) + \frac{\partial}{\partial q_3} (F_3 h_1 h_2) \right].$$

$$(2.6.3)$$

To find the component of $\hat{\mathbf{a}}_1 \cdot \mathbf{curl}\, \mathbf{F}$ of $\mathbf{curl}\, \mathbf{F}$, the definition (2.4.1) is applied to the closed contour PBCD whose normal is $\hat{\mathbf{a}}_1$. The enclosed area is $h_2 h_3 dq_2 dq_3$. The contribution of the edge PB to $\oint_{PBCD} \mathbf{F} \cdot d\mathbf{r}$ is

$$\hat{\mathbf{a}}_2 \cdot \mathbf{F} h_2 dq_2 = F_2 h_2 dq_2,$$

and the contribution from CD is therefore

$$-F_2 h_2 dq_2 - \frac{\partial}{\partial q_3} (F_2 h_2 dq_2) dq_3,$$

yielding a net contribution equal to

$$-\frac{\partial}{\partial q_3} (F_2 h_2) dq_2 dq_3.$$

The contribution from the sides DP and BC is similarly found to be

$$\frac{\partial}{\partial q_2}(F_3 h_3)dq_2 dq_3.$$

Equating the contour integral to $\hat{\mathbf{a}}_1 \cdot \mathbf{curl\,F}\ h_2 h_3 dq_2 dq_3$ we find

$$\hat{\mathbf{a}}_1 \cdot \mathbf{curl\,F} \equiv (\mathbf{curl\,F})_1 = \frac{1}{h_2 h_3}\left[\frac{\partial}{\partial q_2}(F_3 h_3) - \frac{\partial}{\partial q_3}(F_2 h_2)\right].$$

The corresponding expressions for $(\mathbf{curl\,F})_2$ and $(\mathbf{curl\,F})_3$ may be obtained from this expression by permuting the subscripts. The final result may then be cast in the form

$$\mathbf{curl\,F} = \frac{1}{h_1 h_2 h_3}\begin{vmatrix} h_1\hat{\mathbf{a}}_1 & h_2\hat{\mathbf{a}}_2 & h_3\hat{\mathbf{a}}_3 \\ \frac{\partial}{\partial q_1} & \frac{\partial}{\partial q_2} & \frac{\partial}{\partial q_3} \\ h_1 F_1 & h_2 F_2 & h_3 F_3 \end{vmatrix}. \tag{2.6.4}$$

Example   The *cylindrical coordinates* (r, θ, z) of a point are related to the rectangular coordinates (x, y, z) by

$$x = r\cos\theta, \quad y = r\sin\theta.$$

Then

$$ds^2 = dr^2 + r^2 d\theta^2 + dz^2$$

$$h_1 = 1,\ h_2 = r,\ h_3 = 1$$

$$dV = rd\theta dr dz$$

$$\nabla\varphi = \frac{\partial\varphi}{\partial r}\hat{\mathbf{r}} + \frac{1}{r}\frac{\partial\varphi}{\partial\theta}\hat{\boldsymbol{\theta}} + \frac{\partial\varphi}{\partial z}\hat{\mathbf{z}}$$

$$\mathrm{div}\,\mathbf{F} = \frac{1}{r}\frac{\partial(rF_r)}{\partial r} + \frac{1}{r}\frac{\partial F_\theta}{\partial\theta} + \frac{\partial F_z}{\partial z}$$

$$\mathbf{curl\,F} = \frac{1}{r}\left(\frac{\partial F_z}{\partial\theta} - \frac{\partial(rF_\theta)}{\partial z}\right)\hat{\mathbf{r}} + \left(\frac{\partial F_r}{\partial z} - \frac{\partial F_z}{\partial r}\right)\hat{\boldsymbol{\theta}} + \frac{1}{r}\left(\frac{\partial(rF_\theta)}{\partial r} - \frac{\partial F_r}{\partial\theta}\right)\hat{\mathbf{z}}$$

$$\nabla^2\varphi = \frac{1}{r}\frac{\partial}{\partial r}\left(r\frac{\partial\varphi}{\partial r}\right) + \frac{1}{r^2}\frac{\partial^2\varphi}{\partial\theta^2} + \frac{\partial^2\varphi}{\partial z^2}. \tag{2.6.5}$$

Example For *spherical polar coordinates* (r, θ, ϕ)

$$x = r \cos\phi \sin\theta, \quad y = r \sin\phi \sin\theta, \quad z = r \cos\theta \ (0 < \theta < \pi, \ 0 < \phi < 2\pi)$$

Then

$$ds^2 = dr^2 + r^2 d\theta^2 + r^2 \sin^2\theta d\phi^2$$

$$h_1 = 1, \ h_2 = r, \ h_3 = r \sin\theta$$

$$dV = r^2 \sin\theta d\theta d\phi dr$$

$$\nabla\varphi = \frac{\partial\varphi}{\partial r}\hat{\mathbf{r}} + \frac{1}{r}\frac{\partial\varphi}{\partial\theta}\hat{\boldsymbol{\theta}} + \frac{1}{r \sin\theta}\frac{\partial\varphi}{\partial\phi}\hat{\boldsymbol{\phi}}$$

$$\operatorname{div}\mathbf{F} = \frac{1}{r^2}\frac{\partial(r^2 F_r)}{\partial r} + \frac{1}{r \sin\theta}\frac{\partial(\sin\theta F_\theta)}{\partial\theta} + \frac{1}{r \sin\theta}\frac{\partial F_\phi}{\partial\phi}$$

$$\operatorname{\mathbf{curl}}\mathbf{F} = \frac{1}{r \sin\theta}\left(\frac{\partial(\sin\theta F_\phi)}{\partial\theta} - \frac{\partial F_\theta}{\partial\phi}\right)\hat{\mathbf{r}} + \frac{1}{r \sin\theta}\left(\frac{\partial F_r}{\partial\phi} - \frac{\partial(r \sin\theta F_\phi)}{\partial r}\right)\hat{\boldsymbol{\theta}}$$

$$+ \frac{1}{r}\left(\frac{\partial(r F_\theta)}{\partial r} - \frac{\partial F_r}{\partial\theta}\right)\hat{\boldsymbol{\phi}}$$

$$\nabla^2\varphi = \frac{1}{r^2}\frac{\partial}{\partial r}\left(r^2\frac{\partial\varphi}{\partial r}\right) + \frac{1}{r^2 \sin\theta}\frac{\partial}{\partial\theta}\left(\sin\theta\frac{\partial\varphi}{\partial\theta}\right) + \frac{1}{r^2 \sin^2\theta}\frac{\partial^2\varphi}{\partial\phi^2}. \tag{2.6.6}$$

2.7   Evaluation of Line and Surface Integrals

Line integrals Consider line (or contour) integrals of the form

$$\int_C \mathbf{F}{\cdot}d\mathbf{r} \equiv \int_a^b \mathbf{F}{\cdot}d\mathbf{r} = \int_a^b (F_1 dx + F_2 dy + F_3 dz), \tag{2.7.1}$$

along a prescribed path C between the endpoints **a** and **b**. Such integrals are evaluated by introducing a parametric representation of the vector position **r** on C.

Example Evaluate $\int_C \mathbf{F}{\cdot}d\mathbf{r}$ along the section C of the *helix* defined by

$$\mathbf{r} = (\cos t, \sin t, 6t), \quad 0 < t < 4\pi,$$

when $\mathbf{F} = (y, z, x)$.

We have $d\mathbf{r} = (-\sin t, \cos t, 6)dt$ and $\mathbf{F} = (\sin t, 6t, \cos t)$

$$\therefore \quad \int_C \mathbf{F}{\cdot}d\mathbf{r} = \int_0^{4\pi} \left[-\sin^2 t + 6(t+1)\cos t\right]dt = -2\pi.$$

Condition for path independence The line integral (2.7.1) is independent of the path between any two points **a** and **b** provided **F** is the gradient of a scalar *potential* $\varphi(\mathbf{x})$, i.e.

$$\mathbf{F} = \nabla\varphi.$$

For, if $\mathbf{F} = \nabla\varphi$, then $\mathbf{F}\cdot d\mathbf{r}$ is an exact differential, and

$$\int_a^b \mathbf{F}\cdot d\mathbf{r} = \int_a^b \nabla\varphi \cdot d\mathbf{r} = \int_a^b d\varphi = \varphi(\mathbf{b}) - \varphi(\mathbf{a}),$$

depends only on the values of $\varphi(\mathbf{x})$ at the endpoints. On the other hand, if the integral is independent of the path between any two points in the domain $\mathcal{D}$ of definition of **F**, the function

$$\varphi(\mathbf{x}) = \int_a^\mathbf{x} \mathbf{F}\cdot d\mathbf{r},$$

defines a single valued function of **x** in $\mathcal{D}$, and

$$\varphi(\mathbf{x} + \delta\mathbf{x}) - \varphi(\mathbf{x}) = \int_\mathbf{x}^{\mathbf{x}+\delta\mathbf{x}} \mathbf{F}\cdot d\mathbf{r} \approx \mathbf{F}(\mathbf{x}) \cdot \delta\mathbf{x}, \quad \text{as} \quad \delta\mathbf{x} \to 0.$$

By taking in turn $\delta\mathbf{x} = \mathbf{i}\delta x,\ \mathbf{j}\delta y,\ \mathbf{k}\delta z$ it follows that

$$F_1 = \frac{\partial\varphi}{\partial x}, \quad F_2 = \frac{\partial\varphi}{\partial y}, \quad F_3 = \frac{\partial\varphi}{\partial z}, \quad \text{i.e. that} \quad \mathbf{F} = \nabla\varphi.$$

Note that **F** can always be expressed as a gradient if $\mathbf{curl\,F} = \mathbf{0}$. Indeed, the integrals

$$\mathcal{I}_1 = \int_{C_1} \mathbf{F}\cdot d\mathbf{r} \quad \text{and} \quad \mathcal{I}_2 = \int_{C_2} \mathbf{F}\cdot d\mathbf{r},$$

along any two paths C_1 and C_2 between any two points **a** and **b** are always equal, because

$$\mathcal{I}_1 - \mathcal{I}_2 = \oint_{C_{12}} \mathbf{F}\cdot d\mathbf{r} = \int_{S_{12}} \mathbf{curl\,F}\cdot d\mathbf{S} \equiv 0,$$

where C_{12} is the closed contour formed by C_1 traversed from **a** to **b** and C_2 traversed from **b** to **a**, and S_{12} is any open, two-sided surface bounded by C_{12} (Stokes' theorem).

Example Show that $\int_C \sin xy(y\,dx + x\,dy)$ is path independent and find the corresponding potential function.

Path independence is assured because $\mathbf{curl}\,\mathbf{F} = \mathbf{curl}\,(y\sin xy, x\sin xy, 0) \equiv \mathbf{0}$.

Let

$$\frac{\partial\varphi}{\partial x} = y\sin xy, \quad \frac{\partial\varphi}{\partial y} = x\sin xy, \quad \frac{\partial\varphi}{\partial z} = 0.$$

The last equation shows that φ does not depend on z. Integration of the first equation yields $\varphi = -\cos xy + f(y)$ for arbitrary $f(y)$. Substitution into the second equation then implies that $f(y) = $ constant, which may be discarded, because a potential function φ is always undetermined to within an arbitrary constant.

Problems 2E

Evaluate the line integrals $\int_C \mathbf{F}\cdot d\mathbf{r}$:

1. $\mathbf{F} = (3x^4, 3y^6, 0)$ where C is the curve: $x^2 + y^2 = 4$, $z = 0$ from $(2,0,0)$ to $(-2,0,0)$. $[\frac{-192}{5}]$

2. $\mathbf{F} = (z, x, y)$ where C is $\mathbf{r} = (\cos t, \sin t, t)$, $0 < t < 4\pi$. $[6\pi]$

3. $\mathbf{F} = (e^x, e^{4y/x}, e^{2z/y})$ where C is $\mathbf{r} = (t, t^2, t^3)$, $0 < t < 1$. $[\frac{3}{8}e^4 + \frac{3}{4}e^2 + e - \frac{13}{8}]$

Show that the following integrals are path-independent and find the corresponding potential functions:

4. $\int_C [(e^y - ze^x)dx + xe^y dy - e^x dz]$, $[\varphi = xe^y - ze^x]$.

5. $\int_C [y\cos xy\,dx + x\cos xy\,dy - dz]$, $[\varphi = \sin xy - z]$.

6. $\int_C [xe^{2z}dx + x^2 e^{2z}dz]$, $[\varphi = \frac{1}{2}x^2 e^{2z}]$.

7. Show that if $\mathbf{r} = \mathbf{r}(t)$ on C, the *length of arc* between $t = a$ and $t = b$ is given by $\ell = \int_a^b |\dot{\mathbf{r}}(t)|dt$.

8. $\mathbf{r} = \mathbf{i} + t\mathbf{j} + t^2\mathbf{k}$ on C. Show that the length of arc between $t = 0$, T is

$$\ell = \frac{1}{2}T\sqrt{1 + 4T^2} + \frac{1}{4}\ln\{2T + \sqrt{1 + 4T^2}\}.$$

9. $\mathbf{r} = (a\cos\theta,\ a\sin\theta,\ a\theta\tan\alpha)$ on a helix, where a, α are constants. Show that the length of arc measured from $\theta = 0$ is given by $\ell = a\theta\sec\alpha$.

Surface integrals Integrals of the form

$$\int_S \mathbf{F}\cdot d\mathbf{S} = \int_S \mathbf{F}\cdot\mathbf{n}\,dS = \int_S (F_1 n_1 + F_2 n_2 + F_3 n_3)dS, \qquad (2.7.2)$$

over an open or closed surface S, can be evaluated by introducing a representation of a point $\mathbf{r}$ on S of the type

$$\mathbf{r} = x(u, v)\mathbf{i} + y(u, v)\mathbf{j} + z(u, v)\mathbf{k}, \qquad (2.7.3)$$

where u, v are suitable parametric variables.

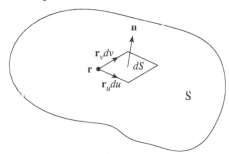

The point $\mathbf{r}(u, v)$ on S and the two neighbouring points

$$\mathbf{r} + \mathbf{r}_u du \equiv \mathbf{r} + \left(\frac{\partial \mathbf{r}}{\partial u}\right)_{(u,v)} du, \quad \text{and} \quad \mathbf{r} + \mathbf{r}_v dv \equiv \mathbf{r} + \left(\frac{\partial \mathbf{r}}{\partial v}\right)_{(u,v)} dv$$

define adjacent sides of a small parallelogram lying on S. The area dS of the parallelogram is $|(\mathbf{r}_u \times \mathbf{r}_v)dudv|$, and the unit normal at $\mathbf{r}$ is

$$\mathbf{n} = \pm\frac{(\mathbf{r}_u \times \mathbf{r}_v)dudv}{|(\mathbf{r}_u \times \mathbf{r}_v)dudv|} = \pm\frac{\mathbf{r}_u \times \mathbf{r}_v}{|\mathbf{r}_u \times \mathbf{r}_v|},$$

where the $\pm$ sign appears because of the ambiguity in the direction of $\mathbf{n}$. Hence, the vector surface element on S can be written

$$d\mathbf{S} = \mathbf{n}dS = \pm(\mathbf{r}_u \times \mathbf{r}_v)dudv, \tag{2.7.4}$$

and the surface integral (2.7.2) becomes

$$\int_S \mathbf{F} \cdot \mathbf{n}\, dS = \pm \int_S \mathbf{F} \cdot \mathbf{r}_u \times \mathbf{r}_v\, dudv. \tag{2.7.5}$$

In applications of this formula the correct sign must be determined from the conditions of the problem.

Example Determine the surface element for a sphere of radius R.
Using spherical polar coordinates (r, θ, ϕ):

$$\mathbf{r} = R(\sin\theta\cos\phi, \sin\theta\sin\phi, \cos\theta)$$

$$\mathbf{r}_\theta = R(\cos\theta\cos\phi, \cos\theta\sin\phi, -\sin\theta)$$

$$\mathbf{r}_\phi = R(-\sin\theta\sin\phi, \sin\theta\cos\phi, 0)$$

$$\therefore \quad \mathbf{r}_\theta \times \mathbf{r}_\phi = R^2(\sin^2\theta\cos\phi, \sin^2\theta\sin\phi, \sin\theta\cos\theta)$$

$$= R^2\sin\theta(\sin\theta\cos\phi, \sin\theta\sin\phi, \cos\theta) = R^2\sin\theta\,\hat{\mathbf{r}}.$$

Hence,

$$\mathbf{n}dS \equiv d\mathbf{S} = \mathbf{r}_\theta \times \mathbf{r}_\phi \, d\theta d\phi = \hat{\mathbf{r}}R^2 \sin\theta d\theta d\phi.$$

Example   Evaluate $\int_S \mathbf{F} \cdot \mathbf{n}dS$ when $\mathbf{F} = z^3(\mathbf{k} - \mathbf{i})$ and S is the surface of the section of the cone

$$\mathbf{r} = u\cos v \, \mathbf{i} + u\sin v \, \mathbf{j} + u \, \mathbf{k}, \quad 0 < u < 5, \ 0 < v < 2\pi,$$

with *outward* normal **n**.

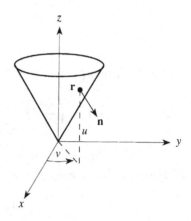

$$\mathbf{r}_u = (\cos v, \sin v, 1)$$

$$\mathbf{r}_v = u(-\sin v, \cos v, 0)$$

$$\mathbf{r}_u \times \mathbf{r}_v = u(-\cos v, -\sin v, 1)$$

$$\mathbf{F} \cdot \mathbf{n}dS = \pm u^4(\cos v + 1)dudv.$$

Choose the negative sign when the normal is outwards from the cone (see figure), to obtain

$$\int_S \mathbf{F} \cdot \mathbf{n}dS = -\int_0^{2\pi} dv \int_0^5 u^4(\cos v + 1)du = -1250\pi.$$

Example   Evaluate $\oint_C \mathbf{F} \cdot d\mathbf{r}$ when $\mathbf{F} = (-4y^3, 4x^3, z^3)$ and C is the contour defined by $x^2 + y^2 + z^2 = R^2$, $z = h$ ($|h| < R$) traversed in the positive direction with respect to the positive z-axis.

Apply Stokes' theorem $\oint_C \mathbf{F} \cdot d\mathbf{r} = \int_S \mathbf{n} \cdot \nabla \times \mathbf{F} \, dS$, where S is the circular surface $x^2 + y^2 \leq R^2 - h^2$, $z = h$. Then

$$\nabla \times \mathbf{F} = \begin{vmatrix} \mathbf{i} & \mathbf{j} & \mathbf{k} \\ \frac{\partial}{\partial x} & \frac{\partial}{\partial y} & \frac{\partial}{\partial z} \\ -4y^3 & 4x^3 & z^3 \end{vmatrix} = (0, 0, 12x^2 + 12y^2),$$

and the unit normal $\mathbf{n} = (0, 0, 1)$ is parallel to the positive z-axis. Transforming to polar coordinates $(x, y) = (r\cos\theta, r\sin\theta)$ to evaluate the surface integral:

$$\int_S \mathbf{n} \cdot \nabla \times \mathbf{F} \, dS = \int_S (0, 0, 1) \cdot (0, 0, 12x^2 + 12y^2)r dr d\theta$$

$$= 24\pi \int_0^{(R^2-h^2)^{\frac{1}{2}}} r^3 dr = 6\pi(R^2 - h^2)^2.$$

Problems 2F

Evaluate $\int_S \mathbf{F} \cdot d\mathbf{S}$:

1. $\mathbf{F} = (2x, 2y, 0)$ where S is the surface: $z = 2x + 3y$, $0 < x < 2$, $-1 < y < 1$.
   $[\pm 16]$

2. $\mathbf{F} = (e^y, -e^z, e^x)$ where S is $x^2 + y^2 = 9$, $x > 0$, $y > 0$, $0 < z < 2$.
   $[\pm(2e^3 - 3e^2 + 1)]$

3. $\mathbf{F} = (1, x^2, xyz)$ where S is $z = xy$, $0 < x < y$, $0 < y < 1$. $\left[\pm \frac{59}{180}\right]$

4. $\mathbf{F} = (z, -xz, y)$ where S is $x^2 + 9y^2 + 4z^2 = 36$, $0 < x$, $0 < y$, $0 < z$. $\left[\pm \frac{53}{2}\right]$

5. $\mathbf{F} = (2xy, x^2, 0)$ where S is $\mathbf{r} = (\cosh u, \sinh u, v)$, $0 < u < 2$, $-3 < v < 3$.
   $[+(2\cosh^3 2 - 2)]$

6. $\mathbf{F} = (x^3, 0, z^3)$ where S is the surface of the cube $|x| < 1$, $|y| < 1$, $|z| < 1$. $[\pm 16]$

7. $\mathbf{F} = (y^2, z^2, x^2 z)$ where S is the surface bounding the region $x^2 + y^2 \leq 4$, $x \geq 0$, $y \geq 0$, $|z| \leq 1$. $[\pm 2\pi]$

Use Stokes' theorem to evaluate $\oint_C \mathbf{F} \cdot d\mathbf{r}$:

8. $\mathbf{F} = xy\mathbf{i} - (2x - y)\mathbf{k}$ where C is the triangle with vertices $(0,0,0)$, $(1,1,0)$, $(1,0,0)$, traversed in this order. $\left[\frac{1}{3}\right]$

9. $\mathbf{F} = x^2 yz\mathbf{j}$ where C is the quadrilateral with vertices $(0,1,0)$, $(1,1,0)$, $(1,0,1)$, $(0,0,1)$ traversed in this order. $\left[-\frac{1}{6}\right]$

10. $\mathbf{F} = x^2 z\mathbf{j}$ where C is the triangle with vertices $(1,-1,0)$, $(1,1,0)$, $(0,0,1)$, traversed in this order. $\left[-\frac{1}{6}\right]$

11. $\mathbf{F} = xyz\mathbf{j}$ where C is the triangle with vertices $(1,0,0)$, $(0,1,0)$, $(0,0,1)$. $[0]$

12. $\mathbf{F} = (2y, z, 3y)$ where C is the circle $x^2 + y^2 + z^2 = 6z$, $z = x + 3$. $[\pm 18\pi\sqrt{2}]$

13. $\mathbf{F} = (-3y, 3x, z)$ where C is $x^2 + y^2 = 4$, $z = 1$. $[\pm 24\pi]$

14. $\mathbf{F} = x^2 z\mathbf{i} + xy^2\mathbf{j} + z^2\mathbf{k}$ where C is $x^2 + y^2 = 9$, $x + y + z = 1$, orientated anticlockwise when viewed from above. $\left[\frac{81}{2}\pi\right]$

15. $\mathbf{F} = 2z\mathbf{i} + 4x\mathbf{j} + 5y\mathbf{k}$ where C is $x^2 + y^2 = 4$, $z - x = 4$, orientated anticlockwise when viewed from above. $[-4\pi]$

2.8 Suffix Notation

Summation convention Algebraic manipulations of vectorial expressions can often be simplified by adopting an explicit suffix notation $\mathbf{F} = (F_1, F_2, F_3)$ for all vectors, including the position vector $\mathbf{x}$. Instead of writing $\mathbf{x} = (x, y, z)$ we write $\mathbf{x} = (x_1, x_2, x_3)$. The components of $\mathbf{x}$ are then x_i, where $i = 1$, 2 or 3, and we can talk about the vector x_i instead of $\mathbf{x}$.

The usual expansion of the scalar product of two vectors a_i and b_i

$$a_1 b_1 + a_2 b_2 + a_3 b_3,$$

Mathematical Methods for Mechanical Sciences

is the sum $\sum_{i=1}^{3} a_i b_i$ over all possible values of the suffix i. In this and more complicated formulae involving summations the expression to be summed is always found to contain the suffix to be summed *twice*. We therefore dispense with the summation sign and adopt the convention that whenever a repeated suffix occurs in a formula it is to be given in turn all possible values ($i = 1, 2, 3$) and the terms are to be added. The shorthand representation of the scalar product is then simply $a_i b_i$.

Differentiation The partial derivative of the jth component of the vector field $\mathbf{F}(\mathbf{x})$ with respect to the ith component of $\mathbf{x}$ is written $\partial F_j / \partial x_i$. When $j = i$ our convention gives

$$\frac{\partial F_i}{\partial x_i} = \frac{\partial F_1}{\partial x_1} + \frac{\partial F_2}{\partial x_2} + \frac{\partial F_3}{\partial x_3} \equiv \text{div } \mathbf{F}. \tag{2.8.1}$$

Similarly, for a scalar field $\varphi(\mathbf{x})$ the derivative $\partial \varphi / \partial x_i$ is just the ith component of $\nabla \varphi$. The directional derivative $\hat{\mathbf{a}} \cdot \nabla \varphi$ of φ in the direction of the unit vector $\hat{\mathbf{a}}$ can therefore be written

$$\hat{\mathbf{a}} \cdot \nabla \varphi = \hat{a}_i \frac{\partial \varphi}{\partial x_i}.$$

Kronecker delta δ_{ij} When $\mathbf{F} \equiv \mathbf{x}$ the partial derivative $\partial F_j / \partial x_i = \partial x_j / \partial x_i$, which is equal to 1 when $i = j$ and 0 if $i \neq j$. The Kronecker delta symbol δ_{ij} is defined to have this property:

$$\delta_{11} = \delta_{22} = \delta_{33} = 1,$$

$$\delta_{12} = \delta_{13} = \delta_{21} = \delta_{23} = \delta_{31} = \delta_{32} = 0,$$

so that

$$\frac{\partial x_j}{\partial x_i} \equiv \frac{\partial x_i}{\partial x_j} = \delta_{ij}. \tag{2.8.2}$$

A quantity like δ_{ij} with two independent suffixes has $3 \times 3 = 9$ different components. It is an example of a *second-order tensor* (The properties of tensors are not considered in this book, but it should be noted that an arbitrary set of nine numbers represented, say, by the symbol A_{ij} must satisfy certain well-defined conditions in order to qualify as a tensor.) A vector F_i is a one-dimensional tensor.

Alternating tensor e_{ijk} Consider the three unit vectors

$$\hat{\mathbf{e}}_1 = (1,0,0), \quad \hat{\mathbf{e}}_2 = (0,1,0), \quad \hat{\mathbf{e}}_3 = (0,0,1).$$

The determinant e_{ijk} formed by the triple scalar product $\hat{\mathbf{e}}_i \cdot \hat{\mathbf{e}}_j \times \hat{\mathbf{e}}_k$ defines the *three-dimensional alternating tensor* with 27 components that satisfy:

1. $e_{ijk} = 0$ if any two of i, j, k are equal;
2. $e_{ijk} = 1$ if i, j, k are all different and in *cyclic* order, i.e. $e_{123} = e_{231} = e_{312} = 1$;
3. $e_{ijk} = -1$ if i, j, k are all different and *not* in *cyclic* order, i.e. $e_{213} = e_{132} = e_{321} = -1$.

For any two vectors a_i, b_i the quantity

$$e_{ijk}a_jb_k \equiv \sum_{j=1}^{3}\sum_{k=1}^{3} e_{ijk}a_jb_k$$

is a *double* sum, because there are two repeated suffixes. For each value of i there are nine terms on the right-hand side. But only two of these are non-zero; for example, when $i = 1$ only the terms $j = 2$, $k = 3$ and $j = 3$, $k = 2$ are non-zero (for which $e_{123} = 1$, $e_{132} = -1$), so that

$$e_{1jk}a_jb_k = a_2b_3 - a_3b_2.$$

Similarly

$$e_{2jk}a_jb_k = a_3b_1 - a_1b_3 \quad \text{and} \quad e_{3jk}a_jb_k = a_1b_2 - a_2b_1.$$

But these are just the three components of the vector product $\mathbf{a} \times \mathbf{b}$, which therefore becomes in suffix notation

$$(\mathbf{a} \times \mathbf{b})_i \equiv e_{ijk}a_jb_k. \tag{2.8.3}$$

We can derive this useful formula directly by putting $\mathbf{a} = (a_1, a_2, a_3) \equiv a_j\hat{\mathbf{e}}_j$ and $\mathbf{b} = (b_1, b_2, b_3) \equiv b_k\hat{\mathbf{e}}_k$. Then

$$(\mathbf{a} \times \mathbf{b})_i = \hat{\mathbf{e}}_i \cdot (a_j\hat{\mathbf{e}}_j \times b_k\hat{\mathbf{e}}_k) = (\hat{\mathbf{e}}_i \cdot \hat{\mathbf{e}}_j \times \hat{\mathbf{e}}_k)a_jb_k = e_{ijk}a_jb_k.$$

Problems 2G

Establish the following formulae:

1. $\nabla^2 \varphi = \frac{\partial^2 \varphi}{\partial x_i \partial x_i} = \frac{\partial^2 \varphi}{\partial x_i^2}$.

2. $(\mathbf{curl\,F})_i = (\nabla \times \mathbf{F})_i \equiv e_{ijk} \frac{\partial F_k}{\partial x_j}$.

3. $\mathbf{a} \cdot \mathbf{b} \times \mathbf{c} = e_{ijk} a_i b_j c_k = \begin{vmatrix} a_1 & a_2 & a_3 \\ b_1 & b_2 & b_3 \\ c_1 & c_2 & c_3 \end{vmatrix}$.

4. $\delta_{ii} = 3$.

5. $\delta_{ij} e_{ijk} = 0$.

6. $e_{ijk} e_{ipq} = \delta_{jp}\delta_{kq} - \delta_{jq}\delta_{kp}$.

7. $e_{ijk} e_{ljk} = 2\delta_{il}$.

8. $e_{ijk} e_{ijk} = 6$.

9. $\left(\mathbf{a} \times (\mathbf{b} \times \mathbf{c})\right)_i \equiv e_{ijk} a_j e_{kpq} b_p c_q = b_i a_j c_j - c_i a_j b_j \equiv b_i \mathbf{a} \cdot \mathbf{c} - c_i \mathbf{a} \cdot \mathbf{b}$.

10. If $\mathbf{n}$ is the unit outward normal to the surface S of a sphere of unit radius, show that

$$\oint_S n_i n_j dS = \frac{4\pi}{3}\delta_{ij}, \qquad \oint_S n_i n_j n_k n_l dS = \frac{4\pi}{15}\left(\delta_{ij}\delta_{kl} + \delta_{ik}\delta_{jl} + \delta_{il}\delta_{jk}\right).$$

3

COMPLEX VARIABLES

3.1 Complex Numbers

A complex number has the form

$$z = x + iy, \quad \text{where} \quad i = \sqrt{-1};$$

x and y are real numbers, respectively called the *real* and *imaginary* parts of z, and sometimes denoted by $x = \operatorname{Re} z$, $y = \operatorname{Im} z$. The *complex conjugate* of z is

$$z^* = x - iy.$$

Algebraic manipulations with complex numbers generally obey the usual rules of algebra with the addition of $i \times i = i^2 = -1$.

Examples Express in the form $x + iy$:

1. $(3 - 2i)(1 + i) = 3 + i - 2i^2 = 5 + i.$

2. $\dfrac{(1 + i)}{(3 - 2i)} = \dfrac{(1 + i)(3 + 2i)}{(3 - 2i)(3 + 2i)} = \dfrac{1 + 5i}{9 + 4} = \dfrac{1}{13} + \dfrac{5i}{13}.$

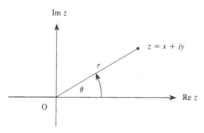

The complex plane, or *Argand diagram*

The complex number $z = x + iy$ can be represented by a point in the *complex plane* or *Argand diagram*, the real and imaginary parts of z being plotted respectively along the conventional x- and y-directions. Thus, z may also be interpreted as a *vector* whose magnitude and direction are specified by polar coordinates (r, θ), in terms of which the *polar form* of z is

$$z = x + iy = r(\cos\theta + i\sin\theta) \equiv re^{i\theta},$$

where $r \equiv |z| = \sqrt{x^2 + y^2} \ (= \sqrt{zz^*})$ is the *modulus* of z, and $\theta = \arctan\left(\frac{y}{x}\right)$ is the *argument*. Because θ is undefined to within a multiple of 2π, we introduce a *principal value* for the argument denoted by $\arg z$ which satisfies

$$-\pi < \arg z \leq \pi.$$

Examples

1. $z = 1 + i\sqrt{3} = 2e^{i(\frac{\pi}{3} + 2n\pi)}; \quad r = 2, \ \arg z = \frac{\pi}{3}.$
2. $z = 3 - 3i = 3\sqrt{2}e^{i(-\frac{\pi}{4} + 2n\pi)}; \quad r = 3\sqrt{2}, \quad \arg z = -\frac{\pi}{4}.$

Complex numbers obey the vector *parallelogram* law of addition

$$z_1 + z_2 = (x_1 + iy_1) + (x_2 + iy_2) = (x_1 + x_2) + i(y_1 + y_2).$$

In the complex plane addition is accomplished by completing the parallelogram whose adjacent sides are represented by z_1 and z_2. This geometrical construction shows that addition satisfies the *triangle inequality*

$$|z_1 + z_2| \leq |z_1| + |z_2|. \qquad (3.1.1)$$

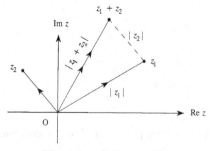

The triangle inequality

Example All complex numbers z satisfying $|z - z_0| = R > 0$, where $z_0 = a + ib$, lie on the circle $(x - a)^2 + (y - b)^2 = R^2$, with centre at $z = z_0$.

The multiplication and division of complex numbers assumes a particularly simple representation in polar form. If

$$z_1 = r_1(\cos\theta_1 + i\sin\theta_1) = r_1 e^{i\theta_1}, \quad z_2 = r_2(\cos\theta_2 + i\sin\theta_2) \equiv r_2 e^{i\theta_2},$$

then

$$z_1 z_2 = r_1 r_2[\cos(\theta_1 + \theta_2) + i\sin(\theta_1 + \theta_2)] = r_1 r_2 e^{i(\theta_1 + \theta_2)},$$

$$\frac{z_1}{z_2} = \frac{r_1}{r_2}\left[\cos(\theta_1 - \theta_2) + i\sin(\theta_1 - \theta_2)\right] = \frac{r_1}{r_2} e^{i(\theta_1 - \theta_2)}.$$

(3.1.2)

These results may be derived from the expansion formulae for trigonometric functions, $\cos(A \pm B) = \cos A \cos B \mp \sin A \sin B$, $\sin(A \pm B) = \sin A \cos B \pm \cos A \sin B$, or otherwise be regarded as obvious from the properties of the exponential function. Repeated application of the multiplication formula for $r_1 = r_2 = 1$ and $\theta_1 = \theta_2 = \theta$ supplies *De Moivre's formula*

$$(\cos\theta + i\sin\theta)^n = \cos n\theta + i\sin n\theta.$$

The nth root of a complex number For a positive integer n and complex number z, the equation

$$w = \sqrt[n]{z},$$

defines n distinct *roots* w.

Let $w = Re^{i\phi}$ and suppose that $z = re^{i\theta}$, where $\theta = \arg z$, i.e. $-\pi < \theta \leq \pi$. Then, because $w^n = z$,

$$R^n e^{in\phi} = re^{i\theta} \equiv re^{i(\theta + 2k\pi)}, \quad k = 0, \pm 1, \pm 2, \ldots,$$

$$\therefore \quad R = r^{\frac{1}{n}}, \quad \phi = \frac{\theta}{n} + \frac{2k}{n}\pi.$$

There are n different values of the argument ϕ that correspond to n distinct nth roots of z, obtained by taking $k = 0, 1, 2, \ldots, n - 1$. The *principal value* of the nth root is obtained by taking $k = 0$.

Example   Find all the values of $z^{\frac{1}{4}} \equiv \sqrt[4]{1 + i\sqrt{3}}$.

$$z = 1 + i\sqrt{3} = 2e^{\frac{i\pi}{3}}$$

$$\therefore \quad z^{\frac{1}{4}} = 2^{\frac{1}{4}} e^{i(\frac{\pi}{12} + \frac{k\pi}{2})} = 2^{\frac{1}{4}} i^k e^{\frac{i\pi}{12}}, \quad k = 0, 1, 2, 3.$$

Example   The cube roots of unity.

$$(1)^{\frac{1}{3}} = \left(e^{2k\pi i}\right)^{\frac{1}{3}} = e^0, \quad e^{\frac{2\pi i}{3}}, \quad e^{\frac{4\pi i}{3}} = 1, \quad -\frac{1}{2} \pm i\frac{\sqrt{3}}{2}.$$

In all cases the sum of the roots, $w_1 + w_2 + w_3 + \cdots + w_n$, vanishes. This is because the equation $w^n - z = 0$ is equivalent to $(w - w_1)(w - w_2)\ldots(w - w_n) = 0$, and the sum of the roots is the coefficient of $-w^{n-1}$ when the terms are multiplied out. The vectors $w_1, w_2, \ldots w_n$ may therefore be regarded as representing a system of forces in equilibrium.

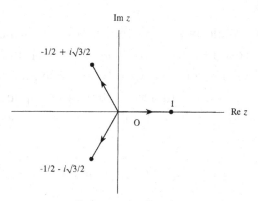

Cube roots of unity

Problems 3A

1. Express in the form $a + ib$:
   (i) $(2 + i)^2 + (2 - i)^2$;   (ii) $\frac{\alpha + \beta i}{\alpha - \beta i} - \frac{\alpha - \beta i}{\alpha + \beta i}$. $[6, \ 4i\alpha\beta/(\alpha^2 + \beta^2)]$

2. Show that
   $$\left|\frac{(2 - 3i)(3 + 4i)}{(6 + 4i)(15 - 8i)}\right| = \frac{5}{34}.$$

3. Find the modulus and principal value of the argument of: $z = -1, i, 3 + 4i$, $-i - \sqrt{3}$.
   $[1, \pi; \ 1, \frac{\pi}{2}; \ 5, 0.927 \text{ radians}; \ 2, -\frac{5\pi}{6}]$

4. Find $\arg z$ for: $z = -10 - i, \ 2 + 2i, \ 3 - 3i$. $[\theta = -3.042, \ \frac{\pi}{4}, \ -\frac{\pi}{4}]$

5. Express $\sqrt{5 + 12i}, \ \sqrt{-5 + 12i}, \ \sqrt{i}$ in the form $a + ib$. $[\pm(3 + 2i), \ \pm(2 + 3i), \ \pm\frac{1}{\sqrt{2}}(1 + i)]$

6. If $(x + iy)^4 = (a + ib)$, show that $a^2 + b^2 = (x^2 + y^2)^4$.

7. Find two real numbers a, b such that $(1+i)a + 2(1-2i)b - 3 = 0$. $[a = 2, b = \frac{1}{2}]$

8. If $z = x + iy$, express z^2, $1/z$, $(z^2+1)/z$ in the form $X + iY$, where X, Y are real functions of x, y. $[X = x^2 - y^2, Y = 2xy; X = x/(x^2 + y^2), Y = -y/(x^2 + y^2); X = x\{1 + 1/(x^2 + y^2)\}, Y = y\{1 - 1/(x^2 + y^2)\}]$

9. If z_1, z_2, z_3 are complex numbers, show that

   (i) $|z_1 + z_2|^2 + |z_1 - z_2|^2 = 2|z_1|^2 + 2|z_2|^2$
   (ii) $|2z_1 - z_2 - z_3|^2 + |2z_2 - z_3 - z_1|^2 + |2z_3 - z_1 - z_2|^2 = 3\{|z_2 - z_3|^2 + |z_3 - z_1|^2 + |z_1 - z_2|^2\}$.

10. Show that when a quadratic equation with real coefficients has a complex root $z = a + ib$, then the other root is the complex conjugate $z^* = a - ib$. If $z = 1 + 3i$ is a root of $x^4 + 16x^2 + 100 = 0$, find all the roots. $[1 + 3i, 1 - 3i, -1 + 3i, 1 - 3i]$

11. Show that
$$\frac{1 + \cos\theta + i\sin\theta}{1 - \cos\theta + i\sin\theta} = \cot\left(\frac{\theta}{2}\right) e^{i(\theta - \frac{\pi}{2})}.$$

Evaluate the roots:

12. $\sqrt{-8i}$. $[\pm 2(1 - i)]$

13. $\sqrt[8]{1}$. $[\pm 1, \pm i, \pm(1 \pm i)/\sqrt{2}]$

14. $\sqrt[3]{1 + i}$. $[2^{\frac{1}{6}} e^{\frac{ik\pi}{12}}, k = 1, 9, 17]$

15. $z^2 - (5 + i)z + 8 + i = 0$. $[z = 3 + 2i, 2 - i]$

16. If $w = 4/z$ and the point representing z in the complex plane describes a circle of unit radius whose centre is at $1 + i$, show that the point representing w in the complex w-plane describes a circle of radius 4.

17. The point $z = x + iy$ moves along a curve in the complex z-plane defined by the equation $f(z) = $ constant. What curves are represented by the equations: $|z - 1| = 2, |z + 1| - |z - 1| = 0, \operatorname{Re}(z^2) = 1$? [Circle of radius 2 with centre at (1,0); straight line $x = 0$; hyperbola $x^2 - y^2 = 1$].

3.2   Functions of a Complex Variable

A complex valued function $f(z)$ of the complex variable z associates a complex number

$$w = f(z),$$

with each given value of z. The function relates corresponding points in the complex z- and w-planes, and is said to provide a transformation (or *map*) between the planes.

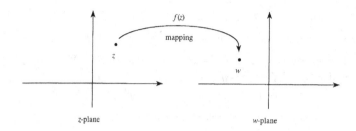

As the point z varies over the whole of the z-plane its *image* will vary over a certain region of the w-plane that may or may not include all possible points of the w-plane. Alternatively, the function $f(z)$ may be defined only over a certain region (its *range* or *domain of definition*) of the z-plane, and then defines a mapping of that region onto a corresponding image domain in the w-plane.

It is frequently useful to split $f(z)$ into its real and imaginary parts:

$$w = f(z) \equiv u(x,y) + iv(x,y).$$

If $u(x,y)$ and $v(x,y)$ are continuous functions of x and y in the usual sense, then $f(z)$ will be a continuous function of z.

Example

$$w = z^3 = (x+iy)^3 = x^3 - 3xy^2 + i(3x^2y - y^3),$$
$$\therefore \quad u(x,y) = x^3 - 3xy^2, \quad v(x,y) = 3x^2y - y^3 \tag{3.2.1}$$

Differentiation $f(z)$ is said to be *differentiable* at z with derivative $f'(z)$ if the following limit exists

$$f'(z) = \lim_{\delta z \to 0} \frac{f(z+\delta z) - f(z)}{\delta z}, \tag{3.2.2}$$

where the complex number $z + \delta z$ may approach z along *any* path as $\delta z \to 0$.

The condition that the limit should exist independently of the path by which $\delta z \to 0$ imposes a severe restriction on the class of functions that are differentiable. Generally speaking ordinary (real or complex valued) differentiable functions $f(x)$ of the real variable x are differentiable when regarded as a function of the complex variable z. But any combination of the form $f(x,y) = u(x,y) + iv(x,y)$, where the real functions $u(x, y)$ and $v(x, y)$ are differentiable with respect x and y, is not necessarily differentiable with respect to z.

Example $f(z) = z^2 = x^2 - y^2 + 2ixy$ is differentiable for all z, because

$$\frac{(z + \delta z)^2 - z^2}{\delta z} = \frac{(2z + \delta z)\delta z}{\delta z} \to 2z, \quad \text{as} \ \ \delta z \to 0,$$

but $f(z) = (z^*)^2 = x^2 - y^2 - 2ixy$ is not differentiable:

$$\frac{f(z + \delta z) - f(z)}{\delta z} = \frac{(2z^* + \delta z^*)\delta z^*}{\delta z}$$

$$= (2z^* + \delta z^*)e^{-2i\theta} \to 2z^* e^{-2i\theta}, \quad \text{where} \ \ \theta = \arg \delta z,$$

depends on the direction θ at which the point z is approached.

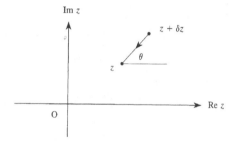

Regular function $f(z)$ is said to be *regular* (or *analytic*) in a domain $\mathcal{D}$ of the complex plane if $f'(z)$ exists at all points of $\mathcal{D}$. A point where $f(z)$ is not regular is called a *singularity*.

Example $f(z) = \frac{1}{z}$ is regular everywhere except for a singularity at $z = 0$, where it is undefined.

The Cauchy–Riemann equations Suppose that $f(z) = u(x, y) + iv(x, y)$ is regular. Introduce the shorthand notation

$$u_x = \frac{\partial u}{\partial x}, \quad u_y = \frac{\partial u}{\partial y} \quad \text{and} \quad v_x = \frac{\partial v}{\partial x}, \quad v_y = \frac{\partial v}{\partial y},$$

then, because $\delta z = \delta x + i\delta y$,

$$f'(z) = \lim_{\delta x, \delta y \to 0} \frac{(u_x + iv_x)\delta x + (u_y + iv_y)\delta y}{(\delta x + i\delta y)}.$$

According to the definition (3.2.2) this limiting operation should not depend on the manner in which $\delta z = \delta x + i\delta y \to 0$. In particular, we can let $\delta z \to 0$ along a direction parallel to the real axis, in which case $\delta y = 0$ and $\delta z = \delta x$; similarly, δz can tend to zero along a direction parallel to the imaginary axis, so that $\delta x = 0$ and $\delta z = i\delta y$. These limits give the following alternative representations of $f'(z)$

$$f'(z) = \frac{\partial u}{\partial x} + i\frac{\partial v}{\partial x} = -i\frac{\partial u}{\partial y} + \frac{\partial v}{\partial y}.$$

Equating the real and imaginary parts of these formulae supplies the **Cauchy–Riemann equations**:

$$\frac{\partial u}{\partial x} = \frac{\partial v}{\partial y}, \quad \frac{\partial u}{\partial y} = -\frac{\partial v}{\partial x}. \tag{3.2.3}$$

These compatibility conditions must be satisfied by the functions $u(x, y)$ and $v(x, y)$ when $f(z) = u(x, y) + iv(x, y)$ is regular.

The *converse* is also true, namely: If $u(x, y), v(x, y)$ satisfy the Cauchy–Riemann equations *then* $f(z) = u(x, y) + iv(x, y)$ is regular.

To prove this we must show that $f'(z) = \lim_{\delta z \to 0} \delta f/\delta z$ exists independently of the way in which $\delta z \to 0$. Indeed, when equations (3.2.3) are satisfied

$$
\begin{aligned}
\delta f &= (u_x + iv_x)\delta x + (u_y + iv_y)\delta y \\
&= (u_x + iv_x)\delta x + (-v_x + iu_x)\delta y \quad \text{(using (3.2.3))} \\
&= (u_x + iv_x)\delta x + i(u_x + iv_x)\delta y \\
&= (u_x + iv_x)(\delta x + i\delta y) \equiv (u_x + iv_x)\delta z,
\end{aligned}
$$

$$\therefore \quad f'(z) = \lim_{\delta z \to 0} \frac{\delta f}{\delta z} = u_x + iv_x \quad \text{exists.}$$

When $f(z) = u(x, y) + iv(x, y)$ is regular, the real and imaginary parts satisfy the Cauchy–Riemann equations (3.2.3). By eliminating in turn v and u between these equations it follows that $u(x, y)$ and $v(x, y)$ are each solutions of **Laplace's equation**:

$$\frac{\partial^2 u}{\partial x^2} + \frac{\partial^2 u}{\partial y^2} = 0, \quad \frac{\partial^2 v}{\partial x^2} + \frac{\partial^2 v}{\partial y^2} = 0,$$

and that the real and imaginary parts of an analytic function are necessarily solutions of the Laplace equation in two dimensions.

Example $w = z^3$ is regular because (see (3.2.1))

$$u_x = v_y = 3x^2 - 3y^2 \quad \text{and} \quad u_y = -v_x = -6xy,$$

where $u_x = \partial u / \partial x$, etc.

Example The exponential function $w = e^z = e^x e^{iy} = e^x [\cos y + i \sin y]$ is an *entire* function, i.e. it is regular for all values of z

Example The complex sine and cosines

$$\cos z = \frac{e^{iz} + e^{-iz}}{2}, \quad \sin z = \frac{e^{iz} - e^{-iz}}{2i},$$

are entire functions, but $\tan z = \sin z / \cos z$ is regular except at $z = (2n+1)\frac{\pi}{2}$ where $\cos z = 0$.

Example The complex hyperbolic sine and cosines

$$\cosh z = \frac{e^z + e^{-z}}{2}, \quad \sinh z = \frac{e^z - e^{-z}}{2},$$

are entire.

Example The logarithmic function $w \equiv u + iv = \operatorname{Ln} z$ is defined by the equation $e^w = z$. If $z = re^{i\theta}$, then $e^w = e^u e^{iv} \equiv r e^{i(\theta + 2n\pi)}$, $n = 0, \pm 1, \pm 2, \ldots$. Hence $\operatorname{Ln} z = \ln r + i(\theta + 2n\pi)$. If $\theta = \arg z$ (the principal value of the argument of z), the *principal value* of $\operatorname{Ln} z$ is denoted by $\ln z$ and defined by

$$\ln z = \ln r + i \arg z.$$

By writing $u = \frac{1}{2} \ln(x^2 + y^2)$, $v = \arctan\left(\frac{y}{x}\right) + 2n\pi$ we can verify that u and v satisfy the Cauchy–Riemann equations except at $z = 0$, and that

$$\frac{d}{dz} \ln z = \frac{1}{z}, \quad z \neq 0.$$

Example We define the complex power a^z by $a^z = e^{z(\ln a + 2n\pi i)}, n = 0, \pm 1, \pm 2, \ldots$, so that there are infinitely many values when a is a complex number not equal to an integer or a rational fraction. The *principal value* of a^z is defined to be $e^{z \ln a}$.

Problems 3B

Express the principal values in the form $x + iy$:

1. $i^{\frac{1}{2}}$. $[(1 + i)/\sqrt{2}]$

2. $(5 - 2i)^{(3 + \pi i)}$. $[-276.2 - 436.0i]$

3. i^i. $[e^{-\frac{\pi}{2}}]$

3.3   Integration in the Complex Plane

Integration in the complex plane between two points $z = a$ and $z = b$ along a contour C joining the points is defined by the limiting operation

$$\int_C f(z)dz = \lim_{n \to \infty} \sum_{j=1}^{n} f_j \delta z_j, \tag{3.3.1}$$

where $a = z_0, z_1, \ldots, z_j, \ldots, z_n = b$ is an ordered sequence of points along the contour from a to b, f_j is the value of $f(z)$ at an arbitrary point z on C within the segment (z_{j-1}, z_j), and the limit is taken in such a manner that $\delta z_j = z_j - z_{j-1} \to 0$ as $n \to \infty$.

By setting $f = u + iv$ and $dz = dx + idy$, the real and imaginary parts of the integral can be represented as conventional line integrals in the xy-plane:

$$\int_C f(z)dz = \int_C \left\{ u(x,y)dx - v(x,y)dy \right\} + i \int_C \left\{ v(x,y)dx + u(x,y)dy \right\}$$

$$\tag{3.3.2}$$

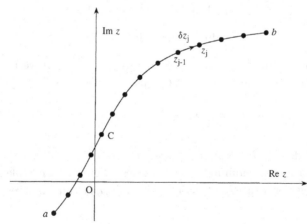

If C is defined parametrically by $z \equiv z(t) = x(t) + iy(t)$, $t_a < t < t_b$, then

$$\int_C f(z)dz = \int_{t_a}^{t_b} \left\{ \Big(u(t)\dot{x}(t) - v(t)\dot{y}(t) \Big) + i \Big(u(t)\dot{y}(t) + v(t)\dot{x}(t) \Big) \right\} dt,$$

$$\tag{3.3.3}$$

where $u(t) = u(x(t), y(t)), v(t) = v(x(t), y(t))$ and $\dot{x}(t) = dx/dt$, $\dot{y}(t) = dy/dt$.

Evidently, when the direction of integration is reversed: $\int_b^a f(z)dz = -\int_a^b f(z)dz$. The precise value of $\int_a^b f(z)dz$ generally depends on the route followed by the path of integration between a and b.

Example $f(z) = \text{Im } z \equiv y$ is not an analytic function (it does not satisfy the Cauchy–Riemann equations). Consider the integral along the curve $y = Y(x)$ over the interval $0 < x < X$, and put $Z = X + iY(X)$:

$$\int_{iY(0)}^{Z} y\,dz = \int_{iY(0)}^{Z} \left[y\,dx + iy\,dy \right]$$

$$= \int_0^X Y(x)dx + \left[\frac{iY(x)^2}{2} \right]_0^X = \int_0^X Y(x)dx + \frac{i}{2}\left[Y(X)^2 - Y(0)^2 \right].$$

The imaginary part $\frac{i}{2}\left[Y(X)^2 - Y(0)^2 \right]$ does not depend on the integration path, but the remaining integral does because it is just equal to the area between the curve $y = Y(x)$ and the x-axis.

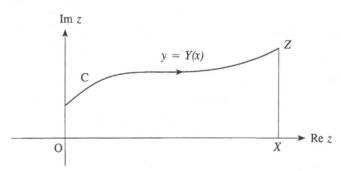

Example Evaluate $\int_C z^2 dz$ along the circular arc $|z| = R$ over (i) $0 < \arg z < \frac{\pi}{2}$, (ii) $0 < \arg z < 2\pi$.

Set $z = Re^{i\theta}$, $dz = iRe^{i\theta}d\theta$, then

(i) $$\int_C z^2 dz = iR^3 \int_0^{\frac{\pi}{2}} e^{3i\theta} d\theta = \frac{R^3}{3}\left[e^{\frac{3\pi i}{2}} - 1 \right] = -\frac{R^3}{3}(1 + i);$$

(ii) $$\int_C z^2 dz = iR^3 \int_0^{2\pi} e^{3i\theta} d\theta = \frac{R^3}{3}\left[e^{6\pi i} - 1 \right] = 0.$$

Integrals of analytic functions   A regular function $f(z)$ always possesses an indefinite integral $F(z)$ which is also regular and satisfies $F'(z) = f(z)$ (see §3.4). In the summation of (3.3.1) we can therefore set

$$f_j \delta z_j = F(z_j) - F(z_{j-1}) \quad \text{as } \delta z_j \to 0,$$

so that the integral becomes

$$\int_C f(z)dz = \Big(F(z_1) - F(a)\Big) + \Big(F(z_2) - F(z_1)\Big)$$

$$+ \cdots + \Big(F(z_{n-1}) - F(z_{n-2})\Big) + \Big(F(b) - F(z_{n-1})\Big)$$
$$= F(b) - F(a).$$

It is evident that the arguments of z_j and z_{j-1} of adjacent points on the integration contour C must ultimately be equal as $\delta z_j \to 0$. Thus, when a definite integral is expressed in the form

$$\int_a^b f(z)dz = F(b) - F(a), \quad \text{where } F'(z) = f(z), \tag{3.3.4}$$

it is implicitly understood that when $F(a)$ has been calculated, the appropriate value of $F(b)$ is determined by taking $\arg b$ to be equal to $\arg a$ plus the smooth increase obtained when the point z translates along C from a to b. This precaution will ensure that the same 'branch' of the function $F(z)$ is used at each end of C.

Example   Evaluate $\int_{-i}^{i} \frac{dz}{z}$ along the contours C_+ and C_-.

$$C_+: \int_{-i}^{i} \frac{dz}{z} = \Big[\ln z\Big]_{-i}^{i} = \Big[\ln r + i\theta\Big]_{\theta=-\frac{\pi}{2}}^{-\frac{\pi}{2}+\pi} = \pi i;$$

$$C_-: \int_{-i}^{i} \frac{dz}{z} = \Big[\ln z\Big]_{-i}^{i} = \Big[\ln r + i\theta\Big]_{\theta=-\frac{\pi}{2}}^{-\frac{\pi}{2}-\pi} = -\pi i.$$

Example   Evaluate $\int_{-i}^{i} \frac{dz}{\sqrt{z}}$ along the contours C_+ and C_-.

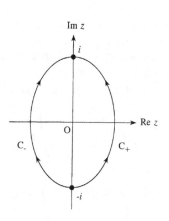

$$C_+: \int_{-i}^{i} \frac{dz}{\sqrt{z}} = \Big[2\sqrt{z}\Big]_{-i}^{i} = \Big[2\sqrt{r}e^{\frac{i\theta}{2}}\Big]_{\theta=-\frac{\pi}{2}}^{-\frac{\pi}{2}+\pi} = 2\sqrt{2}i;$$

$$C_-: \int_{-i}^{i} \frac{dz}{\sqrt{z}} = \Big[2\sqrt{z}\Big]_{-i}^{i} = \Big[2\sqrt{r}e^{\frac{i\theta}{2}}\Big]_{\theta=-\frac{\pi}{2}}^{-\frac{\pi}{2}-\pi} = -2\sqrt{2}.$$

Branch cuts In order to avoid 'ambiguous' results of the kind illustrated by the two previous examples it is often convenient to draw 'barriers' in the complex plane across which the complex variable z cannot pass. For example, the principal values of $\ln z$ and $\sqrt{z}$ are defined by the condition that $-\pi < \arg z \leq \pi$, and for these functions it is usual to make the negative real axis into a barrier. The complex plane is then said to be 'cut' along the negative real axis. In the cut plane $\arg z$ cannot wander outside the range $-\pi < \arg z \leq \pi$, where $\sqrt{z}$ is a single-valued function of position. The difficulty with $\sqrt{z}$, for example, is that it is really a two-valued function of z, namely $\sqrt{z} = \pm\sqrt{r}e^{\frac{i\theta}{2}}$, $-\pi < \theta \leq \pi$. These values are called the two *branches* of $\sqrt{z}$, each of which is one-valued in the cut plane. We can, however, consider *two* separate complex planes P_1 and P_2, each of which is cut along the negative real axis, with the first branch of $\sqrt{z}$ defined on P_1 and the second branch on P_2. We can then imagine that the planes are superposed, and that the upper edge ($\theta = \pi$) of the cut on P_1 is joined to the lower edge of the cut on P_2, and the lower edge ($\theta = -\pi$) of the cut on P_1 is joined to the upper edge of the cut on P_2. Starting on P_1, a complex number z can then be made to follow a continuous closed path encircling the origin twice in, say, the anticlockwise direction. When z first crosses the cut it passes from P_1 onto P_2; the second crossing of the cut brings z back onto P_1 and to its starting point. The double surface formed by joining the two planes along the cut is called a two-sheeted *Riemann surface*, on which both branches of $\sqrt{z}$ define $\sqrt{z}$ as a single-valued function, being equal to its first branch on P_1 and to the second branch on P_2. The point $z = 0$ is called a *branch point*; it is only by going around this point that the two branches of $\sqrt{z}$ can be realised.

The logarithmic function $\operatorname{Ln} z = \ln r + i\Theta$, ($\Theta = \arg z + 2n\pi$, $n = 0, \pm1, \pm2, \ldots$) has *infinitely* many branches; the nth branch is obtained by using the cut to restrict Θ to the range $(2n - 1)\pi < \Theta \leq (2n + 1)\pi$. The Riemann surface for this function has infinitely many sheets; a point encircling in one direction the branch point at $z = 0$ can never return to its starting point.

When a cut is taken along the negative real axis it is not possible to integrate along the contour C_- in the previous figure except by passing onto another sheet of the Riemann surface. This is often undesirable, and the integration contour must then be deformed to pass *around* the branch cut in order to stay on the original Riemann surface and with the original branch of the function; the path C_- is then replaced by C'_- (see the figure below) which passes around the branch point at $z = 0$ in the anticlockwise direction; $\ln z$ and $1/\sqrt{z}$ are defined by their principal values on C'_-. Of course, the corresponding integrals along C_+ and C'_- are equal, because the net change in $\arg z$ along each contour is the same.

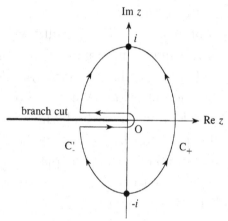

Upper bound for a contour integral (the '*ML* theorem'): Let M be a constant such that $|f(z)| \le M$ along an integration contour. Then, because $|f_j \delta z_j| \equiv |f_j||\delta z_j|$, repeated application of the triangle inequality (3.1.1) shows that

$$\left| \int_C f(z)dz \right| = \lim_{n \to \infty} \left| \sum_{j=1}^{n} f_j \delta z_j \right|$$

$$\le \lim_{n \to \infty} \sum_{j=1}^{n} |f_j||\delta z_j| \le M \lim_{n \to \infty} \sum_{j=1}^{n} |\delta z_j| = ML$$

where L is the length of the contour, i.e.

$$\left| \int_C f(z)dz \right| \le ML. \tag{3.3.5}$$

Example   Find an upper bound for $|\int_C z^4 dz|$ where C is the arc of the quarter circle $|z| = R$, $0 < \arg z < \frac{\pi}{2}$.

On C $|z|^4 = R^4$ and $L = \frac{\pi R}{2}$

$$\therefore \quad \left| \int_C z^4 dz \right| \le \frac{M \pi R}{2} = \frac{\pi R^5}{2}.$$

The exact value of $|\int_C z^4 dz| = \frac{\sqrt{2}R^5}{5} < \frac{\pi R^5}{2}$.

Problems 3C

Evaluate the integrals (closed contours are traversed in the anticlockwise direction):

1. $\int_C (az + b)dz$; a, b are constants and C is the straight line from $-(i + i)$ to $+(1 + i)$. $[2b(1 + i)]$

2. $\int_C \left(3z^3 + \frac{2}{z^2}\right) dz$; C is the unit circle $|z| = 1$. $[0]$

3. $\int_C \text{Re}\{z^2\}dz$; C is the square with corners $0, 1, 1 + i, i$. $[1 + i]$

4. $\int_C \left(\frac{3}{z-2} + \frac{2}{(z-2)^3}\right) dz$; C is the circle $|z| = 3$. $[6\pi i]$

5. $\int_0^{\frac{\pi i}{6}} \sinh 3z \, dz$. $[\frac{-1}{3}]$

6. $\int_{-\pi i}^{\pi i} \cos^2 z \, dz$. $[\pi i + \frac{i}{2} \sinh 2\pi]$

Let C_1 be the straight line path $z = (1 + i)t$, $0 \le t \le 1$; C_2 the quarter circle $z = 1 - \cos t + i \sin t$, $0 \le t \le \frac{\pi}{2}$; and C_3 the path $z = t$, $(0 \le t \le 1)$, $1 + i(t-1)$ $(1 \le t \le 2)$. Show that:

7. $\int_{C_1} z \, dz = \int_{C_2} z \, dz = \int_{C_3} z \, dz = i$.

8. $\int_{C_1} z^* \, dz = 1$, $\int_{C_2} z \, dz = 1 + i(1 - \frac{\pi}{2})$, $\int_{C_3} z \, dz = 1 + i$.

9. From (3.3.4) show that for every path from $z = 0$ to $z = Z$: $\int_0^Z dz = Z$, $\int_0^Z z dz = \frac{1}{2}Z^2$, $\int_0^Z e^z dz = e^Z - 1$. In each case verify the upper bound formula (3.3.5) for a straight line path of integration by taking $M = \max |f(z)|$ on the path.

10. Show that $\int_{C_1} dz/z$ taken along a semi-circular arc from -1 to $+1$ has the value $-\pi i$ or $+\pi i$ according as the arc lies above or below the real axis.

3.4 Cauchy's Theorem

Let C be a *simple closed contour*, i.e. a closed loop that does not intersect itself. We adopt the convention, that as C is traversed in the *positive* direction the *interior* region S is to the left. Cauchy's theorem concerns the integration of a regular function in the positive sense around a closed contour, and it will henceforth be assumed that all curves are to be traversed in this sense unless otherwise indicated.

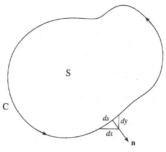

Simple closed curve

Cauchy's theorem   Let $f(z)$ be continuous on a simple closed curve C and regular (analytic) within C, then

$$\oint_C f(z)dz = 0, \qquad\qquad (3.4.1)$$

where the notation $\oint$ implies that the integration is taken around the whole of C.

This is proved by putting $ds = |dz| > 0$, the element of arc length on C, and introducing the outward unit normal **n** on C. Then

$$\mathbf{n}ds = (dy, -dx),$$

and the integral is transformed by means of the two-dimensional divergence theorem (2.3.6) to yield

$$\oint_C f(z)dz \equiv \oint_C f(z)\left(dx + idy\right) = \oint_C \left(if(z), -f(z)\right) \cdot \left(dy, -dx\right)$$

$$= \oint_C \left(if(z), -f(z)\right) \cdot \mathbf{n}ds = \int_S \left(i\frac{\partial f}{\partial x} - \frac{\partial f}{\partial y}\right) dxdy.$$

The final surface integral vanishes because $f'(z) = \partial f(z)/\partial x = \partial f(z)/i\partial y$ for a regular function.

This proof depends on the assumption (implicit in the divergence theorem) that the partial derivatives u_x, u_y, v_x, v_y are *piecewise continuous* within C. This is the case for most functions that arise in applications, but Goursat has shown that it is actually sufficient to assume that $f(z)$ is differentiable everywhere within and on C.

Path independence and deformation The contour integrals $\int_{C_1} f(z)dz, \int_{C_2} f(z)dz$ along two different paths C_1, C_2 between two points a and b are equal provided C_1 can be continuously deformed onto C_2 without encountering any singularities of $f(z)$. This is because Cauchy's theorem implies that the integral

$$\oint_C f(z)dz = \int_{C_1} f(z)dz - \int_{C_2} f(z)dz \equiv 0,$$

where C is the closed contour formed by C_1 traversed in the positive direction and C_2 in the negative direction. Similarly, if $f(z)$ is regular within the annular region formed when a simple closed curve C_2 is

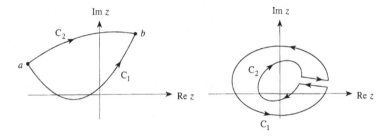

enclosed by a 'larger' simple closed curve C_1, then

$$\oint_{C_1} f(z)dz = \oint_{C_2} f(z)dz, \tag{3.4.2}$$

because Cauchy's theorem implies that the integral $\oint_C f(z)dz = 0$, where C is the simple closed curve formed by C_1 traversed in the positive direction, C_2 and traversed in the negative direction, and the two sides of the 'cut' between C_1 and C_2 illustrated in the figure. The two contributions from the cut are equal and opposite as the width of the cut tends to zero.

In the more general case where C_1 encloses several closed, *non-overlapping* contours $C_2, C_3, \ldots$, etc, the integral around C_1 is clearly equal to the *sum* of the integrals around the interior contours.

Example $\oint_C \frac{dz}{z} = 0$, where C is any closed curve not enclosing the origin $z = 0$.

Example   Evaluate $\oint_{|z|=2} \frac{(z^7 + 3z - 2)dz}{(z-1)(7-z)}$.

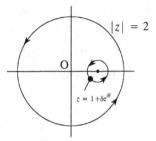

The integrand is regular within the circle $|z| = 2$ except at $z = 1$. By (3.4.2), the integration contour may therefore be collapsed onto a small circle of radius δ that just encloses this point. On this circle $z = 1 + \delta e^{i\theta}, dz = i\delta e^{i\theta} d\theta$, and

$$\oint_{|z|=2} \frac{(z^7 + 3z - 2)dz}{(z-1)(7-z)} = \int_0^{2\pi} \frac{[(1 + \delta e^{i\theta})^7 + 3(1 + \delta e^{i\theta}) - 2]i\delta e^{i\theta}\, d\theta}{\delta e^{i\theta}[7 - (1 + \delta e^{i\theta})]}$$

$$\rightarrow \int_0^{2\pi} \frac{2i\, d\theta}{6} = \frac{2\pi i}{3}, \quad \text{as } \delta \to 0.$$

Example   Show that $\displaystyle\oint_{|z-z_0|>0} (z-z_0)^n dz = \begin{cases} 0 & n \neq -1 \\ 2\pi i & n = -1 \end{cases}$ where n is an integer.

(3.4.3)

Set $z - z_0 = \delta e^{i\theta}$, then

$$\oint_{|z-z_0|>0} (z - z_0)^n dz = \int_0^{2\pi} i\delta^{n+1} e^{i(n+1)\theta} d\theta$$

$$= i\delta^{n+1} \int_0^{2\pi} [\cos(n+1)\theta + i\sin(n+1)\theta]d\theta = \begin{cases} 0 & n \neq -1 \\ 2\pi i & n = -1 \end{cases}.$$

Problems 3D

Evaluate:

1. $\oint_{|z|=2} \frac{dz}{z+i}$. $[2\pi i]$

2. $\oint_{|z|=3} \frac{\cos z\, dz}{z} - \oint_{|z|=1} \frac{\cos z\, dz}{z}$. $[0]$

3. $\oint_C \frac{dz}{z^2 - \frac{1}{4}}$ where C is the square with corners $\pm(1 \pm i)$. $[0]$

4. $\oint_{|z-i|=1} \frac{dz}{(z^2+i)}$. $[\frac{\pi}{\sqrt{2}}(1 - i)]$

5. $\oint_{|z-\frac{\pi}{2}|=1} \frac{\sin z\, dz}{\left(z^2 - \frac{\pi^2}{4}\right)}$. $[2i]$

6. $\oint_{|z|=\frac{1}{2}|a|} \frac{dz}{z-a}$. $[0]$

7. $\oint_{|z|=\frac{3}{2}|a|} \frac{dz}{z-a}$. $\quad [2\pi i]$

8. $\oint_{|z+1|=1} \frac{dz}{1+z^3}$. $\quad [\frac{2\pi i}{3}]$

9. $\oint_{|z|=1} \frac{(2z-3)dz}{z^2-3z}$. $\quad [2\pi i]$

10. $\oint_{|z-3|=1} \frac{(2z-3)dz}{z^2-3z}$. $\quad [2\pi i]$

11. $\oint_{|z|=9} \frac{(2z-3)dz}{z^2-3z}$ $\quad [4\pi i]$

12. Let $f(z)$ be regular in the region bounded by a simple closed contour C. Show that $\oint_C f(z)dz/(z-a) = 2\pi i f(a)$ or $= 0$ according as a is in the interior of C or outside C.

Indefinite integral of a regular function

Let $f(z)$ be regular within a region $\mathcal{D}$ bounded by a simple closed curve (a 'simply connected region'). The integral

$$F(z) = \int_a^z f(\zeta)\,d\zeta,$$

along any contour in $\mathcal{D}$ from a fixed point a in $\mathcal{D}$ to an arbitrary point z in $\mathcal{D}$ defines a function $F(z)$ that is regular (analytic) in $\mathcal{D}$, and $F'(z) = f(z)$.

Cauchy's theorem (3.4.1) shows that $F(z)$ is a single-valued function of z in $\mathcal{D}$, because the integral is the same for any path in $\mathcal{D}$ between a and z. Also,

$$\frac{F(z+\delta z) - F(z)}{\delta z} = \frac{1}{\delta z}\left[\int_a^{z+\delta z} f(\zeta)d\zeta - \int_a^z f(\zeta)d\zeta\right]$$

$$= \frac{1}{\delta z}\int_z^{z+\delta z} f(\zeta)d\zeta \to f(z), \quad \text{as } \delta z \to 0,$$

i.e. $F'(z) = f(z)$.

3.5   Cauchy's Integral Formula

Suppose $f(z)$ is continuous on a simple closed contour C and regular within C. Then for z within C

$$f(z) = \frac{1}{2\pi i} \oint_C \frac{f(\zeta)d\zeta}{\zeta - z}. \tag{3.5.1}$$

Indeed, $f(\zeta)/(\zeta - z)$ is regular (as a function of ζ) in C except at $\zeta = z$, so that the contour may be collapsed onto a small circle $\zeta - z = \delta e^{i\theta}$, and the integration becomes equivalent to

$$\frac{1}{2\pi i} \int_0^{2\pi} \frac{f(z + \delta e^{i\theta})\, i\delta e^{i\theta}\, d\theta}{\delta e^{i\theta}}$$

$$= \frac{1}{2\pi i} \int_0^{2\pi} i f(z + \delta e^{i\theta})\, d\theta \;\rightarrow\; \frac{f(z)}{2\pi} \int_0^{2\pi} d\theta = f(z), \quad \text{as } \delta \rightarrow 0.$$

Cauchy's integral formula shows that the value of $f(z)$ at any point z is determined by its values on *any* boundary within which $f(z)$ is regular. It also shows that a regular function $f(z)$ is infinitely differentiable inside C, because

$$\frac{f(z + \delta z) - f(z)}{\delta z} = \frac{1}{2\pi i} \oint_C \frac{f(\zeta)}{\delta z} \left(\frac{1}{\zeta - z - \delta z} - \frac{1}{\zeta - z} \right) d\zeta$$

$$= \frac{1}{2\pi i} \oint_C \frac{f(\zeta)d\zeta}{(\zeta - z - \delta z)(\zeta - z)}$$

$$\rightarrow \frac{1}{2\pi i} \oint_C \frac{f(\zeta)d\zeta}{(\zeta - z)^2} \quad \text{as } \delta z \rightarrow 0.$$

$$\therefore \quad f'(z) = \frac{1}{2\pi i} \oint_C \frac{f(\zeta)d\zeta}{(\zeta - z)^2} \quad \text{for } z \text{ inside C},$$

and by repeated application of this procedure we deduce

$$f^{(n)}(z) = \frac{n!}{2\pi i} \oint_C \frac{f(\zeta)d\zeta}{(\zeta - z)^{n+1}} \quad \text{for } z \text{ inside C.} \qquad (3.5.2)$$

Example Morera'a Theorem If $f(z)$ is continuous in a certain region $\mathcal{D}$ and $\oint_C f(z)dz = 0$ for any closed curve C in $\mathcal{D}$, then $f(z)$ is regular in $\mathcal{D}$. Because,

$F(z) = \int_{z_0}^{z} f(\zeta)d\zeta$, for some fixed z_0, is independent of the path of integration, and

$$\frac{F(z+\delta z) - F(z)}{\delta z} = \frac{1}{\delta z} \int_{z}^{z+\delta z} f(\zeta)d\zeta \to f(z) \quad \text{as } \delta z \to 0.$$

Thus $F(z)$ is regular, and so therefore is its derivative $f(z)$.

Example Liouville's Theorem If $f(z)$ is regular and $|f(z)| < M =$ constant for all z, then $f(z) =$ constant.

Suppose z_1 and z_2 are two points inside the circle C: $|z| = R$. Then

$$f(z_1) - f(z_2) = \frac{1}{2\pi i} \oint_{C} \left(\frac{1}{z - z_1} - \frac{1}{z - z_2} \right) f(z)dz$$

and

$$|f(z_1) - f(z_2)| = \frac{1}{2\pi} \left| \oint_{C} \frac{(z_1 - z_2)f(z)dz}{(z - z_1)(z - z_2)} \right| < \frac{1}{2\pi} \frac{|z_1 - z_2|M 2\pi R}{(R - |z_1|)(R - |z_2|)} \to 0 \quad \text{as } R \to \infty.$$

Hence, $f(z_1) = f(z_2)$, i.e. $f(z)$ is constant.

If $|f(z)| \le A|z|^k$ as $|z| \to \infty$ ($A =$ constant, $k > 0$), apply the same argument using (3.5.2) with $n =$ the largest integer $\le k$ to show that $f(z)$ is a polynomial of degree $\le k$.

3.6 Taylor's Theorem

Suppose $f(z)$ is regular inside the circle $|z - z_0| < R$ (and continuous on $|z - z_0| = R$), then

$$f(z) = \sum_{m=0}^{\infty} a_m(z - z_0)^m \equiv \sum_{m=0}^{\infty} \frac{f^{(m)}(z_0)}{m!}(z - z_0)^m, \quad \text{for } |z - z_0| < R.$$

$$(3.6.1)$$

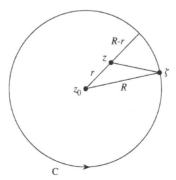

This follows from (3.5.1) and the exact formula

$$\frac{1}{1-X} = 1 + X + X^2 + X^3 + \cdots + X^{n-1} + \frac{X^n}{1-X},$$

by setting $X = (z - z_0)/(\zeta - z_0)$ in

$$\frac{1}{\zeta - z} \equiv \frac{1}{(\zeta - z_0) - (z - z_0)} = \frac{1}{(\zeta - z_0)\left[1 - \frac{z - z_0}{\zeta - z_0}\right]}$$

$$= \sum_{m=0}^{n-1} \frac{(z - z_0)^m}{(\zeta - z_0)^{m+1}} + \left(\frac{z - z_0}{\zeta - z_0}\right)^n \frac{1}{(\zeta - z)}. \qquad (3.6.2)$$

Substituting into (3.5.1) and using the formula (3.5.2) (with z replaced by z_0) we find

$$f(z) = \sum_{m=0}^{n-1} \frac{f^{(m)}(z_0)}{m!}(z - z_0)^m + R_n,$$

where R_n is the remainder after n terms, given by

$$R_n = \frac{1}{2\pi i} \int_C \frac{(z - z_0)^n f(\zeta) d\zeta}{(\zeta - z_0)^n (\zeta - z)}, \quad \text{for which } |R_n| \leq \frac{1}{2\pi}\left(\frac{r}{R}\right)^n \frac{M 2\pi R}{(R - r)},$$

where $|f(\zeta)| < M$ on C, $r = |z - z_0| < |\zeta - z_0| = R$, and $R - r$ is the smallest value of $|\zeta - z|$. This proves Taylor's theorem, because $R_n \to 0$ as $n \to \infty$.

This calculation shows that the *radius of convergence* R of Taylor's series (3.6.1) is equal to the distance from z_0 to the nearest singularity.

Example The radius of convergence of the Taylor expansion

$$\frac{1}{1 + z^2} = 1 - z^2 + z^4 - z^6 + z^8 - \cdots$$

is $R = 1$, the singularities being at $z = \pm i$.

3.7 Laurent's Expansion

This is a generalisation of Taylor's theorem to situations where $f(z)$ has singularities within the circle of integration C.

Suppose that $f(z)$ is regular within the annular region between the concentric circles C of radius R, and C' of radius $R' < R$, with centre z_0. Then

$$f(z) = \sum_{m=-\infty}^{\infty} a_m(z - z_0)^m \quad \text{for } R' < |z - z_0| < R,$$

$$a_m = \frac{1}{2\pi i} \oint \frac{f(\zeta)d\zeta}{(\zeta - z_0)^{m+1}}, \tag{3.7.1}$$

where the integration is around any simple closed contour within the annulus enclosing C'.

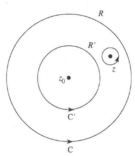

If z lies within the annulus, then $f(\zeta)/(\zeta - z)$ is regular as a function of ζ in the annulus except at $\zeta = z$. Thus

$$\frac{1}{2\pi i} \oint_C \frac{f(\zeta)d\zeta}{\zeta - z} = f(z) + \frac{1}{2\pi i} \oint_{C'} \frac{f(\zeta)d\zeta}{\zeta - z},$$

where the right-hand side is obtained by shrinking the contour C onto a small circle enclosing the singularity at $\zeta = z$ and onto the inner contour C'. Hence,

$$f(z) = \frac{1}{2\pi i} \oint_C \frac{f(\zeta)d\zeta}{\zeta - z} - \frac{1}{2\pi i} \oint_{C'} \frac{f(\zeta)d\zeta}{\zeta - z}. \tag{3.7.2}$$

The denominators in the integrands are now expanded as in the derivation of Taylor's theorem. For the integration around C (on which $|z - z_0| < |\zeta - z_0|$), we write (as in (3.6.2))

$$\frac{1}{\zeta - z} = \frac{1}{\zeta - z_0} + \frac{(z - z_0)}{(\zeta - z_0)^2} + \cdots + \frac{(z - z_0)^n}{(\zeta - z_0)^{n+1}} + \cdots,$$

and for the integration around C' (on which $|\zeta - z_0| < |z - z_0|$), we write

$$\frac{-1}{\zeta - z} = \frac{1}{z - z_0} + \frac{(\zeta - z_0)}{(z - z_0)^2} + \cdots + \frac{(\zeta - z_0)^n}{(z - z_0)^{n+1}} + \cdots .$$

By substituting these expansions respectively into the first and second integrals on the right of (3.7.2) we obtain the first line of (3.7.1), but with

$$a_m = \frac{1}{2\pi i} \oint_C \frac{f(\zeta)d\zeta}{(\zeta - z_0)^{m+1}}, \quad \text{for } m \geq 0$$

$$= \frac{1}{2\pi i} \oint_{C'} \frac{f(\zeta)d\zeta}{(\zeta - z_0)^{m+1}}, \quad \text{for } m \leq -1.$$

However, the integrands are regular in $R' < |z - z_0| < R$, which means that any path encircling C' within the annulus may be used to evaluate the coefficients a_m, and (3.7.1) is therefore proved.

3.8   Poles and Essential Singularities

A point where $f(z)$ is not regular is called a *singularity*. If $f(z)$ is regular near a point z_0, but singular at z_0, then z_0 is called an *isolated singularity*. If $f(z)$ is regular in a region $\mathcal{D}$ except at an isolated singularity z_0, we can draw two concentric circles centred on z_0 of radii $r_1, r_2 (r_1 < r_2)$ both lying within $\mathcal{D}$. In $r_1 < |z - z_0| < r_2$ $f(z)$ has the Laurent expansion

$$f(z) = \sum_{n=0}^{\infty} a_n(z - z_0)^n + \sum_{n=1}^{\infty} \frac{b_n}{(z - z_0)^n}. \tag{3.8.1}$$

This series converges for $0 < |z - z_0| < R$, where R is the largest value of r_2 for which the larger circle lies entirely within $\mathcal{D}$. The second term on the right of (3.8.1) is called the *principal part* of $f(z)$ at z_0. If $b_m \neq 0$, but $b_{m+1} = b_{m+2} = \cdots = 0$, there are m terms in the principal part

$$\frac{b_1}{z - z_0} + \frac{b_2}{(z - z_0)^2} + \cdots + \frac{b_m}{(z - z_0)^m},$$

and the isolated singularity is called a *pole of order m* of $f(z)$. A pole of order one is called a *simple pole*. The coefficient b_1 plays a particularly important role in applications of contour integration, and is called the *residue* of $f(z)$ at z_0. For a simple pole, the residue

$$b_1 = \lim_{z \to z_0} (z - z_0)f(z); \tag{3.8.2}$$

at a pole of order m the residue is

$$b_1 = \frac{1}{(m-1)!} \left(\frac{\partial^{m-1}}{\partial z^{m-1}} \left\{ (z - z_0)^m f(z) \right\} \right)_{z=z_0}. \tag{3.8.3}$$

In applications, however, the use of these general formulae is not necessarily the best way to proceed.

The point $z = z_0$ is called an *essential singularity* when the principal part contains an infinite number of terms.

Example $e^{\frac{1}{z}} = 1 + \frac{1}{z} + \frac{1}{2!z^2} + \frac{1}{3!z^3} + \frac{1}{4!z^4} + \cdots$, has an essential singularity at $z = 0$, with residue $b_1 = 1$.

Example $f(z) = \frac{z^5+1}{(z-1)^2} \equiv \frac{2}{(z-1)^2} + \frac{5}{(z-1)} + 10 + 10(z - 1) + 5(z - 1)^2 + (z - 1)^3$ is regular everywhere except for a double pole at $z = 1$ with residue 5.

3.9   Cauchy's Residue Theorem

Let $f(z)$ be continuous on a simple closed contour C and regular inside C except for a finite number of singularities at $z_1, z_2, \ldots, z_n$. Then

$$\oint_C f(z)dz = 2\pi i \sum_{m=1}^{n} \mathcal{R}_m, \tag{3.9.1}$$

where $\mathcal{R}_m$ is the residue of $f(z)$ at $z = z_m$ (i.e. $\sum_{m=1}^{n} \mathcal{R}_m$ is the sum of all the residues *inside* C).

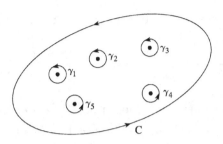

The proof follows from the principle of path deformation deduced from Cauchy's theorem in §3.4. Let $\gamma_1, \gamma_2, \ldots, \gamma_n$ be small, non-overlapping circles of radius δ with centres $z_1, z_2, \ldots, z_n$ contained entirely within C. Then $f(z)$ is regular in the region between C and the circles, so that the integration contour may be collapsed onto the circles to obtain

$$\oint_C f(z)dz = \oint_{\gamma_1} f(z)dz + \oint_{\gamma_2} f(z)dz + \cdots + \oint_{\gamma_n} f(z)dz.$$

On the mth circle $f(z)$ is given by a Laurent expansion of the form (3.8.1)

$$f(z) = \sum_{k=0}^{\infty} a_k(z - z_m)^k + \sum_{k=1}^{\infty} \frac{b_k}{(z - z_m)^k}.$$

Therefore, integrating around γ_m, making use of equation (3.4.3), we see that $\oint_{\gamma_m} f(z)dz = 2\pi i b_1 \equiv 2\pi i \mathcal{R}_m$. This proves the theorem.

Calculation of residues

(i) Direct calculation from the Laurent expansion.

Example   Find the residue of $z^5 e^{\frac{2}{z}}$ at $z = 0$.

$$z^5 e^{\frac{2}{z}} = z^5 \left\{ 1 + \frac{2}{z} + \frac{2^2}{2!z^2} + \frac{2^3}{3!z^3} + \frac{2^4}{4!z^4} + \frac{2^5}{5!z^5} + \frac{2^6}{6!z^6} + \frac{2^7}{7!z^7} + \cdots \right\}$$

$$= z^5 + 2z^4 + \frac{2^2 z^3}{2!} + \frac{2^3 z^2}{3!} + \frac{2^4 z}{4!} + \frac{2^5}{5!} + \frac{2^6}{6!z} + \frac{2^7}{7!z^2} + \cdots,$$

$$\therefore \quad \text{residue} = \frac{2^6}{6!} = \frac{4}{45}.$$

Example $$\oint_{|z|=1} z^5 e^{\frac{2}{z}} dz = 2\pi i \times \left(\text{Residue at } z = 0 \right) = 2\pi i \left(\frac{4}{45} \right) = \frac{8\pi i}{45}.$$

Example Find the residue of $z^3/(z-1)^2$ at $z = 1$.
Set $\zeta = z - 1$, then

$$\frac{z^3}{(z-1)^2} = \frac{(1+\zeta)^3}{\zeta^2} = \frac{1 + 3\zeta + 3\zeta^2 + \zeta^3}{\zeta^2} = \frac{1}{\zeta^2} + \frac{3}{\zeta} + 3 + \zeta$$

$$\equiv \frac{1}{(z-1)^2} + \frac{3}{(z-1)} + 3 + (z-1)$$

$$\therefore \qquad \text{residue} = 3.$$

Example $$\oint\limits_{|z-1|>0} \frac{z^3 dz}{(z-1)^2} = 2\pi i \times \left(\text{Residue at } z = 1 \right) = 6\pi i.$$

(iia) Residue at a simple pole when

$$f(z) = \frac{P(z)}{(z - z_0)},$$

where $P(z)$ is regular at $z = z_0$:

$$\text{Residue at } z_0 = P(z_0), \qquad (3.9.2)$$

(cf. formula (3.8.2)).
(iib) Residue at a simple pole when

$$f(z) = \frac{P(z)}{Q(z)},$$

where $P(z)$ and $Q(z)$ are regular at $z = z_0$, such that $Q(z)$ has a *simple* zero there but $P(z_0) \neq 0$, so that $f(z)$ has a simple pole at $z = z_0$. This means that near $z = z_0$, $Q(z) = (z - z_0)Q'(z_0) + \frac{1}{2!}(z - z_0)^2 Q''(z_0) + \cdots$ where $Q'(z_0) \neq 0$. Hence, using (3.8.2)

$$\text{Residue at } z_0 = \lim_{z \to z_0} \frac{(z - z_0)P(z)}{Q(z)}$$

$$= \lim_{z \to z_0} \frac{(z - z_0)[P(z_0) + (z - z_0)P'(z_0) + \cdots]}{(z - z_0)Q'(z_0) + (z - z_0)^2 \frac{Q''(z_0)}{2!} + \cdots}$$

$$= \frac{P(z_0)}{Q'(z_0)}. \qquad (3.9.3)$$

Example   Find the residue of $\cot z$ at $z = n\pi, n = 0, \pm 1, \pm 2, \ldots$.

$$\cot z = \frac{\cos z}{\sin z} \quad \text{has a simple pole at } z = n\pi$$

$$\therefore \quad \text{residue} = \left(\frac{\cos z}{\frac{d(\sin z)}{dz}} \right)_{z=n\pi} = \left(\frac{\cos z}{\cos z} \right)_{z=n\pi} = 1.$$

Example $\displaystyle\oint_{|z|=1} \cot z \, dz = 2\pi i.$

(iii) Residue at a double pole when

$$f(z) = \frac{P(z)}{(z - z_0)^2},$$

where $P(z)$ is regular at $z = z_0$.

Use equation (3.8.3) with $m = 2$:

$$\text{residue at } z_0 = \left(\frac{\partial}{\partial z} \left\{ (z - z_0)^2 f(z) \right\} \right)_{z=z_0} = P'(z_0).$$

Example   Find the residues of $f(z) = 1/(z^2 + 1)^2$.

This has double poles at $z = \pm i$.

Write $f(z) = \dfrac{1}{(z - i)^2 (z + i)^2}.$

At $z = i$:   residue $= \left(\dfrac{\partial}{\partial z} \left\{ (z - i)^2 f(z) \right\} \right)_{z=i}$

$$= \left(\frac{\partial}{\partial z} \left\{ \frac{1}{(z + i)^2} \right\} \right)_{z=i} = \left(\frac{-2}{(z + i)^3} \right)_{z=i} = -\frac{i}{4}.$$

At $z = -i$:   residue $= \left(\dfrac{\partial}{\partial z} \left\{ \dfrac{1}{(z - i)^2} \right\} \right)_{z=-i} = \dfrac{i}{4}.$

Example $\displaystyle\oint_{|z|>1} \frac{dz}{(z^2 + 1)^2} = 0; \quad \oint_{|z-i|<1} \frac{dz}{(z^2 + 1)^2} = \frac{\pi}{2}.$

Problems 3E

Evaluate the residues of:

1. $z \cosh(3/z)$. $[\frac{9}{2}$ at $z = 0]$
2. $(z + 2)/(z^2 - 3z)$. $[-\frac{2}{3}$ at $z = 0; \frac{5}{3}$ at $z = 3]$
3. $z^4/(z^2 + 1)$. $[-\frac{i}{2}$ at $z = i; \frac{i}{2}$ at $z = -i]$

4. $1/z^3(z-1)^2$.   [3 at $z = 0$; -3 at $z = 1$]

5. $\frac{\sin z}{z^4}$. $\left[-\frac{1}{6} \text{ at } z = 0\right]$

6. $\frac{1}{z^4+1}$. $\left[\mp\frac{1}{4}e^{\pm i\pi/4} \text{ at } z = \pm e^{\pm i\pi/4}\right]$

7. $z^2 e^{\frac{1}{z}}$. $\left[\frac{1}{6} \text{ at } z = 0\right]$

8. $e^{\frac{1}{z}}/(z-1)^2$.   [e at $z = 0$; $-$e at $z = 1$]

9. $(\cosh 2z)/z^5$. $\left[\frac{2}{3} \text{ at } z = 0\right]$

10. $z/\sin z$. $[(-1)^n n\pi \text{ at } z = n\pi, n = \pm 1, \pm 2, \pm 3, \ldots]$.

Evaluate by the residue theorem (contours are traversed in the anticlockwise sense):

11. $\oint_{|z|=1} \frac{z^6+7}{z^2-2z} dz$. $[-7\pi i]$

12. $\oint_{|z|=1} \frac{e^{-z^2}}{\sin 2z} dz$. $[\pi i]$

13. $\oint_{|z|=1} \frac{\sinh z}{4z^2+1} dz$. $\left[\pi i \sin\left(\frac{1}{2}\right)\right]$

14. $\oint_{|z|=1} \frac{\tan \pi z}{z^3} dz$. $[0]$

15. $\oint_{|z-\frac{i}{2}|=1} \frac{z^4}{z^2+1} dz$. $[\pi]$

16. $\oint_{|z|=20} z^2 e^{\frac{1}{z}} dz$. $[\pi i/3]$

17. $\oint_{|z-\frac{i}{2}|=1} \frac{1}{z^3(z-1)^2} dz$. $[6\pi i]$

18. $\oint_{|z|=\frac{1}{2}} e^{\frac{1}{z}}/(z-1)^2 dz$. $[2\pi i e]$

19. $\oint_{|z|=\frac{3}{2}} e^{\frac{1}{z}}/(z-1)^2 dz$. $[0]$

20. $\oint_{|z|=2} \frac{\sinh z}{(1+z^2)^2} dz$. $[\pi i(\sin 1 - \cos 1)]$

21. $\oint_{|z|>|a|} (z-a)^n dz$, n is an integer. $[2\pi i, \ n = -1; 0, \ n \neq -1]$

22. $\oint_{|z|>0} \frac{\cos z}{z} dz$. $[2\pi i]$

23. $\oint_{|z|>1} \frac{z}{(z+i)(z^2+1)} dz$. $[0]$

24. $\oint_{|z|>0} \frac{\sin z}{z} dz$. $[0]$

25. $\oint_{|z|>0} \frac{\cosh 2z}{z^5} dz$. $\left[\frac{4\pi i}{3}\right]$

3.10   Applications of the Residue Theorem to Evaluate Real Integrals

Type I: Integration around the unit circle

Consider

$$I = \int\limits_{0}^{2\pi} F(\cos\theta, \sin\theta)d\theta, \qquad (3.10.1)$$

where $F(\cos\theta, \sin\theta)$ is a *rational function* (i.e. a ratio of polynomials in $\sin\theta$ and $\cos\theta$).

Set $z = e^{i\theta}$, then

$$\cos\theta = \frac{1}{2}\left(e^{i\theta} + e^{-i\theta}\right) = \frac{1}{2}\left(z + \frac{1}{z}\right),$$

$$\sin\theta = \frac{1}{2i}\left(e^{i\theta} - e^{-i\theta}\right) = \frac{1}{2i}\left(z - \frac{1}{z}\right), \quad d\theta = \frac{dz}{iz}.$$

Therefore,

$$I = \int_0^{2\pi} F(\cos\theta, \sin\theta)d\theta$$

$$= \oint_{|z|=1} F\left(\frac{1}{2}\left(z + \frac{1}{z}\right), \frac{1}{2i}\left(z - \frac{1}{z}\right)\right)\frac{dz}{iz} = 2\pi i \sum \mathcal{R},$$

where $\sum \mathcal{R}$ denotes the sum of the residues inside the unit circle $|z| = 1$.

Example

$$\int_0^{2\pi} \frac{d\theta}{\frac{5}{4} + \cos\theta} = \oint_{|z|=1} \frac{dz}{iz\left[\frac{5}{4} + \frac{1}{2}\left(z + \frac{1}{z}\right)\right]} = \oint_{|z|=1} \frac{-2i\,dz}{\left(z + \frac{1}{2}\right)(z + 2)}$$

$$= 2\pi i \times \left(\text{residue at } z = -\frac{1}{2}\right) = 2\pi i \left(\frac{-2i}{z+2}\right)_{z=-\frac{1}{2}} = \frac{8\pi}{3}.$$

Type II

$$I = \int_{-\infty}^{\infty} f(x)dx = 2\pi i \sum \mathcal{R}^+ \qquad (3.10.2)$$

where $f(z)$ has only isolated singularities in the upper half-plane, $zf(z) \to 0$ ('uniformly') as $|z| \to \infty$, and $\sum \mathcal{R}^+$ is the sum of the residues of $f(z)$ in the upper half-plane.

Consider the semi-circular contour of radius R shown in the figure, where R is large enough that all the singularities are within the contour. Then

$$\int\limits_{-R}^{R} f(x)dx + \int\limits_{\gamma} f(z)dz = 2\pi i \sum \mathcal{R}^+.$$

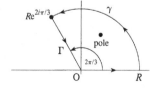

As $R \to \infty$ the 'ML' upper bound (3.3.5) shows that $\int_{\gamma} f(z)dz \to 0$, because

$$\left| \int\limits_{\gamma} f(z)dz \right| \equiv \left| \int\limits_{\gamma} \frac{zf(z)dz}{z} \right| \leq \frac{M \times \pi R}{R} = \pi M \to 0 \quad \text{as } R \to \infty,$$

where M is the maximum value of $|zf(z)|$ on γ, which $\to 0$ as $|z| \to \infty$. This proves (3.10.2).

Example Evaluate $\int_0^\infty \frac{dx}{1+x^4}$. The integrand has simple poles where $z^4 = e^{i\pi}$, i.e. at $z = \pm e^{\frac{i\pi}{4}}$, $\pm e^{\frac{3i\pi}{4}}$ and $M \approx 1/R^3 \to 0$ as $R \to \infty$. The poles at $z = e^{\frac{i\pi}{4}}$ and $e^{\frac{3i\pi}{4}}$ are in the upper half-plane, so that (evaluating the residues by (3.9.3))

$$\int\limits_0^\infty \frac{dx}{1+x^4} = \frac{1}{2} \int\limits_{-\infty}^{\infty} \frac{dx}{1+x^4} = \frac{1}{2} \times 2\pi i \left(\text{residues at } z = e^{\frac{i\pi}{4}}, \ e^{\frac{3i\pi}{4}} \right)$$

$$= \frac{\pi i}{4} \left(e^{-\frac{3i\pi}{4}} + e^{-\frac{9i\pi}{4}} \right) = \frac{\pi}{4} \left(e^{-\frac{i\pi}{4}} + e^{\frac{i\pi}{4}} \right) = \frac{\pi}{2\sqrt{2}}.$$

Example Evaluate $\int_0^\infty \frac{dx}{1+x^3}$. The procedure of the previous example does not work because the integrand is not an even function. However, setting $z = re^{i\theta}$, let us choose θ in the range $0 < \theta < \pi$ such that $1 + z^3 = 1 + r^3$. This requires $3\theta = 2\pi$, i.e. $\theta = \frac{2\pi}{3}$. We now consider the integral around the contour shown in the figure

$$\oint \frac{dz}{1+z^3} = \int\limits_{\Gamma} \frac{dz}{1+z^3} + \int\limits_0^R \frac{dx}{1+x^3} + \int\limits_{\gamma} \frac{dz}{1+z^3}.$$

On Γ, $z = re^{\frac{2i\pi}{3}}$ and $dz = e^{\frac{2i\pi}{3}} dr$

$$\therefore \quad \int\limits_{\Gamma} \frac{dz}{1+z^3} = \int\limits_R^0 \frac{e^{\frac{2i\pi}{3}} dr}{1+r^3} = -e^{\frac{2i\pi}{3}} \int\limits_0^R \frac{dx}{1+x^3}.$$

Also $\left|\int_\gamma \frac{dz}{1+z^3}\right| \approx \frac{2\pi}{3R^2} \to 0$ as $R \to \infty$. There is one pole within the contour at $z = e^{\frac{i\pi}{3}}$, with residue $\frac{1}{3}e^{-\frac{2i\pi}{3}}$. Hence

$$\int_\Gamma \frac{dz}{1+z^3} + \int_0^\infty \frac{dx}{1+x^3} \equiv \left(1 - e^{\frac{2i\pi}{3}}\right) \int_0^\infty \frac{dx}{1+x^3} = \frac{2\pi i e^{-\frac{2i\pi}{3}}}{3}$$

$$\therefore \int_0^\infty \frac{dx}{1+x^3} = \frac{2\pi i e^{-\frac{2i\pi}{3}}}{3\left(1 - e^{\frac{2i\pi}{3}}\right)} = \frac{2\pi i}{3\left(e^{\frac{i\pi}{3}} - e^{\frac{-i\pi}{3}}\right)} = \frac{\pi}{3\sin\left(\frac{\pi}{3}\right)} = \frac{2\pi}{3\sqrt{3}}.$$

Type III: Fourier integrals

$$I = \int_{-\infty}^\infty f(x)e^{\pm ikx}\,dx = \pm 2\pi i \sum \mathcal{R}^\pm, \quad k > 0, \tag{3.10.3}$$

where $f(z)$ has only isolated singularities in the upper/lower half-plane, $f(z) \to 0$ ('uniformly') as $|z| \to \infty$, and $\sum \mathcal{R}^\pm$ is the sum of the residues of $f(z)e^{\pm ikz}$ in the upper/lower half-plane. Observe that (for $k > 0$) $e^{\pm ikz} \to 0$ as $\mathrm{Im}(z) \to \pm\infty$.

Consider the '+' case. Because $e^{+ikz} \to 0$ as $\mathrm{Im}(z) \to +\infty$, we integrate around a large semi-circular contour in the upper half-plane of radius R (as for the Type II integral). Then, by the usual procedure

$$\oint f(z)e^{ikz}\,dz = \int_{-R}^R f(x)e^{ikx}\,dx + \int_\gamma f(z)e^{ikz}\,dz = 2\pi i \sum \mathcal{R}^+.$$

(3.10.3) will be true provided the integral around γ tends to zero as $R \to \infty$. The fact that this is so is a consequence of *Jordan's lemma*, which is proved as follows. On γ, where $z = Re^{i\theta}(0 < \theta < \pi)$,

$$\left|\int_\gamma f(z)e^{ikz}\,dz\right| = \left|\int_0^\pi f(z)e^{ikz}\,Re^{i\theta}id\theta\right|$$

$$< M\int_0^\pi e^{-kR\sin\theta}Rd\theta = 2M\int_0^{\frac{\pi}{2}} e^{-kR\sin\theta}Rd\theta,$$

where $|f(z)| < M$ on γ (and $M \to 0$ as $R \to \infty$). But

$$\frac{2\theta}{\pi} \leq \sin\theta$$

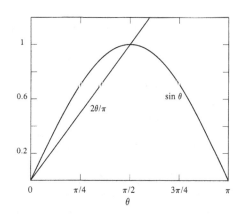

$$\therefore \quad \mathrm{e}^{-kR\sin\theta} \leq \mathrm{e}^{-2kR\theta/\pi}$$

$$\text{for } 0 \leq \theta \leq \frac{\pi}{2}.$$

Thus,

$$\left| \int_\gamma f(z)\mathrm{e}^{ikz}\,dz \right| < 2M \int_0^{\frac{\pi}{2}} \mathrm{e}^{-2kR\theta/\pi} R\,d\theta$$

$$= \frac{\pi M}{k}\left(1 - \mathrm{e}^{-kR}\right) < \frac{\pi M}{k} \to 0, \quad \text{as } R \to \infty,$$

because $M \to 0$ in this limit. This proves the lemma and (3.10.3) for the '+' case.

The argument is easily modified for the '$-$' case, by taking γ to be a semi-circle in the lower half-plane. This time, however, the contour is traversed in the 'negative' direction, so the sign of the residues must be reversed. Note that Jordan's lemma is *not* required if $|f(z)| \sim 1/R^{1+\sigma}$ on γ, where $\sigma > 0$, because the simpler '*ML*' method used for Type II integrals is then applicable.

Example Evaluate $\displaystyle\int_{-\infty}^{\infty} \frac{\cos(\alpha x)dx}{1+x^4} \equiv \int_{-\infty}^{\infty} \frac{\mathrm{e}^{i\alpha x}dx}{1+x^4}, \quad \alpha > 0.$

Integrate around a large semi-circular contour in the upper half-plane, noting that $|1/(1+z^4)| \approx 1/R^4 \to 0$ as $R \to \infty$ on γ. There are simple poles within the contour at $z_1 = \mathrm{e}^{\frac{i\pi}{4}} \equiv (1+i)/\sqrt{2}$, $z_2 = \mathrm{e}^{\frac{3i\pi}{4}} \equiv (-1+i)/\sqrt{2}$, with residues

$$\mathcal{R}_1 = \frac{\mathrm{e}^{i\alpha z_1}}{4z_1^3} = \frac{z_1 \mathrm{e}^{i\alpha z_1}}{-4} = \frac{-\mathrm{e}^{\frac{-\alpha}{\sqrt{2}}+i\left(\frac{\alpha}{\sqrt{2}}+\frac{\pi}{4}\right)}}{4},$$

and

$$\mathcal{R}_2 = \frac{e^{i\alpha z_2}}{4z_2^3} = \frac{z_2 e^{i\alpha z_2}}{-4} = \frac{e^{\frac{-\alpha}{\sqrt{2}} - i\left(\frac{\alpha}{\sqrt{2}} + \frac{\pi}{4}\right)}}{4}.$$

Hence,

$$\int_{-\infty}^{\infty} \frac{\cos(\alpha x)dx}{1 + x^4} = 2\pi i(\mathcal{R}_1 + \mathcal{R}_2)$$

$$= \frac{\pi i e^{\frac{-\alpha}{\sqrt{2}}}}{2}\left[-e^{i\left(\frac{\alpha}{\sqrt{2}} + \frac{\pi}{4}\right)} + e^{-i\left(\frac{\alpha}{\sqrt{2}} + \frac{\pi}{4}\right)}\right] = \pi \sin\left(\frac{\alpha}{\sqrt{2}} + \frac{\pi}{4}\right)e^{\frac{-\alpha}{\sqrt{2}}}.$$

Type IV: Integrals of many-valued functions

$$I = \int_0^{\infty} x^{\alpha-1}f(x)dx = \frac{2\pi i \sum \mathcal{R}}{1 - e^{2\pi i\alpha}},$$

where $|z^{\alpha}f(z)| \to 0$ as $|z| \to \infty$ and $|z| \to 0$,   (3.10.4)

and $f(z)$ is a rational function with no poles on the positive real axis, $\alpha > 0$ is not an integer, and $\sum \mathcal{R}$ is the sum of the residues of $z^{\alpha-1}f(z)$ in the complex plane *cut* along the positive real axis.

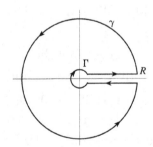

When α is not an integer, z^{α} is a many-valued function, whose argument increases by $2\pi\alpha$ when z traverses a closed contour encircling the origin. Integrals of this kind can be evaluated by 'cutting' the complex plane along the positive real axis, from $x = 0$ to $x = \infty$, and considering the integral around the contour C shown in the figure, which includes a large circle γ of radius R, a small circle Γ of radius δ enclosing the origin, and the upper and lower 'sides' of the positive real axis. Then z^{α} is regular and single valued within C, and the residue theorem is applicable. The contribution from γ vanishes as $R \to \infty$ in the usual way (as for Type II integrals). Similarly, on Γ, $|z^{\alpha-1}f(z)| < M/\delta$ where $M \to 0$ as $|z| \to 0$. Hence

$$\left| \int_\Gamma z^{\alpha-1}f(z)dz \right| < \frac{2\pi\delta M}{\delta} \to 0 \quad \text{as } \delta \to 0.$$

Taking the principal value of z^α within and on C, we now have respectively from the upper and lower sides of the positive real axis (as $R \to \infty$)

$$\int_0^\infty x^{\alpha-1}f(x)dx + e^{2\pi i(\alpha-1)}\int_\infty^0 x^{\alpha-1}f(x)dx = 2\pi i\sum\mathcal{R}.$$

The formula (3.10.4) is now obtained by reversing the sign of the second integral after interchanging the limits of integration.

Example Evaluate the Type IV integral: $\int_0^\infty \frac{x^\mu\,dx}{(x+1)(x+2)}$, where $-1 < \mu < 1$.
    There are two simple poles at $z = -1, -2$ with residues $e^{i\pi\mu}$ and $-2^\mu e^{i\pi\mu}$, respectively. Hence, (3.10.4) supplies

$$\int_0^\infty \frac{x^\mu\,dx}{(x+1)(x+2)} = \frac{2\pi i}{1 - e^{2\pi i\mu}}\left(e^{i\pi\mu} - 2^\mu e^{i\pi\mu} \right) = \frac{\pi(2^\mu - 1)}{\sin(\pi\mu)}. \tag{3.10.5}$$

The value of the integral when $\mu = 0$ is easily evaluated by elementary means, using partial fractions, and must coincide with the limit $\mu \to 0$ of (3.10.5),

$$\int_0^\infty \frac{dx}{(x+1)(x+2)} = \lim_{\mu\to 0}\frac{\pi(e^{\mu\ln 2} - 1)}{\sin(\pi\mu)} = \lim_{\mu\to 0}\frac{\pi\mu\ln 2}{\pi\mu} = \ln 2.$$

Type V: Principal value integrals

$$I = \fint_{-\infty}^\infty \frac{f(x)dx}{x - a} = \pm 2\pi i\sum\mathcal{R}^\pm \pm \pi i f(a), \tag{3.10.6}$$

where a is real, and $f(z)/(z-a)$ is a Type II or Type III integrand. The 'upper' and 'lower' signs go together, and $\sum\mathcal{R}^\pm$ are residues respectively in the upper and lower half-planes, the choice being determined for Type III integrals by the sign in the exponential on the left of (3.10.3). There is a pole on the real axis, and the notation $\fint$ implies that the contribution to the integral from the neighbourhood of this pole is to be interpreted as a 'principal value',

$$\fint_{-\infty}^{\infty} \frac{f(x)dx}{x-a} = \lim_{\epsilon \to +0} \left[\int_{-\infty}^{a-\epsilon} \frac{f(x)dx}{x-a} + \int_{a+\epsilon}^{\infty} \frac{f(x)dx}{x-a} \right].$$

The principal value integral can be converted to a Type II or Type III integral by indenting the contour to pass above or below the pole at $z = a$, by adjoining a semi-circular arc Γ of radius ϵ connecting the two halves $(-\infty, a - \epsilon)$ and $(a + \epsilon, \infty)$ of the real axis. The choice of arc is a matter of convenience. When a closed integration contour C is to be formed by introducing a large semi-circle γ (radius R) in the upper half-plane, it is convenient to take Γ to pass *above* the pole at a.

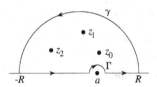

Then

$$\oint_{C} \frac{f(z)dz}{z-a} \equiv \int_{-R}^{a-\epsilon} \frac{f(x)dx}{x-a} + \int_{\Gamma} \frac{f(z)dz}{z-a} + \int_{a+\epsilon}^{R} \frac{f(x)dx}{x-a} + \int_{\gamma} \frac{f(z)dz}{z-a}$$

$$= 2\pi i \sum \mathcal{R}^+,$$

where the residues are from poles in the upper half-plane inside C. By the usual argument, $\int_{\gamma} \frac{f(z)dz}{z-a} \to 0$ as $R \to \infty$. On Γ, $z = a + \epsilon e^{i\theta}$, and

$$\int_{\Gamma} \frac{f(z)dz}{z-a} = \int_{\pi}^{0} \frac{f(a + \epsilon e^{i\theta}) i\epsilon e^{i\theta} \, d\theta}{\epsilon e^{i\theta}} \to -\pi i f(a) \quad \text{as } \epsilon \to 0.$$

Hence, as $\epsilon \to 0$

$$\oint_{C} \frac{f(z)dz}{z-a} \to \fint_{-\infty}^{\infty} \frac{f(x)dx}{x-a} - \pi i f(a) = 2\pi i \sum \mathcal{R}^+,$$

which proves (3.10.6) for the case in which the contour is closed in the upper half-plane.

Example
$$\oint_{-\infty}^{\infty} \frac{e^{ikx}\, dx}{x-a} = \pi i\, \mathrm{sgn}(\mathrm{k}) e^{ika}, \text{ where } k \text{ is real,}$$

because the semi-circular contour must be taken in the upper half-plane for $k > 0$ and the lower half-plane for $k < 0$.

Example Consider $\oint_C \frac{z^{\alpha-1}\, dz}{z-1}, -1 < \alpha < 0$, about the Type IV integration contour. The pole at $z = 1$ is avoided by the integration paths along the upper and lower positive x-axis by semi-circular indentations, respectively into the upper and lower half-planes. Hence, because there are no poles within C,

$$\oint_0^\infty \frac{x^{\alpha-1}\, dx}{x-1} - \pi i + \int_\infty^0 \frac{e^{2\pi i\alpha} x^{\alpha-1}\, dx}{x-1} - \pi i e^{2\pi i\alpha} = 0,$$

$$\therefore \quad \oint_0^\infty \frac{x^{\alpha-1}\, dx}{x-1} = \frac{\pi i(1 + e^{2\pi i\alpha})}{1 - e^{2\pi i\alpha}} \equiv -\pi \cot(\pi\alpha).$$

Example Consider $I_n = \int_0^{2\pi} \frac{\cos n\theta\, d\theta}{\cos\theta - \cos\phi}$, $n = 0, 1, 2, \ldots$. Using the substitution $z = e^{i\theta}$ (as for a Type I integral) and setting $\zeta = e^{i\phi}$ we find

$$I_n = \mathrm{Re}\left\{ -2i \oint_{|z|=1} \frac{z^n\, dz}{(z-\zeta)(z-1/\zeta)} \right\},$$

where the principal value contour integral is taken around the unit circle $|z| = 1$ in the positive sense. A 'closed' contour C can be constructed (with the poles at $z = \zeta$ and $z = 1/\zeta$ outside C) from the sectional principal value path of integration by adding small semi-circular arcs (on which $|z| < 1$) centred on each of the poles at $z = \zeta$ and $z = 1/\zeta$. Then

$$\oint_{|z|=1} \frac{z^n\, dz}{(z-\zeta)(z-1/\zeta)} = \oint_C \frac{z^n\, dz}{(z-\zeta)(z-1/\zeta)} + \pi i \left(\frac{\zeta^n}{\zeta - 1/\zeta} + \frac{\zeta^{-n}}{1/\zeta - \zeta} \right),$$

$$\therefore \quad I_n = \frac{2\pi \sin n\phi}{\sin \phi},$$

because $\oint_C \frac{z^n\, dz}{(z-\zeta)(z-1/\zeta)} \equiv 0$.

Problems 3F

Evaluate by the residue theorem:

1. $\int_0^{2\pi} \frac{d\theta}{1+a\sin\theta}$, $|a| < 1$. $\quad\left[\frac{2\pi}{\sqrt{1-a^2}}\right]$

2. $\int_0^{\pi} \frac{d\theta}{a^2+\sin^2\theta}$, $a > 0$. $\quad\left[\frac{\pi}{a\sqrt{1+a^2}}\right]$

3. $\int_0^{2\pi} \frac{\cos 2\theta\, d\theta}{1-2p\cos\theta+p^2}$, $|p| < 1$. $\quad[\frac{2\pi p^2}{1-p^2}]$

4. $\int_0^{\frac{\pi}{2}} \cos^6\theta\, d\theta$. $\quad[\frac{5\pi}{32}]$

5. $\int_0^{\pi} \frac{\sin^4\theta\, d\theta}{a+\cos\theta}$, $a > 1$. $\quad\left[\pi\left(\frac{3}{2}a - a^3 + (a^2-1)^{\frac{3}{2}}\right)\right]$

6. $\int_0^{2\pi} \frac{\sin^2\theta\, d\theta}{2+\cos\theta}$. $\quad\left[2\pi\left(2 - \sqrt{3}\right)\right]$

7. $\int_0^{\infty} \frac{dx}{(1+x^2)(4+x^2)}$. $\quad\left[\frac{\pi}{12}\right]$

8. $\int_0^{\infty} \frac{dx}{1+x^4}$. $\quad\left[\frac{\pi}{2\sqrt{2}}\right]$

9. $\int_0^{\infty} \frac{dx}{1+x^{2n}}$. $\quad\left[\frac{\pi}{2n\sin\left(\frac{\pi}{2n}\right)}\right]$

10. $\int_0^{\infty} \frac{x^{\alpha-1}\, dx}{1+x}$, $0 < \alpha < 1$. $\quad\left[\frac{\pi}{\sin(\alpha\pi)}\right]$

11. $\int_0^{\infty} \frac{dx}{(1+x^2)^2}$. $\quad\left[\frac{\pi}{4}\right]$

12. $\int_0^{\infty} \frac{dx}{\sqrt{x}(1+x+x^2)}$. $\quad\left[\frac{\pi}{\sqrt{3}}\right]$

13. $\int_0^{\infty} \frac{dx}{1+x^{3/2}}$. $\quad\left[\frac{4\pi}{3\sqrt{3}}\right]$

14. $\int_{-\infty}^{\infty} \frac{\sin x\, dx}{x}$. $\quad[\pi]$

15. $\int_0^{\infty} \frac{x\sin 2x\, dx}{1+x^2}$. $\quad\left[\frac{\pi}{2e^2}\right]$

16. $\int_{-\infty}^{\infty} \frac{\cos kx\, dx}{x^2+2x+5}$, ($k$ real). $\quad\left[\frac{\pi}{2}\cos k\, e^{-2|k|}\right]$

17. $\fint_{-\infty}^{\infty} \frac{\cos kx\, dx}{x-a}$, ($a$, k real, $k > 0$). $\quad[-\pi\sin ka]$

18. $\fint_{-\infty}^{\infty} \frac{x^4\, dx}{x^6-a^6}$, ($a > 0$). $\quad\left[\frac{\pi}{\sqrt{3}a}\right]$

19. $\fint_{-\infty}^{\infty} \frac{dx}{x(x-ai)}$, ($a > 0$). $\quad\left[\frac{\pi}{a}\right]$

20. $\fint_0^{2\pi} \frac{\cos^2\theta\, d\theta}{\cos\theta-\cos\phi}$. $\quad[2\pi\cos\phi]$

21. $\int_0^{\infty} \frac{\ln x\, dx}{4+x^2}$. $\quad\left[\frac{\pi\ln 2}{4}\right]$

22. $\int_0^{\infty} \frac{dx}{1+x^5}$. $\quad\left[\frac{\pi}{5\sin(\pi/5)}\right]$

23. $\int_0^\infty \frac{\sin^3 x}{x} dx.$ $\left[\frac{\pi}{4}. \text{ Use the formula } 4\sin^3 x = 3\sin x - \sin 3x\right]$

24. $\int_0^\infty \frac{\sin^3 x}{x^3} dx.$ $\left[\frac{3\pi}{8}\right]$

25. $\int_0^\infty \frac{\sin^2 x}{x^2} dx.$ $\left[\frac{\pi}{2}\right]$

26. $\int_0^\infty \frac{\cos x}{1+x^2} dx.$ $\left[\frac{\pi}{2e}\right]$

27. $\int_{-\infty}^\infty \frac{\sin(x+1)\sin(x-1)}{x^2-1} dx.$ $\left[\frac{\pi}{2}\sin 2\right]$

28. $\int_0^\infty \frac{\sin^2 x}{x^2(1+x^2)} dx.$ $\left[\frac{\pi}{4}\left(1+e^{-2}\right)\right]$

3.11   Contour Integration Applied to the Summation of Series

Consider the evaluation of

$$S = \sum_{n=-\infty}^{\infty} f(n),$$

where $f(z)$ is regular except for *simple* poles at a finite number of points $z = a_1, a_2, \ldots, a_m$ where the residues are $b_1, b_2, \ldots, b_m$ (a 'meromorphic' function), and $|zf(z)| \to 0$ as $|z| \to \infty$.

The function $\pi \cot \pi z$ has simple poles on the real axis at $z = 0, \pm 1, \pm 2, \ldots$, each of residue 1. Hence, by integrating $F(z) = \pi \cot \pi z\, f(z)$ around a contour C_N that includes all the poles of $f(z)$ and the poles of $\pi \cot \pi z$ at $z = 0, \pm 1, \pm 2, \ldots, \pm N$ we find

$$\oint_{C_N} F(z)dz = 2\pi i \sum_{n=-N}^{N} f(n) + 2\pi i \sum_{k=1}^{m} b_k \pi \cot \pi a_k. \qquad (3.11.1)$$

As N increases the contour C_N must increase in size. If C_N can be chosen such that $|zF(z)| \to 0$ on C_N as $N \to \infty$, the usual argument (see §3.10) implies that the contour integral $\to 0$, so that

$$\sum_{n=-\infty}^{\infty} f(n) = -\pi \sum_{k=1}^{m} b_k \cot \pi a_k. \qquad (3.11.2)$$

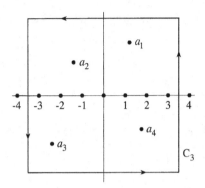

The square contour C_N with corners at $\pm(N+\frac{1}{2})(1\pm i)$ satisfies our requirements, because on C_N

$$|\cot \pi z|^2 = \frac{\cos^2 \pi x + \sinh^2 \pi y}{\sin^2 \pi x + \sinh^2 \pi y} < \frac{3}{2}, \quad \text{for } N \geq 0, \quad (z = x + iy).$$

Example Take
$$f(z) = \frac{1}{z^2 + a^2}, \quad a \text{ real},$$
with simple poles at $z = \pm ai$ with residues $\pm 1/2ai$. Then

$$\frac{1}{a^2} + 2\sum_{n=1}^{\infty} \frac{1}{n^2 + a^2} = -\pi \left(\frac{\cot(\pi ai)}{2ai} - \frac{\cot(-\pi ai)}{2ai} \right) = \frac{\pi \coth \pi a}{a}.$$

Hence,

$$\sum_{n=1}^{\infty} \frac{1}{n^2 + a^2} = \frac{1}{2} \left\{ \frac{\pi \coth \pi a}{a} - \frac{1}{a^2} \right\},$$

$$\therefore \quad \sum_{n=1}^{\infty} \frac{1}{n^2} = \lim_{a \to 0} \frac{1}{2} \left\{ \frac{\pi \coth \pi a}{a} - \frac{1}{a^2} \right\} = \frac{\pi^2}{6}.$$

Example Evaluate $\displaystyle\sum_{n=0}^{\infty} \frac{(-1)^n}{n^2 + a^2}.$

The function $\pi/\sin \pi z$ has simple poles with residues $(-1)^n$ at $z = n\pi, n = 0, \pm 1, \pm 2, \ldots,$ and

$$\frac{1}{|\sin \pi z|^2} = \frac{1}{\sin^2 \pi x + \sinh^2 \pi y} < 1 \quad \text{on } C_N.$$

We therefore consider $\int_{C_N} \frac{\pi f(z) dz}{\sin \pi z}$ to obtain

$$\sum_{n=0}^{\infty} \frac{(-1)^n}{n^2 + a^2} = \frac{1}{2a^2}\left\{1 + \frac{\pi a}{\sinh \pi a}\right\} \quad \text{and} \quad \sum_{n=1}^{\infty} \frac{(-1)^{n+1}}{n^2} = \frac{\pi^2}{12}.$$

Problems 3G

Evaluate by the residue theorem:

1. $\sum_{n=-\infty}^{\infty} \frac{1}{n^4+1}.$ $\left[\frac{\pi}{\sqrt{2}} \frac{\sinh(\pi\sqrt{2})+\sin(\pi\sqrt{2})}{\cosh(\pi\sqrt{2})-\cos(\pi\sqrt{2})}\right]$

2. $\sum_{n=1}^{\infty} \frac{1}{n^4-a^4}.$ $\left[\frac{1}{2a^4}\left\{1 - \frac{\pi a}{2}\left(\coth \pi a + \cot \pi a\right)\right\}\right]$

3. $\sum_{n=1}^{\infty} \frac{1}{n^2-a^2}.$ $\left[\frac{1}{2a^2}\left\{1 - \pi a \cot \pi a\right\}\right]$

4. $\sum_{n=1}^{\infty} \frac{(-1)^{n+1}}{n^2-a^2}.$ $\left[-\frac{1}{2a^2}\left\{1 - \frac{\pi a}{\sin \pi a}\right\}\right]$

5. $\sum_{n=-\infty}^{\infty} \frac{(-1)^n}{n^4-a^4}.$ $\left[-\frac{\pi}{2a^3}\left\{\frac{1}{\sin \pi a} + \frac{1}{\sinh \pi a}\right\}\right]$

6. $\sum_{n=-\infty}^{\infty} \frac{1}{(n+a)^2}.$ $\left[\frac{\pi^2}{\sin^2 \pi a}\right]$

7. By taking $F(z) = \frac{\pi \sin \frac{\pi z}{2}}{z^3 \sin \pi z}$ and considering the integral $\oint_{C_N} F(z)dz$, show that
$\sum_{n=0}^{\infty} \frac{(-1)^n}{(2n+1)^3} = \frac{\pi^3}{32}.$

8. By taking $F(z) = \frac{\pi \sin \frac{\pi z}{2}}{z \sin \pi z}$ and considering the integral $\oint_{C_N} F(z)dz$, show that
$\sum_{n=0}^{\infty} \frac{(-1)^n}{(2n+1)} = \frac{\pi}{4}.$

3.12   Conformal Representation

It has already been pointed out that a complex function $f(z)$ defines a transformation between points in the z-plane and points $Z \equiv X+iY = f(z)$ in the Z-plane. We now consider the *geometrical* properties of this transformation when $f(z)$ is regular in a region $\mathcal{D}$ of the z-plane.

Consider three neighbouring points z_0, z_1, z_2 in $\mathcal{D}$ and their corresponding *images* Z_0, Z_1, Z_2 in the Z-plane. When z_1, z_2 are *very close* to z_0 we can write

$$Z_1 - Z_0 = f'(z_0)(z_1 - z_0), \quad Z_2 - Z_0 = f'(z_0)(z_2 - z_0), \quad \text{provided } f'(z_0) \neq 0.$$

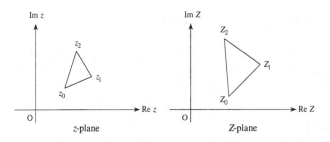

Thus, from the first of equations (3.1.2),

$$|Z_1 - Z_0| = |f'(z_0)||z_1 - z_0|, \quad |Z_2 - Z_0| = |f'(z_0)||z_2 - z_0|,$$

which shows that *small* distances between points in the z-plane in the vicinity of z_0 are all *magnified* by a factor $|f'(z_0)|$ in the Z-plane. Also,

$$\arg(Z_1 - Z_0) = \arg\{f'(z_0)(z_1 - z_0)\} = \arg\{f'(z_0)\} + \arg(z_1 - z_0)$$

$$\arg(Z_2 - Z_0) = \arg\{f'(z_0)(z_2 - z_0)\} = \arg\{f'(z_0)\} + \arg(z_2 - z_0)$$

which means that the angle between the rays $Z_1 - Z_0$ and $Z_2 - Z_0$ has the *same magnitude and sense* as the angle between the rays $z_1 - z_0$ and $z_2 - z_0$.

The effect of the transformation is to *rotate* all small straight lines in the neighbourhood of z_0 (such as the sides of the triangle in the figure) through the same angle $\arg\{f'(z_0)\}$ (in the anticlockwise or clockwise direction according as $\arg\{f'(z_0)\} \gtrless 0$), and to change their lengths by a factor $|f'(z_0)|$. The *area* of the triangle is therefore increased by this factor squared: $|f'(z_0)|^2$. It is also clear that when two curves intersect in the z-plane, their images in the Z-plane will *intersect at the same angle*. A transformation with this property is said to be 'conformal'.

The transformation is *not* conformal at $z = z_0$ if $f'(z_0) = 0$. The point $z = z_0$ is then called a *critical point* of the transformation, about which the Taylor series expansion of $f(z)$ has the form

$$f(z) = f(z_0) + a(z - z_0)^n + \cdots, \quad a \neq 0, \quad n \geq 2.$$

When $|z_1 - z_0|$ and $|z_2 - z_0|$ are small

$$Z_1 - Z_0 = a(z_1 - z_0)^n, \quad \arg(Z_1 - Z_0) = \arg a + n \arg(z_1 - z_0)$$

$$Z_2 - Z_0 = a(z_2 - z_0)^n, \quad \arg(Z_2 - Z_0) = \arg a + n \arg(z_2 - z_0)$$

Hence, the angle between the rays is n times larger in the Z-plane, and the magnification factor is zero.

It can also be shown, that if $f(z)$ is regular and single valued in $\mathcal{D}$, and the boundary $\mathcal{C}$ of $\mathcal{D}$ is a simple closed curve, then the image $\mathcal{D}'$ of $\mathcal{D}$ in the Z-plane is bounded by the image $\mathcal{C}'$ of $\mathcal{C}$. When z moves along $\mathcal{C}$ in the positive direction relative to $\mathcal{D}$ (with the 'interior' of $\mathcal{D}$ on the 'left') the image Z moves along $\mathcal{C}'$ also in the positive direction relative to $\mathcal{D}'$. The result remains true when one or more sections of the boundaries $\mathcal{C}, \mathcal{C}'$ are at infinity.

Example 1 Show that the transformation $Z = f(z) \equiv z^2$ maps the first quadrant $\mathcal{D}(x > 0, y > 0)$ of the z-plane onto the upper half (Im $Z = Y > 0$) of the Z-plane.

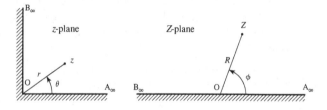

Let $Z = Re^{i\phi}$, then

$$Z = Re^{i\phi} \equiv r^2 e^{2i\theta}, \quad 0 < \theta < \frac{\pi}{2},$$

$$\therefore \quad R = r^2, \quad \phi = 2\theta, \quad 0 < \phi < \pi.$$

Because $0 < \phi < \pi$, the image of every point in $\mathcal{D}$ lies in the upper half of the Z-plane. The transformation is conformal except at $z = 0$ where $f'(z) = 2z = 0$. Each 'ray' inclined at angle θ to the x-axis is rotated to an image ray in the Z-plane inclined at angle 2θ to the x-axis. The positive x-axis transforms into the positive x-axis, and the imaginary axis OB_∞ ($\theta = \frac{\pi}{2}$) in the z-plane maps onto the negative x-axis.

Example 2 Show that the transformation $Z = f(z) \equiv i\sqrt{z}$, where the square root is the *principal value* $\sqrt{r}e^{i\frac{\theta}{2}}$ ($z = re^{i\theta}, -\pi < \theta < \pi$), maps the infinite region $\mathcal{D}$ consisting of the z-plane *cut along the negative real axis* onto the upper half (Im $Z = Y > 0$) of the Z-plane.

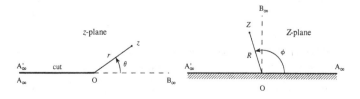

Let $Z = Re^{i\phi}$, then

$$Z = Re^{i\phi} \equiv \sqrt{r}e^{i(\frac{\theta}{2} + \frac{\pi}{2})}, \quad -\pi < \theta < \pi,$$

$$\therefore \quad R = \sqrt{r}, \quad \phi = \frac{\theta}{2} + \frac{\pi}{2}, \quad 0 < \phi < \pi.$$

Because $0 < \phi < \pi$, the image of every point in $\mathcal{D}$ lies in the upper half of the Z-plane. The transformation is conformal except at $z = 0$ where $f(z)$ ceases to be regular. The cut is necessary to ensure that each point of $\mathcal{D}$ corresponds to precisely one image point in the Z-plane. Thus, any path in $\mathcal{D}$ connecting any two points z_1 and z_2 cannot cross the negative real axis, thereby forcing $\theta = \arg z$ to lie in the range $(-\pi, \pi)$. The 'lower' side $A_\infty O$ of the cut (where $\theta \to -\pi$) is transformed into the positive real axis in the Z-plane; the upper side OA'_∞ ($\theta \to \pi$) is mapped into the negative real axis, and the positive real axis ($\theta = 0$) transforms into the positive imaginary axis.

It is important to note that different domains $\mathcal{D}$ correspond to different orientations of the cut in the z-plane. For example, if the cut were taken along the positive imaginary axis, the argument θ would be restricted to the range $(-\frac{3\pi}{2}, \frac{\pi}{2})$, and the image points $Z = Re^{i\phi}$ would occupy $-\frac{\pi}{4} < \phi < \frac{3\pi}{4}$, i.e. the cut plane would be mapped onto the half-plane consisting of all points to the right of the line $Y = -X$ in the Z-plane. In the absence of a cut the argument θ of z is unrestricted, and because any complex number has two square roots, each point in the z-plane would correspond to *two* image points $Z = \pm\sqrt{r}e^{i(\frac{\theta}{2} + \frac{\pi}{2})}$.

Example 3 Show that the transformation $Z = e^z$ maps the 'strip' $\mathcal{D}: -\infty < x < \infty, 0 < y < \pi$ of the $z = x + iy$ plane onto the half-plane Im $Z > 0$.

The real axis in the z-plane (the lower boundary of the strip) transforms into the positive real axis, and the upper boundary ($y = \pi$) maps onto $Z = e^{i\pi + x} = -e^x$, the negative real axis. Straight lines $y = \Theta = $ constant $(0 < \Theta < \pi)$ map into rays $Z = e^{i\Theta + x}$ radiating from the origin in the Z-plane; the image of the line $x = C = $ constant $(-\infty < C < \infty)$ is the semi-circular arc $Z = e^{C + iy}, 0 < y < \pi$. The rays cut the circles at right angles because the mapping is conformal and the original curves $y = $ constant, $x = $ constant are orthogonal. Corresponding points are illustrated in the figure.

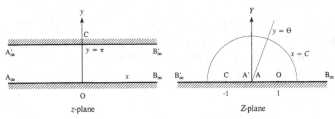

z-plane   Z-plane

Example 4 Show that the transformation $Z = z + \sqrt{z^2 - 1}$ maps the z-plane cut along the x-axis between $x = -1$ and $x = +1$ onto the exterior of the circle $|Z| = 1$. The square root is defined to be real and positive when $x > 1$ on the real axis.

Referring to the figure

$$\sqrt{z^2 - 1} = \sqrt{z - 1}\sqrt{z + 1} = \sqrt{r_1 r_2}\, e^{i\left(\frac{\theta_1}{2} + \frac{\theta_2}{2}\right)}, \qquad (3.12.1)$$

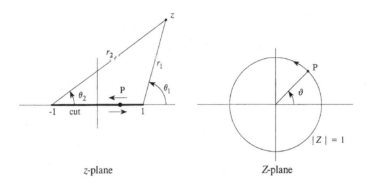

z-plane   Z-plane

The angles θ_1 and θ_2 both increase by 2π when z makes one complete traverse around any simple closed path surrounding the cut, and therefore $\sqrt{z^2 - 1}$ is unchanged in value after this traverse. This ensures that each point in the cut z-plane corresponds to a unique image point in the Z-plane. When z moves along the 'upper' side of the cut (where $\theta_1 = \pi, \theta_2 = 0$) from $x = 1$ to $x = -1$ we can set $z = \cos\vartheta, 0 < \vartheta < \pi$; the image $Z = \cos\vartheta + i\sin\vartheta \equiv e^{i\vartheta}$ accordingly travels along the semi-circular arc $|Z| = 1$ from right to left in the upper Z-plane. The 'lower' side of the cut ($\theta_1 = \pi, \theta_2 = 2\pi$) is traversed from left to right when ϑ increases from π to 2π (or equivalently from $-\pi$ to π). In this case Z travels along the arc of the semi-circle in the lower Z-plane from left to right.

Thus the upper and lower edges of the cut map onto the circle $|Z| = 1$. Also $Z \sim 2z$ as $|z| \to \infty$; this means that distant parts of the z- and Z-planes correspond and therefore that the image of an arbitrary point in the cut z-plane lies in the region $|Z| > 1$.

Problems 3H

Verify that:

1. The transformation $Z = e^{i\alpha}z$ (α real) rotates all points of the z-plane through an angle α about the origin.

2. $Z = iz$ maps the region $x > 0$ onto $Y > 0$.

3. $Z = z^2$ maps the half-plane $\operatorname{Im} z \geq 0$ onto the Z-plane cut along the positive real axis.

4. $Z = iz^3$ maps the wedge $0 < \arg z < \frac{\pi}{3}$ onto the left half-plane $X < 0$.

5. $Z = z^{\frac{\pi}{\alpha}}$ ($0 < \alpha < \pi$) maps the infinite wedge shaped region $0 < \arg z < \alpha$ onto the upper half of the Z-plane.

6. $Z = z + 1/z$ maps points on the circle $|z| = c \neq 1$ onto the ellipse

$$\frac{X^2}{(c+1/c)^2} + \frac{Y^2}{(c-1/c)^2} = 1, \quad Z = X + iY.$$

7. $Z = z^2 + 1/z^2$ maps the region $\mathcal{D}$: $|z| \geq 1, x \geq 0, y \geq 0$ (in the *first quadrant*) onto the upper half of the Z-plane. Where in $\mathcal{D}$ does the transformation cease to be conformal? $[z = 1, i]$

8. $Z = \ln z$ maps the wedge $\alpha < \arg z < \beta$ onto the infinite strip: $-\infty < X < \infty$, $\alpha < Y < \beta$.

9. The region $2 \leq |z| \leq 3, \frac{\pi}{4} \leq \arg z \leq \frac{\pi}{2}$ is mapped onto the rectangle $\ln 2 \leq X \leq \ln 3, \frac{\pi}{4} \leq Y \leq \frac{\pi}{2}$ by the transformation $Z = \ln z$.

10. The transformation $Z = z - \sqrt{z^2 - 1}$, where $\sqrt{z^2 - 1}$ is defined as in Example 4, maps the z-plane cut along the x-axis between $x = \pm 1$ onto the *interior* of the circle $|Z| = 1$.

11. The rectangular region $0 \leq x \leq 1, 0 \leq y \leq \frac{\pi}{2}$ is mapped onto $1 \leq |Z| \leq e$, $0 \leq \arg Z < \frac{\pi}{2}$ by the transformation $Z = e^z$.

12. $Z = e^{5z}$ maps the strip $0 < y < \frac{\pi}{5}$ onto $Y > 0$.

13. $Z = e^z$ maps the semi-infinite strip $x < 0, 0 < y < \pi$ onto $|Z| < 1, \operatorname{Im} Z > 0$.

14. $Z = 1/z$ maps $|z| < R$ onto $|Z| > 1/R$.

15. The transformation defined for real and positive c by $\frac{z-c}{z+c} = ie^{iZ}$ maps the strip $0 < X < \pi, -\infty < Y < \infty$ into the circle $|z| \leq c$.

16. $Z = z + \alpha^2/z$ (α real and positive) maps the region $|z| > \alpha$ *in the upper half-plane* onto the upper half of the Z-plane.

17. $Z = \cosh z$ maps the semi-infinite strip $x \geq 0, 0 \leq y \leq \pi$ onto the upper half of the Z-plane.

18. $Z = -\cos \pi z$ maps the semi-infinite strip $0 < x < 1, y > 0$ onto $Y > 0$.

19. $Z = \sin z$ maps the semi-infinite strip $0 < x < \frac{\pi}{2}, y > 0$ onto the first quadrant $X > 0, Y > 0$.

20. Find the points where the mapping $Z = \sin z$ is *not* conformal. $[z = \pm(n + \frac{1}{2})\pi, n = 0, 1, 2, \ldots]$

3.13   Laplace's Equation in Two Dimensions

If

$$w(z) = \varphi(x, y) + i\psi(x, y), \quad z = x + iy,$$

is regular in a region $\mathcal{D}$ of the z-plane, the real and imaginary parts $\varphi(x, y)$ and $\psi(x, y)$ satisfy Laplace's equation in $\mathcal{D}$, i.e.

$$\frac{\partial^2 \varphi}{\partial x^2} + \frac{\partial^2 \varphi}{\partial y^2} = 0, \quad \frac{\partial^2 \psi}{\partial x^2} + \frac{\partial^2 \psi}{\partial y^2} = 0, \quad \text{in } \mathcal{D}. \tag{3.13.1}$$

Now $z = x + iy$ and $z' = x - iy$, so that

$$\frac{\partial}{\partial x} = \frac{\partial z}{\partial x}\frac{\partial}{\partial z} + \frac{\partial z^*}{\partial x}\frac{\partial}{\partial z^*} = \frac{\partial}{\partial z} + \frac{\partial}{\partial z^*}$$

$$\frac{\partial}{\partial y} = \frac{\partial z}{\partial y}\frac{\partial}{\partial z} + \frac{\partial z^*}{\partial y}\frac{\partial}{\partial z^*} = i\left(\frac{\partial}{\partial z} - \frac{\partial}{\partial z^*}\right),$$

and

$$2\frac{\partial}{\partial z} = \frac{\partial}{\partial x} - i\frac{\partial}{\partial y}, \quad 2\frac{\partial}{\partial z^*} = \frac{\partial}{\partial x} + i\frac{\partial}{\partial y}.$$

Hence,

$$\frac{\partial^2}{\partial x^2} + \frac{\partial^2}{\partial y^2} = \left(\frac{\partial}{\partial x} + i\frac{\partial}{\partial y}\right)\left(\frac{\partial}{\partial x} - i\frac{\partial}{\partial y}\right) = 4\frac{\partial^2}{\partial z^*\partial z}$$

and equations (3.13.1) can also be written

$$\frac{\partial^2 \varphi}{\partial z^*\partial z} = 0, \quad \frac{\partial^2 \psi}{\partial z^*\partial z} = 0.$$

Therefore the most general solution of Laplace's equation in two dimensions is

$$f_1(z) + f_2(z^*) \equiv f_1(x + iy) + f_2(x - iy),$$

where f_1 and f_2 are arbitrary functions. The most general *real* solution is $F(z) + F^*(z^*) \equiv F(z) + \{F(z)\}^*$, for an arbitrary function F.

The function $w(z)$ cannot depend on z^* if it is regular, but it also satisfies Laplace's equation because

$$2\frac{\partial w(z)}{\partial z^*} = \left(\frac{\partial}{\partial x} + i\frac{\partial}{\partial y}\right)w(x + iy) \equiv (1 - 1)\, w'(z) = 0. \tag{3.13.2}$$

Now let $f(z)$ be regular in $\mathcal{D}$, and define a conformal transformation $Z = f(z)$ of $\mathcal{D}$ into a region Δ in the Z-plane. Let $\mathcal{W}(Z)$ be regular in Δ with real and imaginary parts $\Phi(X, Y)$, $\Psi(X, Y)$. Then

$$\frac{\partial^2 \Phi}{\partial X^2} + \frac{\partial^2 \Phi}{\partial Y^2} = 0, \quad \frac{\partial^2 \Psi}{\partial X^2} + \frac{\partial^2 \Psi}{\partial Y^2} = 0, \quad \text{in } \Delta.$$

The transformation $Z = f(z)$ permits us to define a corresponding function $w(z) \equiv \varphi(x, y) + i\psi(x, y) = \mathcal{W}(f(z))$ which is regular in $\mathcal{D}$, with derivative $w'(z) = f'(z)\mathcal{W}'(f(z))$. For corresponding points in $\mathcal{D}$ and Δ we have

$$\varphi(x, y) = \Phi\left(X(x, y), Y(x, y)\right), \quad \psi(x, y) = \Psi\left(X(x, y), Y(x, y)\right).$$

In other words: The solutions Φ and Ψ of Laplace's equation in Δ are also solutions of Laplace's equation in $\mathcal{D}$.

The explicit transformation of Laplace's equation from Δ to $\mathcal{D}$ is effected as follows:

$$\frac{\partial^2 \Phi}{\partial X^2} + \frac{\partial^2 \Phi}{\partial Y^2} = 4\frac{\partial^2 \Phi}{\partial Z^* \partial Z} = \frac{4}{\{f'(z)\}^*} \frac{\partial}{\partial z^*} \left[\frac{1}{f'(z)} \frac{\partial}{\partial z}\right] \Phi\left(X(x, y), Y(x, y)\right)$$

$$= \frac{4}{|f'(z)|^2} \frac{\partial^2 \varphi}{\partial z^* \partial z} = \frac{1}{|f'(z)|^2} \left[\frac{\partial^2 \varphi}{\partial x^2} + \frac{\partial^2 \varphi}{\partial y^2}\right], \qquad (3.13.3)$$

where, in passing from the first to the second line, we have used (3.13.2)

$$\frac{\partial}{\partial z^*}\left[\frac{1}{f'(z)}\right] \equiv 0,$$

which is valid because $1/f'(z)$ is regular provided $f'(z) \neq 0$ in $\mathcal{D}$.

These results have the following significance: The solution of Laplace's equation within a given two-dimensional bounded region $\mathcal{D}$ is equivalent to the solution of Laplace's equation within the transformed region Δ. If it is possible to solve the latter problem, the solution to the original problem in $\mathcal{D}$ can be found by transforming back to the z-plane. Problems may arise at isolated points where $f'(z) = 0$ or if there are points where $f(z)$ ceases to be regular, but these can usually be dealt with by considering the detailed behaviour of the transformation near these points.

3.14   Applications to Hydrodynamics

Irrotational motion of an ideal, incompressible fluid in two dimensions (in planes parallel to the xy-plane) can be investigated by introducing the *complex potential* $w(z) = \varphi(x,y) + i\psi(x,y)$, which is a regular function of $z = x + iy$. The fluid velocity $\mathbf{v}$ is determined in terms of the *velocity potential* $\varphi(x,y)$ by $\mathbf{v} = \nabla\varphi = (\partial\varphi/\partial x, \partial\varphi/\partial y)$ at the point (x,y). The function ψ is called the *stream function*. For *steady* motion the velocity at any fixed point (x,y) does not change with time, and the fluid particles travel along a fixed system of streamlines each of which is a member of the family of curves $\psi(x,y) = $ constant.

Both $\varphi(x,y)$ and $\psi(x,y)$ are solutions of Laplace's equation that satisfy the Cauchy–Riemann equations (3.2.3):

$$\frac{\partial\varphi}{\partial x} = \frac{\partial\psi}{\partial y}, \quad \frac{\partial\varphi}{\partial y} = -\frac{\partial\psi}{\partial x},$$

which imply that $\nabla\varphi \cdot \nabla\psi = 0$, i.e. that the streamlines intersect the 'equipotentials' $\varphi = $ constant at right angles. In the usual notation of theoretical fluid mechanics we write $\mathbf{v} = (u,v)$. The *complex velocity*

$$w'(z) = \frac{\partial\varphi}{\partial x} - i\frac{\partial\varphi}{\partial y} \equiv u - iv$$

is also regular, with Cauchy–Riemann equations

$$\frac{\partial u}{\partial x} + \frac{\partial v}{\partial y} = 0, \quad \frac{\partial v}{\partial x} - \frac{\partial u}{\partial y} = 0,$$

the first of which is just the equation of continuity div $\mathbf{v} = 0$ for incompressible flow. The expression on the left of the second equation is the *vorticity* (in the $\mathbf{k}$-direction) which vanishes for irrotational flow.

The fact that $w(z)$ is a regular function of z can greatly simplify the solution of many problems. This will be illustrated by consideration of two methods based on the theory of complex variables.

Method 1   The real and imaginary parts of every regular function $w(z)$ determine the velocity potential and stream function of a possible flow. A catalogue of flows can therefore be constructed by studying the properties of arbitrarily selected $w(z)$.

Example $w = Uz$, $U =$ real constant:

$$\varphi = Ux, \quad \psi = Uy, \quad \mathbf{v} = (U, 0).$$

The motion is uniform at speed U along streamlines parallel to the x-direction.

Example

$$w = U\left(z + \frac{1}{z}\right), \quad U = \text{ real constant}, \quad |z| > 1. \tag{3.14.1}$$

At large distances from the origin $w \to Uz$, and the motion becomes uniform at speed U parallel to the x-axis. In terms of the polar form $z = re^{i\theta}$,

$$w = U\left(re^{i\theta} + \frac{e^{-i\theta}}{r}\right), \quad \varphi = U\cos\theta\left(r + \frac{1}{r}\right).$$

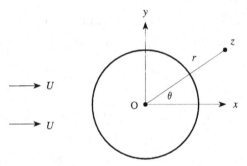

The radial component of velocity

$$\frac{\partial\varphi}{\partial r} = U\cos\theta\left(1 - \frac{1}{r^2}\right),$$

vanishes at $r = 1$. The motion therefore represents steady flow in the x-direction past a rigid cylinder of unit radius with centre at the origin. (This problem is treated by the method of separation of variables in §4.4.)

Example

$$w = -iU\left(z - \frac{1}{z}\right), \quad U = \text{ real constant}, \quad |z| > 1, \quad \left(\varphi = U\sin\theta\left(r + \frac{1}{r}\right)\right),$$

$$\tag{3.14.2}$$

describes potential flow in the y-direction past a rigid cylinder of unit radius with centre at the origin.

Example The function

$$w = \frac{1}{2\pi}\ln z, \quad \left(\varphi = \frac{1}{2\pi}\ln r, \ \psi = \frac{\theta}{2\pi}, \ z = re^{i\theta}\right),$$

is regular except at $z = 0$. The flow is radially outwards from the origin along streamlines $\theta = $ constant, at speed $\partial\varphi/\partial r = 1/2\pi r$. The origin is a singularity of the flow where fluid is *created* at a rate equal to $\oint_C \nabla\varphi \cdot \mathbf{n} ds$, where C is any simple closed curve enclosing the origin with outward normal $\mathbf{n}$, and ds is the element of arc length on C. In particular, taking C to be a circle of radius r,

$$\oint_C \nabla\varphi \cdot \mathbf{n} ds = \int_0^{2\pi} \frac{\partial\varphi}{\partial r} r d\theta = 1.$$

The origin is therefore a simple source of *unit* strength.

When the source situated at $z_0 = x_0 + iy_0$

$$w = \frac{1}{2\pi} \ln(z - z_0), \quad \left(\varphi = \frac{1}{2\pi} \ln|z - z_0| = \frac{1}{2\pi} \ln \sqrt{(x - x_0)^2 + (y - y_0)^2} \right).$$

Example The function

$$w = \frac{1}{2\pi} \left(\ln(z - z_0) + \ln(z - z_0^*) \right), \quad \left(\varphi = \frac{1}{2\pi} \left(\ln r_1 + \ln r_2 \right) \right),$$

represents the flow produced by two unit point sources located at $z_0 = x_0 + iy_0$ and $z_0^* = x_0 - iy_0$. The motion is symmetric with respect to the x-axis, and $\partial\varphi/\partial y = 0$ on $y = 0$. Therefore, in the region $y > 0$ the potential also describes the flow produced by a point source at z_0 *adjacent to a rigid wall* at $y = 0$ (the presence of the wall is said to be accounted for by an 'image' source).

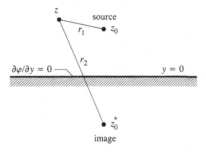

Method 2 The flow past a system of rigid boundaries in the z-plane is represented by means of a conformal transformation $Z = f(z)$ by an equivalent flow in the Z-plane. The transformation is usually chosen to simplify the boundary conditions, thereby permitting the solution in the Z-plane to be found in a relatively straightforward manner. Point source singularities of the flow are *preserved* under the transformation. Indeed, if $Z = Z_0$ is the image of a unit point source at $z = z_0$, the

complex potential in the neighbourhood of Z_0 is determined by

$$\mathcal{W}(Z) = w(z) = \frac{1}{2\pi} \ln(z - z_0) + \text{terms finite at } z_0$$

$$= \frac{1}{2\pi} \ln \left(\frac{Z - Z_0}{f'(z_0)} \right) + \text{terms finite at } Z_0$$

$$= \frac{1}{2\pi} \ln(Z - Z_0) + \text{terms finite at } Z_0.$$

The point source in the z-plane therefore maps onto an equal point source at the image point in the Z-plane.

Example Derive the following formula for the velocity potential of irrotational flow around the edge of the rigid half-plane $x < 0$, $y = 0$ in terms of polar coordinates (r, θ):

$$\varphi = \alpha \sqrt{r} \sin \frac{\theta}{2}, \quad \alpha = \text{a real constant,}$$

and make a plot of the streamlines.

The transformation $Z = i\sqrt{z}$ maps the z-plane cut along the negative real axis onto the upper half of the Z-plane (see §3.12, Example 2). The complex potential of flow in the positive X-direction parallel to the boundary $Y = 0$ in Z-plane corresponds to flow around the edge of the half-plane in the clockwise sense, and has the general representation $\mathcal{W} = UZ$, where U is real. In the z-plane this becomes

$$w = iU\sqrt{z} \equiv -U\sqrt{r} \sin \left(\frac{\theta}{2} \right) + iU\sqrt{r} \cos \left(\frac{\theta}{2} \right), \quad -\pi < \theta < \pi.$$

The polar representation of the velocity is therefore

$$\mathbf{v} = (v_r, v_\theta) = \left(\frac{\partial \varphi}{\partial r}, \frac{1}{r} \frac{\partial \varphi}{\partial \theta} \right) = \frac{-U}{2\sqrt{r}} \left(\sin \frac{\theta}{2}, \cos \frac{\theta}{2} \right).$$

This satisfies the rigid wall condition on the half-plane (where $\theta = \pm\pi$) because the component of velocity normal to the wall is v_θ, which vanishes at $\theta = \pm\pi$. The streamlines of the flow are the parabolas

$$\sqrt{r} \cos \left(\frac{\theta}{2} \right) = \text{constant,} \quad \text{i.e.} \quad y = \pm 2\beta \sqrt{1 - \frac{x}{\beta}},$$

where $x < \beta$, β being a positive constant.

When $U > 0$ fluid particles travel along the parabolic streamlines around the edge in the *clockwise* direction. The streamline for $\beta = 0$ corresponds to the upper and lower surfaces of the half-plane, which maps into the streamline $Y = 0$ on the surface of the wall in the Z-plane. The flow velocity becomes infinite like $1/\sqrt{r}$ as $r \to 0$ at the sharp edge.

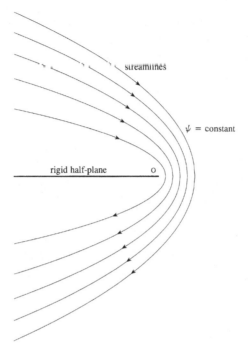

Example Calculate the irrotational flow past a flat rigid plate that lies on the x-axis between $x = \pm 1$, given that the flow at large distances from the plate is at speed U in the y-direction.

The transformation $Z = z + \sqrt{z^2 - 1}$ maps the z-plane cut along the x-axis between $x = \pm 1$ onto the region of the Z-plane outside the circular cylinder $|Z| = 1$ (§3.12, Example 4, $\sqrt{z^2 - 1}$ being defined as in (3.12.1)), with the plate mapping onto the cylinder. At large distances from the plate $Z \sim 2z$, so that the distant parts of the z- and Z-planes have the same orientations. From (3.14.2), the complex potential

$$\mathcal{W}(Z) = -iU'\left(Z - \frac{1}{Z}\right), \quad U' = \text{real constant},$$

represents flow past the cylinder that ultimately has speed U' in the Y-direction at large distances. Hence, the required potential in the z-plane has the form

$$w(z) = \mathcal{W}(Z(z)) = -iU'\left(z + \sqrt{z^2 - 1} - \frac{1}{z + \sqrt{z^2 - 1}}\right)$$

$$= -iU'\left(z + \sqrt{z^2 - 1} - \frac{z - \sqrt{z^2 - 1}}{(z + \sqrt{z^2 - 1})(z - \sqrt{z^2 - 1})}\right)$$

$$= -iU'\left(z + \sqrt{z^2 - 1} - (z - \sqrt{z^2 - 1})\right)$$

$$= -2iU'\sqrt{z^2 - 1}.$$

This implies that $w \approx -2iU'z = 2U'y - 2iU'x$ as $|z| \to \infty$, and we must therefore take $U' = \frac{1}{2}U$ to have the correct flow speed at infinity.

Thus, in the notation of Example 4, §3.12,

$$w = -iU\sqrt{z^2 - 1},$$

and

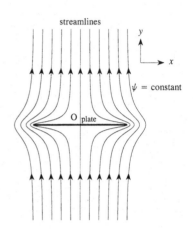

streamlines

$\psi = \text{constant}$

$$\varphi = U\sqrt{r_1 r_2}\sin\left(\frac{\theta_1 + \theta_2}{2}\right).$$

The streamlines are shown in the figure, and correspond to the family of curves

$$\psi \equiv -U\sqrt{r_1 r_2}\cos\left(\frac{\theta_1 + \theta_2}{2}\right) = \text{constant},$$

which is equivalent to

$$x = \pm\beta\sqrt{1 + \frac{1}{\beta^2 + y^2}}, \quad \beta = \text{constant} > 0.$$

Problems 3I

1. Show that the complex potential $w(z) = Ue^{-i\alpha}z$ represents uniform flow at speed U in a direction making an angle α with the x-axis.

2. According to (3.14.1), the complex potential $W(Z) = U(Z + 1/Z)$ describes flow at speed U in the X-direction past a cylinder of unit radius with centre at $Z = 0$. Use the rotational mapping $Z = e^{-i\alpha}z$ to deduce that the velocity potential of flow past the cylinder in a direction making an angle α with the x-axis is given by

$$\varphi = U\cos(\theta - \alpha)\left[r + \frac{1}{r}\right].$$

3. Two sources of unit strength are placed at the points $z = \pm a$. Show that the complex potential of the induced flow is

$$w(z) = \frac{1}{2\pi}\ln\{z^2 - a^2\}.$$

Deduce that the maximum flow speed on the plane of symmetry $x = 0$ is $1/2\pi a$.

4. Derive the following expression for the complex potential of the flow past a rigid plate $|x| < 1$, $y = 0$,

$$w(z) = U\left[z\cos\alpha - i\sin\alpha\sqrt{z^2 - 1}\right].$$

given that the flow at large distances from the plate is at speed U in a direction making an angle α with the x-axis.

5. Verify that the complex potential

$$w(z) = \frac{1}{2\pi} \left[\ln(z - R) + \ln\left(z - \frac{a^2}{R}\right) - \ln z \right]$$

describes the flow produced by a unit point source located at $z = R$ in the presence of a rigid cylinder of radius a with centre at the origin, where $R > a$.

6. By using the transformation $Z = i\sqrt{z}$ (§3.12, Example 2), show that the complex potential of the flow produced by a point source at $z = z_0$ in the presence of the rigid half-plane $x < 0$, $y = 0$ can be taken in the form

$$w(z) = \frac{1}{2\pi} \left[\ln\left(\sqrt{z} - \sqrt{z_0}\right) + \ln\left(\sqrt{z} + \{\sqrt{z_0}\}^*\right) \right],$$

where the square roots are principal values (i.e. $\sqrt{z} = \sqrt{r}e^{\frac{i\theta}{2}}$, $-\pi < \theta < \pi$).

7. Show that the velocity potential $w = U\sqrt{z^2 + a^2}$ represents the flow parallel to a rigid wall at $y = 0$ in the presence of a thin obstacle of length a projecting perpendicularly from the wall at $x = 0$.

8. Show that the transformation $Z = e^{\frac{i\pi}{3}} z^{\frac{2}{3}}$ maps the region $-\frac{\pi}{2} < \arg z < \pi$ onto the upper half of the Z-plane. Hence deduce that the streamlines of potential flow over the right-angled step with top $x < 0$, $y = 0$ and vertical face $x = 0$, $y < 0$ are given by the polar equation

$$r = \frac{\beta}{\sin^{\frac{3}{2}}\left(\frac{2\theta + \pi}{3}\right)}, \quad 0 \le \beta = \text{constant}, \quad -\frac{\pi}{2} < \theta < \pi.$$

9. A unit point source is placed at the point $z = \frac{i\pi}{2}$ within the infinite, rigid duct: $-\infty < x < \infty$, $0 < y < \pi$ of §3.12, Example 3. Under the transformation $Z = e^z$ the source maps onto an equal point source at $Z = i$. One half of the fluid created by this source must be absorbed by a 'sink' (negative source) placed at the origin $A'A$ in the Z-plane, because this point corresponds to the left-hand end of the duct $A_\infty A'_\infty$ in the z-plane, and the flow must be symmetric with respect to $x = 0$. Deduce that the complex potential can be taken in the form

$$w(z) = \frac{1}{2\pi} \ln\{\cosh z\}.$$

10. The transformation $Z = \cosh z$ maps the interior of the semi-infinite duct: $x > 0$, $0 < y < \pi$ onto the upper half of the Z-plane. Deduce that the velocity potential of the motion generated by a unit point source at the corner $x = 0$, $y = 0$ is given by

$$w(z) = \frac{2}{\pi} \ln\left(\sinh \frac{z}{2}\right).$$

4

PARTIAL DIFFERENTIAL EQUATIONS

4.1 Classification of Second-Order Equations

The principal second-order, linear partial differential equations are:

Laplace's equation: $\qquad\qquad\qquad \nabla^2 u = 0$

Wave equation: $\qquad \nabla^2 u - \dfrac{1}{c^2}\dfrac{\partial^2 u}{\partial t^2} = 0 \quad c = \text{constant} > 0$

Diffusion equation: $\qquad \nabla^2 u - \dfrac{1}{\kappa}\dfrac{\partial u}{\partial t} = 0 \quad \kappa = \text{constant} > 0$

The Laplace equation occurs in the theory of time-independent electrical and thermal phenomena, and of irrotational motion of an 'ideal' fluid. The wave equation governs the propagation of 'non-dispersive' waves (at speed c), such as sound waves in still air, electromagnetic waves in free space, and flexural waves on strings and membranes. Heat conduction in a homogeneous body and 'vorticity' diffusion in fluid at low *Reynolds number* satisfy the diffusion equation.

The equations are also encountered in inhomogeneous form, with 'sources' on the right-hand side. The inhomogeneous Laplace equation

$$\nabla^2 u = f(\mathbf{x}, t)$$

is also called the *Poisson* equation.

The *two-dimensional* equation, with independent variables x, y, has the general form

$$au_{xx} + 2bu_{xy} + cu_{yy} = f(x, y, u_x, u_y), \qquad (4.1.1)$$

where $u_{xx} = \partial^2 u/\partial x^2$, etc, and a, b, c are real coefficients that generally depend on x, y and possibly also on u_x, u_y. Historically, second-order partial differential equations of this type are classified by consideration of *Cauchy's problem*, in which a solution is to be deduced when the values of $u(x, y)$ and its normal derivative $\partial u(x, y)/\partial n$ are prescribed on a curve Γ in the xy-plane.

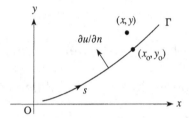

When $u(x, y)$ is known on Γ we can calculate $\partial u(x, y)/\partial s$, where s denotes arc length along Γ. Then u_x, u_y can be determined at all points on Γ by using also the prescribed value of $\partial u(x, y)/\partial n$, and we shall denote them by

$$u_x = p(s), \quad u_y = q(s). \qquad (4.1.2)$$

We can now attempt to solve the differential equation at a point (x, y) close to Γ by developing $u(x, y)$ in a Taylor series expansion about a nearby point x_0, y_0 on Γ:

$$u(x, y) = u(x_0, y_0) + u_x(x - x_0) + u_y(y - y_0)$$
$$+ \frac{1}{2}\Big\{u_{xx}(x - x_0)^2 + 2u_{xy}(y - y_0)(x - x_0)$$
$$+ u_{yy}(y - y_0)^2\Big\} + \cdots, \qquad (4.1.3)$$

where $u_x, u_y, u_{xx}, u_{xy}, u_{yy}$, etc., on the right-hand side are evaluated at (x_0, y_0). The first three terms on the right-hand side are known from the initial data. The differential equation (4.1.1) and equations (4.1.2) can be used to work out the coefficients u_{xx}, u_{xy}, u_{yy} of the quadratic terms in the expansion. To do this, differentiate equations (4.1.2) *along* Γ to obtain

$$u_{xx}\frac{dx}{ds} + u_{xy}\frac{dy}{ds} = \frac{dp}{ds}, \quad u_{xy}\frac{dx}{ds} + u_{yy}\frac{dy}{ds} = \frac{dq}{ds}. \qquad (4.1.4)$$

The three linear equations (4.1.1) and (4.1.4) determine u_{xx}, u_{xy}, u_{yy} uniquely *except* when the determinant of the coefficients vanishes, i.e. when

$$\begin{vmatrix} a & 2b & c \\ \frac{dx}{ds} & \frac{dy}{ds} & 0 \\ 0 & \frac{dx}{ds} & \frac{dy}{ds} \end{vmatrix} \equiv a\left(\frac{dy}{ds}\right)^2 - 2b\frac{dx}{ds}\frac{dy}{ds} + c\left(\frac{dx}{ds}\right)^2 = 0 \quad \text{on } \Gamma.$$

The curves that satisfy this condition are called *characteristics*. By writing the condition in the form

$$a\,dy^2 - 2b\,dx\,dy + c\,dx^2 = 0, \qquad (4.1.5)$$

it can be seen that there are generally two families of characteristics with slopes

$$\frac{dy}{dx} = \frac{b \pm \sqrt{b^2 - ac}}{a}.$$

The characteristics are straight lines when a, b, c are constants; more generally a, b, c are functions of x and y and the characteristics are curved; if a, b, c are also functions of u_x, u_y, the characteristics depend on the solution of the differential equation. When $b^2 - ac < 0$ they have *complex conjugate* slopes.

By drawing an analogy between (4.1.5) and the equation for the asymptotes of a conic section, three classes of equations (4.1.1) can be identified:

$b^2 - ac > 0$ *hyperbolic type* e.g. the wave equation $\quad u_{xx} - u_{yy} = 0$
$b^2 - ac = 0$ *parabolic type* e.g. the diffusion equation $\quad u_{xx} - u_y = 0$
$b^2 - ac < 0$ *elliptic type* e.g. Laplace's equation $\quad u_{xx} + u_{yy} = 0.$

It may be verified that the higher-order terms in the expansion (4.1.3) can also be found by the above procedure provided Γ is not a characteristic. Thus, in particular, it can be asserted that Cauchy's problem is always soluble by this method for an elliptic equation, which has complex characteristics. Hyperbolic and parabolic equations have respectively two families and one family of real characteristics, and a unique solution cannot normally be derived from data specified on a characteristic. If the coefficients a, b, c vary with position or depend on the first derivatives u_x, u_y, the differential equation may be of different type in different parts of the xy-plane.

Example   When equations (4.1.1) and (4.1.4) are solved for u_{xy} in the *homogeneous* case ($f = 0$) the result may be written

$$u_{xy}[a\,dy^2 - 2b\,dx\,dy + c\,dx^2] = a\,dp\,dy - c\,dq\,dx, \qquad (4.1.6)$$

where the differentials represent changes along Γ. This is valid for any solution of (4.1.1), not just one specified by initial conditions on Γ. In this general case u_{xy} is a well-defined quantity, even on a characteristic, and (4.1.6) can therefore be satisfied only if the right-hand side also vanishes on a characteristic, i.e.

$$a\,dp\,dy - c\,dq\,dx = 0 \quad \text{on } \Gamma. \qquad (4.1.7)$$

For the wave equation $u_{xx} - u_{yy} = 0$ ($a = 1, b = 0, c = -1$) the characteristics are the straight lines $x \pm y = $ constant on which $dx = \mp dy$, and we therefore have

$$dp - dq = 0 \quad \therefore \quad p - q = 2\Phi'(x - y) \quad \text{on} \quad x - y = \text{constant}$$

$$dp + dq = 0 \quad \therefore \quad p + q = 2\Psi'(x + y) \quad \text{on} \quad x + y = \text{constant}.$$

for some functions Φ and Ψ, where the primes denote differentiation with respect to the argument. Hence, solving for $p = u_x$ and $q = u_y$, we find

$$u_x = \Phi'(x - y) + \Psi'(x + y), \quad u_y = -\Phi'(x - y) + \Psi'(x + y),$$

which together imply that the solution of the wave equation has the general form

$$u(x, y) = \Phi(x - y) + \Psi(x + y). \qquad (4.1.8)$$

Example   The general solution (4.1.8) of the wave equation $u_{xx} - u_{yy} = 0$ can also be derived by first transforming to the *characteristic coordinates*

$$\xi = x - y, \quad \eta = x + y,$$

in terms of which

$$\frac{\partial}{\partial x} = \frac{\partial \xi}{\partial x}\frac{\partial}{\partial \xi} + \frac{\partial \eta}{\partial x}\frac{\partial}{\partial \eta} = \frac{\partial}{\partial \xi} + \frac{\partial}{\partial \eta}, \quad \text{and} \quad \frac{\partial}{\partial y} = \frac{\partial \xi}{\partial y}\frac{\partial}{\partial \xi} + \frac{\partial \eta}{\partial y}\frac{\partial}{\partial \eta} = -\frac{\partial}{\partial \xi} + \frac{\partial}{\partial \eta}.$$

Then

$$\frac{\partial^2 u}{\partial x^2} - \frac{\partial^2 u}{\partial y^2} = \left[\left(\frac{\partial}{\partial \xi} + \frac{\partial}{\partial \eta}\right)^2 - \left(-\frac{\partial}{\partial \xi} + \frac{\partial}{\partial \eta}\right)^2\right] u \equiv 4\frac{\partial^2 u}{\partial \xi \partial \eta} = 0.$$

$$\therefore \quad \text{integrating with respect to } \xi : \frac{\partial u}{\partial \eta} = \Psi'(\eta),$$

and integrating next with respect to $\eta : u = \Phi(\xi) + \Psi(\eta),$

$$\text{i.e.} \quad u = \Phi(x - y) + \Psi(x + y).$$

D'Alembert's solution of the wave equation This is the solution of the initial value (*Cauchy*) problem:

$$\frac{\partial^2 u}{\partial x^2} - \frac{1}{c^2}\frac{\partial^2 u}{\partial t^2} = 0, \quad -\infty < x < \infty, \ t > 0,$$

$$\text{when } u = f(x) \quad \text{and} \quad \frac{\partial u}{\partial t} = g(x) \quad \text{at } t = 0.$$

From (4.1.8) $\quad u(x, t) = \Phi(x - ct) + \Psi(x + ct),$

$$\therefore \quad \text{at } t = 0 : \quad \Phi(x) + \Psi(x) = f(x)$$

and $\quad -c\Phi'(x) + c\Psi'(x) = g(x).$

$$\therefore \quad -\Phi(x) + \Psi(x) = \frac{1}{c}\int^{x} g(\eta)\, d\eta + 2\mathcal{C} \quad (\mathcal{C} = \text{constant}).$$

Hence, solving for Φ and Ψ:

$$\Phi(x) = \frac{1}{2}f(x) - \frac{1}{2c}\int^{x} g(\eta)\, d\eta - \mathcal{C},$$

$$\Psi(x) = \frac{1}{2}f(x) + \frac{1}{2c}\int^{x} g(\eta)\, d\eta + \mathcal{C}$$

$$\therefore \quad u(x, t) = \Phi(x - ct) + \Psi(x + ct)$$

$$= \frac{1}{2}\Big[f(x - ct) + f(x + ct)\Big] + \frac{1}{2c}\int_{x-ct}^{x+ct} g(\eta)\, d\eta. \qquad (4.1.9)$$

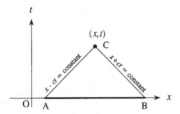

The solution consists of 'waves' propagating at speed c in the positive and negative x-directions. The solution at the point C:(x, t) in the figure is determined entirely by the initial conditions within the interval AB of the x-axis between the points of intersection of the x-axis with the two characteristics through C. AB is called the *domain of dependence* of C.

4.2   Boundary Conditions for Well-Posed Problems

Cauchy's problem is always soluble when the initial data are not given on a characteristic, but this does not guarantee that the solution is acceptable physically. The Taylor series expansion (4.1.3) is valid only for $\sqrt{(x - x_0)^2 + (y - y_0)^2} < R$, where R is the distance from (x_0, y_0) to the nearest singularity of u, which may be complex and very close to Γ. In applications to physical problems Cauchy-type data are not usually associated with elliptic equations like the Laplace equation, because the problem turns out not to be *well posed*. A boundary value problem is said to be well posed provided (i) the solution is *unique* and (ii) the solution changes smoothly with correspondingly smooth changes in the boundary conditions. The latter condition requires that small changes in the boundary data should produce small changes in the solution. This is a vital consideration when a problem is to be solved numerically: a mathematical model of a physical system would have to be rejected if small numerical errors in the boundary data produced large changes in the solution.

Example   *Dirichlet's problem* for the circle.

The real and imaginary parts of a regular function $f(z) = u + iv$ of $z = x + iy$ are solutions of Laplace's equation (§3.2). It might therefore be expected from Cauchy's integral formula (3.5.1) that a well-posed problem for the Laplace equation requires the imposition of only *one* condition on the boundary. To determine $u = \text{Re } f(z)$

within a bounded domain $\mathcal{D}$, for example, the value of u can be prescribed on the boundary; this is known as *Dirichlet's problem*. However, we could also prescribe $\partial u/\partial n$ on the boundary. Both of these problems are well posed (the Cauchy–Riemann equations show (i) that specifying $\partial u/\partial n$ on the boundary is equivalent to specifying v, so that the solution in this case determines $v = \operatorname{Im} f(z)$ and, (ii) that consistency requires $\oint_C (\partial u/\partial n)ds = 0$ where C is the boundary of $\mathcal{D}$).

Suppose that $u(x, y)$ satisfies Laplace's equation in the circle $x^2 + y^2 < R^2$, and is given by $u = u_0(\theta), 0 < \theta < 2\pi$, at (R, θ) on the boundary C. For z inside C, Cauchy's integral formula, with $\zeta = Re^{i\theta}$ and $d\zeta = i\zeta\, d\theta$, gives

$$f(z) = \frac{1}{2\pi} \int_0^{2\pi} \frac{f(\zeta)\zeta\, d\theta}{\zeta - z}.$$

The point R^2/z^* (the 'inverse point' of z with respect to the circle) lies on the same ray from the origin, but *outside* C.

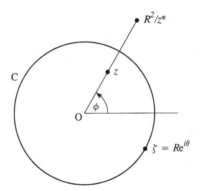

The Cauchy integral is zero at this point, i.e.

$$0 = \frac{1}{2\pi} \int_0^{2\pi} \frac{f(\zeta)\zeta\, d\theta}{\zeta - R^2/z^*}.$$

Now $\zeta^* = R^2/\zeta$ when ζ lies on C, and therefore

$$\frac{\zeta}{\zeta - R^2/z^*} = \frac{z^*}{z^* - R^2/\zeta} = -\frac{z^*}{\zeta^* - z^*}.$$

Subtracting the two integrals:

$$f(z) = \frac{1}{2\pi} \int_0^{2\pi} f(\zeta) \left\{ \frac{\zeta}{\zeta - z} + \frac{z^*}{\zeta^* - z^*} \right\} d\theta \equiv \frac{1}{2\pi} \int_0^{2\pi} f(\zeta) \left(\frac{R^2 - |z|^2}{|\zeta - z|^2} \right) d\theta.$$

$$(4.2.1)$$

The term in the large brackets in the final integral is real. Hence, setting $z = re^{i\phi}$, we find $|\zeta - z|^2 = R^2 - 2rR\cos(\theta - \phi) + r^2$, and by taking the real part we deduce

Poisson's formula for the solution of Dirichlet's problem for the circle

$$u(r, \phi) = \frac{1}{2\pi} \int_0^{2\pi} \frac{(R^2 - r^2) u_0(\theta) \, d\theta}{R^2 - 2rR\cos(\theta - \phi) + r^2}. \qquad (4.2.2)$$

Example The *Cauchy* problem for Laplace's equation in the half-plane $y > 0$:

$$u_{xx} + u_{yy} = 0, \quad \text{given} \quad u = x^3 \quad \text{and} \quad u_y = 0 \quad \text{at} \quad y = 0, \qquad (4.2.3)$$

has the exact solution $u = x^3 - 3xy^2$ (i.e. $u = \text{Re}(z^3)$, $z = x + iy$), but is not well posed. Consider a small change in the boundary conditions to

$$u = x^3 + \epsilon \sin\left(\frac{x}{\epsilon}\right), \quad u_y = 0, \quad \text{on } y = 0,$$

where ϵ is an arbitrarily small parameter. The solution now becomes

$$u = \text{Re}\left[z^3 + \epsilon \sin\left(\frac{z}{\epsilon}\right)\right] = x^3 - 3xy^2 + \epsilon \sin\left(\frac{x}{\epsilon}\right) \cosh\left(\frac{y}{\epsilon}\right).$$

As $\epsilon \to 0$ the boundary data for this problem are the same as for (4.2.3), but $|u| \to \infty$ for any value of $y > 0$. If the condition $u_y = 0$ at $y = 0$ is removed the problem becomes well posed with solution $u = x^3 - 3xy^2 + \epsilon \sin\left(\frac{x}{\epsilon}\right) e^{-\frac{y}{\epsilon}}$.

Four principal kinds of boundary value problems occur frequently in well-posed models of physical systems, with data specified on a boundary S as in Table 4.1.

Table 4.1 Boundary value problems

Problem	Data on S	Equation type
Dirichlet	u	Elliptic, parabolic
Neumann	$\dfrac{\partial u}{\partial n}$	Elliptic, parabolic
Mixed	u and $\dfrac{\partial u}{\partial n}$ given on different parts of S	Elliptic, parabolic
Cauchy	u and $\dfrac{\partial u}{\partial n}$	Hyperbolic

When an equation is to be solved in an unbounded region it is usually necessary to impose additional conditions 'at infinity' to make the solution unique. For example, to solve the wave equation for the sound generated in an infinite fluid by a vibrating body, a 'radiation condition' is applied that requires waves at large distances from the body to be propagating *away* from the body. In practice, the appropriate conditions at infinity are usually clear from the physics of the problem. Thus, when calculating the diffusion of heat from a hot body into an unbounded medium one would require the solution to be bounded or to tend to zero at large distances from the body.

4.3 Method of Separation of Variables

Henceforth attention is confined principally to *homogeneous* partial differential equations that are linear in the dependent variable and all of its derivatives. These satisfy the principle of superposition, i.e. that a solution can be formed by a linear combination of particular solutions. In the method of separation of variables, particular solutions, *eigenfunctions*, are found by assuming the dependent variable to be a product of functions each of which is only dependent on one of the independent variables. A linear combination of the eigenfunctions is then posited as the general solution of the boundary value problem governed by the differential equation. The procedure and the general problem of eigenfunctions and eigenfunction expansions were considered briefly in §1.9. That discussion is extended here by a consideration of several examples. In particular, the partial analysis in §1.9 of waves on a stretched string will be completed.

Oscillations of a hanging chain Small amplitude oscillations about the vertical of a hanging chain of length ℓ satisfy the equation

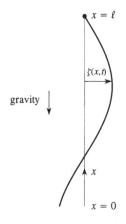

$$\frac{\partial}{\partial x}\left(x\frac{\partial \zeta}{\partial x}\right) - \frac{1}{g}\frac{\partial^2 \zeta}{\partial t^2} = 0, \quad 0 < x < \ell,$$

where $\zeta(x,t)$ is the horizontal displacement of the chain at distance x above the undisturbed equilibrium position of the free end, and g is the acceleration due to gravity. Let us solve this equation subject to the initial conditions

$$\zeta = f(x), \quad \frac{\partial \zeta}{\partial t} = 0, \quad \text{at } t = 0.$$

To derive a particular solution, we set

$$\zeta = X(x)T(t),$$

where $X(x)$ and $T(t)$ are, respectively, functions of x and t alone. Substitute into the differential equation, and divide through by $X(x)T(t)$

$$\frac{1}{X}\frac{d}{dx}\left(x\frac{dX}{dx}\right) = \frac{1}{gT}\frac{d^2T}{dt^2}.$$

The left-hand side of this equation is a function of x alone, and the right-hand side is a function of t alone. This is only possible if both sides are *constant*. This constant must be *negative* if the solution is to be oscillatory in time, i.e.

$$\frac{1}{X}\frac{d}{dx}\left(x\frac{dX}{dx}\right) = \frac{1}{gT}\frac{d^2T}{dt^2} = -k^2 = \text{ constant.}$$

Hence, $$T = A\cos(k\sqrt{g}t) + B\sin(k\sqrt{g}t),$$

where A, B are constants. The function X satisfies the Sturm–Liouville problem:

$$\frac{d}{dx}\left(x\frac{dX}{dx}\right) + k^2 X = 0, \quad \text{where } X(\ell) = 0 \quad \text{and} \quad X(0) \text{ is finite.}$$

$$(4.3.1)$$

Because $p(x) \equiv x = 0$ at $x = 0$, where $p(x)$ is the function in the general Sturm–Liouville equation (1.9.1), the eigenfunctions $X_n(x)$, say, satisfy the orthogonality relation $\int_0^\ell X_n(x)X_m(x)dx = 0, n \neq m$. Equation (4.3.1) can be solved by making the substitution

$$x = \frac{z^2}{4k^2} \quad \text{which gives} \quad \frac{d}{dx} = \frac{2k^2}{z}\frac{d}{dz},$$

and transforms (4.3.1) into Bessel's equation of zero order (§1.8)

$$\frac{d^2X}{dz^2} + \frac{1}{z}\frac{dX}{dz} + X = 0,$$

so that
$$X = CJ_0(2k\sqrt{x}) + DY_0(2k\sqrt{x}),$$

where C, D are constants. However, we must set $D = 0$ to ensure that the motion remains finite at $x = 0$ (where $Y_0(2k\sqrt{x}) \to -\infty$). The condition that the chain is fixed at its upper end $x = \ell$ then yields the eigenvalue equation satisfied by admissible values of k

$$J_0(2k\sqrt{\ell}) = 0, \quad \therefore \quad k = k_n, \ n = 1, 2, \ldots,$$

where $2k_n\sqrt{\ell}$ is the nth positive zero of $J_0(z)$ (an even function of z), and we can therefore take $X_n = J_0(2k_n\sqrt{x})$.

The most general solution of the hanging chain equation is therefore

$$\zeta = \sum_{n=1}^{\infty} \left\{ A_n \cos(k_n\sqrt{g}t) + B_n \sin(k_n\sqrt{g}t) \right\} J_0(2k_n\sqrt{x}), \quad 0 < x < \ell.$$

$$(4.3.2)$$

By setting $B_n = 0$, we satisfy the initial condition $\partial\zeta/\partial t = 0$ at $t = 0$. The coefficients A_n are determined by the initial shape of the chain from the equation

$$\sum_{n=1}^{\infty} A_n J_0(2k_n\sqrt{x}) = f(x), \quad 0 < x < \ell. \tag{4.3.3}$$

To do this we use the orthogonality relations between the eigenfunctions $X_n = J_0(2k_n\sqrt{x})$, namely (see §1.9 and the following example):

$$\int_0^{\ell} J_0(2k_n\sqrt{x})J_0(2k_m\sqrt{x})dx = \begin{cases} 0, & n \neq m, \\ \ell J_1^2(2k_n\sqrt{\ell}), & n = m \end{cases} \tag{4.3.4}$$

The method described in detail in §1.9 (see equation (1.9.6)) now gives

$$\zeta = \sum_{n=1}^{\infty} A_n \cos(k_n\sqrt{g}t)J_0(2k_n\sqrt{x}), \quad 0 < x < \ell.$$

where
$$A_n = \frac{\int_0^{\ell} f(x)J_0(2k_n\sqrt{x})dx}{\int_0^{\ell} J_0^2(2k_n\sqrt{x})dx} = \frac{\int_0^{\ell} f(x)J_0(2k_n\sqrt{x})dx}{\ell J_1^2(2k_n\sqrt{\ell})}.$$

It is sometimes convenient to replace the eigenfunctions $J_0(2k_n\sqrt{x})$ by the *orthonormal* set

$$\varphi_n(x) = \frac{J_0(2k_n\sqrt{x})}{\sqrt{\int\limits_0^\ell J_0^2(2k_n\sqrt{x})dx}} \equiv \frac{J_0(2k_n\sqrt{x})}{\sqrt{\ell}J_1(2k_n\sqrt{\ell})},$$

which satisfy

$$\int\limits_0^\ell \varphi_n(x)\varphi_m(x)dx = \begin{cases} 0, & n \neq m, \\ 1, & n = m, \end{cases} \tag{4.3.5}$$

in which case the solution assumes the compact form

$$\zeta = \sum_{n=1}^\infty A_n' \cos(k_n\sqrt{g}t)\varphi_n(x), \quad \text{where } A_n' = \int\limits_0^\ell f(x)\varphi_n(x)dx.$$

The characteristic frequencies of the chain are given by

$$\omega_n = k_n\sqrt{g} = \frac{z_n}{2}\sqrt{\frac{g}{\ell}}$$

where z_n is the nth positive zero of $J_0(z)$ (≈ 2.4048, 5.5201, 8.6537, $11.7915, \ldots$). These are not harmonically related (multiples of ω_1), in contrast to the eigenfrequencies of the simple stretched string considered in §1.9.

Example Derive the orthogonality formula (4.3.4) for the hanging chain.

This is a special kind of Sturm–Liouville problem, because at the lower end $x = 0$ of the chain it is merely required that ζ should be *bounded*, rather than satisfy a condition of the form (1.9.2). However, the eigenfunctions are still orthogonal. To show this, set $\psi_n(x) = J_0(2k_n\sqrt{x})$, then, according to (4.3.1)

$$\frac{d}{dx}\left(x\frac{d\psi_n}{dx}\right) + k_n^2\psi_n = 0, \quad \frac{d}{dx}\left(x\frac{d\psi_m}{dx}\right) + k_m^2\psi_m = 0.$$

Multiply these equations respectively by ψ_m, ψ_n, subtract, and integrate over the length of the chain to obtain

$$(k_n^2 - k_m^2)\int\limits_0^\ell \psi_n\psi_m\, dx = \int\limits_0^\ell \left[\psi_n\frac{d}{dx}\left(x\frac{d\psi_m}{dx}\right) - \psi_m\frac{d}{dx}\left(x\frac{d\psi_n}{dx}\right)\right]dx$$

$$\equiv \int\limits_0^\ell \frac{d}{dx}\left[x\psi_n\frac{d\psi_m}{dx} - x\psi_m\frac{d\psi_n}{dx}\right]dx$$

$$= \left[x\psi_n \frac{d\psi_m}{dx} - x\psi_m \frac{d\psi_n}{dx} \right]_0^\ell \tag{4.3.6}$$

$$= 0, \quad \text{for} \quad n \neq m.$$

This proves the first case of (4.3.4).

To obtain the result for $n = m$, we first note that the relation (4.3.6) is valid for any values of k_n, k_m, whether or not they are eigenvalues. Thus, let us take $k_m = k_n + \epsilon$ where $\epsilon \ll 1$ and k_n is an eigenvalue, then

$$\int_0^\ell \psi_n^2 \, dx = \frac{1}{2k_n} \lim_{\epsilon \to 0} \left[\frac{\ell[\psi_{n+\epsilon}(\ell)\psi_n'(\ell) - \psi_n(\ell)\psi_{n+\epsilon}'(\ell)]}{\epsilon} \right], \tag{4.3.7}$$

where the prime denotes differentiation with respect to x. But, $\psi_n(\ell) = 0$ because k_n is an eigenvalue. Also

$$\psi_{n+\epsilon}(\ell) \equiv J_0\left(2(k_n + \epsilon)\sqrt{\ell} \right) \approx J_0(2k_n\sqrt{\ell}) + 2\epsilon\sqrt{\ell}J_0'(2k_n\sqrt{\ell}) \equiv 2\epsilon\sqrt{\ell}J_0'(2k_n\sqrt{\ell})$$

and

$$\psi_n'(\ell) = \left(\frac{d}{dx} J_0(2k_n\sqrt{x}) \right)_{x=\ell} = \frac{k_n}{\sqrt{\ell}}J_0'(2k_n\sqrt{\ell}).$$

Hence, (4.3.7) becomes

$$\int_0^\ell \psi_n^2 \, dx = \frac{1}{2k_n} \lim_{\epsilon \to 0} \frac{\ell 2\epsilon\sqrt{\ell}J_0'(2k_n\sqrt{\ell})(k_n/\sqrt{\ell})J_0'(2k_n\sqrt{\ell})}{\epsilon}$$

$$= \ell\left(J_0'(2k_n\sqrt{\ell}) \right)^2 \equiv \ell J_1^2(2k_n\sqrt{\ell}),$$

where the relation $J_0'(x) = -J_1(x)$ has been used (§1.8). This proves the second of (4.3.4).

Example Apply the method of separation of variables to derive the solution (1.9.5) of the wave equation

$$\frac{\partial^2 \zeta}{\partial x^2} - \frac{\partial^2 \zeta}{\partial t^2} = 0, \quad 0 < x < 1; \text{ where } \zeta = 0 \text{ at } x = 0, \ 1,$$

with the initial conditions $\zeta = f(x)$, $\partial\zeta/\partial t = 0$ at $t = 0$ for $0 < x < 1$.

Set $\zeta = X(x)T(t)$ and substitute into the wave equation to obtain

$$\frac{1}{X}\frac{d^2 X}{dx^2} = \frac{1}{T}\frac{d^2 T}{dt^2} = -k^2 = \text{constant},$$

where the constant $-k^2$ is taken to be negative to ensure that the solution is oscillatory in time. Hence X is the solution of the Sturm–Liouville problem:

$$X'' + k^2 X = 0, \quad \text{where } X = 0 \text{ at } x = 0, \ 1.$$

Therefore
$$X(x) = A \cos kx + B \sin kx,$$

and the conditions at $x = 0, 1$ give

$$A = 0, \quad \sin(k) = 0 \quad (\text{for } B \neq 0),$$

$$\therefore \quad k = n\pi, \quad n = 0, 1, 2, \text{etc.}$$

The eigenfunctions of the Sturm–Liouville problem are therefore

$$X_n(x) = \sin(n\pi x), \quad n \geq 1, \quad \text{and} \quad \int_0^1 \sin(n\pi x) \sin(m\pi x)\,dx = \begin{cases} 0, & n \neq m, \\ \dfrac{1}{2}, & n = m. \end{cases}$$

Next,

$$T'' + k^2 T = 0 \quad (k = n\pi) \quad \therefore \quad T = C \cos(n\pi t) + D \sin(n\pi t), \quad C, \ D = \text{constant},$$

We can now form the general solution by combining the different separable solutions $\zeta = X(x)T(t)$ for all $n \geq 1$. To do this first set $C_n = BC$, $D_n = BD$ in the nth separable solution, then

$$\zeta = \sum_{n=1}^{\infty} \left(C_n \cos(n\pi t) + D_n \sin(n\pi t) \right) \sin(n\pi x).$$

The initial conditions yield $D_n = 0$, $n \geq 1$, and the values of C_n are found by using the orthogonality relation for the eigenfunctions. Hence,

$$\zeta = \sum_{n=1}^{\infty} C_n \cos(n\pi t) \sin(n\pi x), \quad \text{where } C_n = 2 \int_0^1 f(x) \sin(n\pi x)\,dx, \quad n \geq 1, \quad t \geq 0.$$

By using the trigonometric formula $2 \sin A \cos B = \sin(A+B) + \sin(A-B)$, this result can be written

$$\zeta = \frac{1}{2} \sum_{n=1}^{\infty} C_n \sin[n\pi(x+t)] + \frac{1}{2} \sum_{n=1}^{\infty} C_n \sin[n\pi(x-t)] \equiv \frac{1}{2}\left(f(x+t) + f(x-t) \right),$$

which expresses the result in D'Alembert's form (4.1.9). It shows that the displacement of the string consists of two equal waves propagating respectively to the left and right whose waveforms are identical in shape but of half the amplitude of the initial displacement of the string. The two waves may be regarded as propagating along an *infinite* string (in $-\infty < x < \infty$), whose initial displacement has period 2 in x, such that $\zeta(x,0) = f(x)$ in $0 < x < 1$ and $\zeta(x,0) = -f(-x)$ in $-1 < x < 0$, i.e. $f(x)$ is imagined to be extended as an *odd* function of x with respect to *both* ends $x = 0$ and $x = 1$ of the original, finite string.

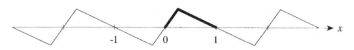

$f(x)$ extended as an odd function of x of period 2

Example Solve by the method of separation of variables the boundary value problem

$$\frac{\partial^2 u}{\partial x^2} = \frac{1}{\kappa} \frac{\partial u}{\partial t}, \quad 0 < x < \ell, \tag{4.3.8}$$

where $u = 0$ at $x = 0$, ℓ, and u satisfies the initial condition $u = f(x)$ at $t = 0$ for $0 < x < \ell$.

This problem represents the diffusion of heat from the ends $x = 0$, ℓ of a bar of thermal diffusivity κ and length ℓ whose sides are thermally insulated, when the initial distribution of temperature is specified by the function $f(x)$. There is only one initial condition at $t = 0$ because the diffusion equation is first order in the time derivative.

Set $u = X(x)T(t)$ and substitute into the differential equation to obtain

$$\frac{1}{X} \frac{d^2 X}{dx^2} = \frac{1}{\kappa T} \frac{dT}{dt} = -k^2 = \text{constant},$$

where the constant $-k^2$ is taken to be negative to ensure that the solution $T = Ce^{-k^2 \kappa t}$ ($C = $ constant) is *bounded* for $t > 0$. Then,

$$X'' + k^2 X = 0, \quad 0 < x < \ell,$$
$$\text{where } X = 0 \quad \text{at } x = 0 \quad \text{and} \quad x = \ell$$

This is a Sturm–Liouville problem, where

$$X(x) = A \cos kx + B \sin kx,$$

and the conditions at $x = 0, \ell$ give

$$A = 0, \quad \sin(k\ell) = 0 \quad (\text{for } B \neq 0),$$

$$\therefore \quad k\ell = n\pi, \quad \text{i.e. } k = \frac{n\pi}{\ell}, \quad n = 0, 1, 2, \text{ etc.}$$

The eigenfunctions of the Sturm–Liouville problem are therefore

$$X_n(x) = \sin\left(\frac{n\pi x}{\ell}\right), \ n \geq 1, \ \text{and} \ \int_0^\ell \sin\left(\frac{n\pi x}{\ell}\right) \sin\left(\frac{m\pi x}{\ell}\right) dx = \begin{cases} 0, & n \neq m, \\ \ell/2 & n = m. \end{cases}$$

$$\tag{4.3.9}$$

We can now form the general solution by combining the different separable solutions $u = X(x)T(t)$ for all $n \geq 1$. To do this first set $A_n = BC$ in the nth separable solution, then

$$u(x,t) = \sum_{n=1}^{\infty} A_n \sin\left(\frac{n\pi x}{\ell}\right) e^{-\left(\frac{n\pi}{\ell}\right)^2 \kappa t}.$$

To satisfy the initial condition at $t = 0$ the coefficients A_n must be chosen to ensure that

$$u(x,0) \equiv f(x) = \sum_{n=1}^{\infty} A_n \sin\left(\frac{n\pi x}{\ell}\right).$$

Hence, using the orthogonality relations (4.3.9) we obtain the solution of the boundary value problem in the form

$$u = \sum_{n=1}^{\infty} A_n \sin\left(\frac{n\pi x}{\ell}\right) e^{-\left(\frac{n\pi}{\ell}\right)^2 \kappa t}, \quad t > 0, \quad \text{where} \quad A_n = \frac{2}{\ell} \int_0^{\ell} f(x) \sin\left(\frac{n\pi x}{\ell}\right) dx.$$

As $t \to \infty$ the temperature $u \to 0$ as the bar tends to thermal equilibrium with its surroundings.

Problems 4A

1. If

$$f(x) = \begin{cases} x, & 0 < x < \ell/2, \\ \ell - x, & \ell/2 < x < \ell, \end{cases}$$

in the boundary value problem for the diffusion equation (4.3.8), show that

$$u = \frac{4\ell}{\pi^2} \sum_{n=0}^{\infty} \frac{(-1)^n}{(2n+1)^2} \sin\left(\frac{(2n+1)\pi x}{\ell}\right) \exp\left(\frac{-(2n+1)^2 \pi^2 \kappa t}{\ell^2}\right), \quad t > 0.$$

2. When the boundary conditions for the heat equation (4.3.8) are

$$\frac{\partial u}{\partial x} = 0 \quad \text{at} \quad x = 0, \ \ell, \ t > 0,$$

and $u = f(x)$ at $t = 0$, show that

$$u = \sum_{n=0}^{\infty} A_n \cos\left(\frac{n\pi x}{\ell}\right) e^{-\left(\frac{n\pi}{\ell}\right)^2 \kappa t}, \quad t > 0,$$

where $A_0 = \frac{1}{\ell} \int_0^{\ell} f(x)dx, \quad A_n = \frac{2}{\ell} \int_0^{\ell} f(x) \cos\left(\frac{n\pi x}{\ell}\right) dx \ (n \geq 1).$

The final equilibrium temperature is $u = A_0$.

3. When $f(x) = x$ in problem 2, deduce that

$$u = \frac{\ell}{2} - \frac{4\ell}{\pi^2} \sum_{n=0}^{\infty} \frac{\cos\left(\frac{(2n+1)\pi x}{\ell}\right) \exp\left(\frac{-(2n+1)^2 \pi^2 \kappa t}{\ell^2}\right)}{(2n+1)^2}, \quad t > 0.$$

4. The displacement $\zeta(x,t)$ of a stretched string of unit length satisfies the wave equation

$$\frac{\partial^2 \zeta}{\partial x^2} - \frac{\partial^2 \zeta}{\partial t^2} = 0, \quad 0 < x < 1; \text{ where } \zeta = 0 \text{ at } x = 0, 1.$$

Show that if the string is pulled aside a distance h at its midpoint and released from rest, the displacement after time t is given by

$$\zeta(x,t) = \sum_{n=0}^{\infty} \frac{8h(-1)^n}{(2n+1)^2\pi^2} \sin[(2n+1)\pi x] \cos[(2n+1)\pi t].$$

5. Show that the solution of the *Dirichlet* problem for u:

$$\frac{\partial^2 u}{\partial x^2} + \frac{\partial^2 u}{\partial y^2} = 0$$

in the rectangular domain $0 < x < a, 0 < y < b$, where $u = 0$ on each side of the rectangle *except* that along the x-axis, where $u = f(x)$ $(0 < x < a, y = 0)$, can be expressed in the form

$$u = \sum_{n=1}^{\infty} A_n \sin\left(\frac{n\pi x}{a}\right) \sinh\left(\frac{n\pi(b-y)}{a}\right),$$

$$\text{where } A_n = \frac{2}{a} \int_0^a \frac{f(x)\sin(n\pi x/a)dx}{\sinh(n\pi b/a)}.$$

6. Show that if $u(x,y)$ satisfies

$$\frac{\partial^2 u}{\partial x^2} + \frac{\partial^2 u}{\partial y^2} = 0, \quad 0 < x < a,\ 0 < y < b,$$

where $u(0,y) = u(a,y) = 0$, $u(x,0) = u(x,b) = 2x(x-a)$, then

$$u = \sum_{n=0}^{\infty} A_n \sin\left(\frac{(2n+1)\pi x}{a}\right)$$

$$\times \left\{ \sinh\left(\frac{(2n+1)\pi(b-y)}{a}\right) + \sinh\left(\frac{(2n+1)\pi y}{a}\right) \right\},$$

$$A_n = \frac{-16a^2}{\pi^3(2n+1)^3 \sinh[(2n+1)\pi b/a]}.$$

7. Obtain all solutions of

$$\frac{\partial^2 u}{\partial x^2} - \frac{\partial u}{\partial y} = u$$

of the form $u(x,y) = (A\cos\lambda x + B\sin\lambda x)f(y)$, where A, B, λ are constants. Show that the particular solution that satisfies $u(0,y) = 0, u(\pi,y) = 0, u(x,1) = x$ $(0 < x < \pi)$ is given by

$$u = -2\sum_{n=1}^{\infty} \frac{(-1)^n}{n} e^{(1+n^2)(1-y)} \sin nx.$$

8. Show that

$$u(x,t) = (A\cos\Omega t + B\sin\Omega t)\sqrt{\frac{x}{a}}\sin\left\{\frac{n\pi}{\ln 2}\ln\left(\frac{x}{a}\right)\right\},$$

$$\Omega = \sqrt{\frac{n^2\pi^2}{(\ln 2)^2} + \frac{1}{4}}, \quad n = 1, 2, 3, \ldots,$$

is a particular solution of the problem

$$x^2\frac{\partial^2 u}{\partial x^2} = \frac{\partial^2 u}{\partial t^2}, \quad u(a,t) = 0, \ u(2a,t) = 0.$$

9. Show that

$$u(x,t) = \frac{1}{2}a - \frac{4a}{\pi^2}\sum_{n=0}^{\infty}\frac{\cos\{(2n+1)\pi x/a\}}{(2n+1)^2}e^{-t[(2n+1)\pi/a]^2}$$

satisfies

$$\frac{\partial^2 u}{\partial x^2} = \frac{\partial u}{\partial t}, \quad \frac{\partial u}{\partial x}(\pm a,t) = 0, \quad u(x,0) = |x|, \quad |x| < a.$$

10. By noting that $u = x$ is a particular solution of

$$\frac{\partial^2 u}{\partial x^2} = \frac{\partial u}{\partial t},$$

show that the solution $u(x,t)$ that satisfies the conditions (i) $u(0,t) = 0$ and $u(1,t) = 1$ for $t > 0$, (ii) $u(x,0) = 0$ for $0 \le x \le 1$, can be written

$$u(x,t) = x + \frac{2}{\pi}\sum_{n=1}^{\infty}\frac{(-1)^n\sin n\pi x}{n}e^{-n^2\pi^2 t}.$$

11. Show that when the conditions of Problem 10 are replaced by (i) $u(0,t) = 1$ and $u(1,t) = 0$ for $t > 0$, (ii) $u(x,0) = \cos(\frac{\pi}{2}x)$ for $0 \le x \le 1$,

$$u(x,t) = 1 - x + \frac{2}{\pi}\sum_{n=1}^{\infty}\frac{\sin n\pi x}{n(4n^2 - 1)}e^{-n^2\pi^2 t}.$$

12. Show that

$$u(x,t) = \frac{2}{\pi} - \frac{4}{\pi}\sum_{n=1}^{\infty}\frac{\cos(2n\pi x/\ell)}{4n^2 - 1}e^{-4\pi^2 n^2 t/\ell^2}, \quad 0 < x < \ell, \ t > 0,$$

satisfies

$$\frac{\partial^2 u}{\partial x^2} = \frac{\partial u}{\partial t}, \quad \text{where}\ \left(\frac{\partial u}{\partial x}\right)_{x=0,\ \ell} = 0, \quad \text{and}\ \ u(x,0) = \sin\left(\frac{\pi x}{\ell}\right), \quad 0 < x < \ell.$$

13. Show that the solution $u(x,y)$ of

$$\frac{\partial^2 u}{\partial x^2} + \frac{\partial^2 u}{\partial y^2} = 0, \quad 0 < x < \pi, \ \ 0 < y < \pi,$$

where $\partial u/\partial x = 0$ at $x = 0$ and $x = \pi$, $u(x, \pi) = 0$, and $u(x, 0) = f(x)$, is given by

$$u = A_0 \frac{(\pi - y)}{2\pi} + \sum_{n=1}^{\infty} A_n \frac{\cos nx \sinh n(\pi - y)}{\sinh n\pi}, \quad A_n = \frac{2}{\pi} \int_0^{\pi} f(x) \cos nx \, dx.$$

14. If $u = 0$ for $y > 0$ and $x = 0$ and $x = \ell$, and $u = 1$ when $y = 0$ and $0 < x < \ell$, and if

$$\frac{\partial^2 u}{\partial x^2} + \frac{\partial^2 u}{\partial y^2} = 0, \quad 0 < x < \ell, \ y > 0,$$

show that

$$u(x, y) = \frac{4}{\pi} \sum_{n=0}^{\infty} \frac{\sin[(2n + 1)\pi x/\ell]}{2n + 1} \left\{ \cosh \left(\frac{(2n + 1)\pi y}{\ell} \right) - \sinh \left(\frac{(2n + 1)\pi y}{\ell} \right) \right\}.$$

15. If $u(x, t)$ satisfies

$$\frac{\partial^2 u}{\partial x^2} = \frac{\partial u}{\partial t}, \quad 0 < x < a, \ t > 0, \quad \text{subject to } u(0, t) = 0,$$

$$\frac{\partial u}{\partial x}(a, t) = 0, \ u(x, 0) = \sin^3 \left(\frac{\pi x}{2a} \right),$$

show that

$$u(x, t) = \frac{3}{4} \sin \left(\frac{\pi x}{2a} \right) e^{-\pi^2 t/4a^2} - \frac{1}{4} \sin \left(\frac{3\pi x}{2a} \right) e^{-9\pi^2 t/4a^2}.$$

4.4   Problems with Cylindrical Boundaries

Laplace's equation in polar coordinates (r, θ)

$$\nabla^2 u \equiv \frac{\partial^2 u}{\partial r^2} + \frac{1}{r} \frac{\partial u}{\partial r} + \frac{1}{r^2} \frac{\partial^2 u}{\partial \theta^2} = 0, \quad 0 \le \theta \le 2\pi,$$

has separable solutions $u = R(r)\Theta(\theta)$ that satisfy

$$\frac{r^2}{R} \frac{d^2 R}{dr^2} + \frac{r}{R} \frac{dR}{dr} = -\frac{1}{\Theta} \frac{d^2\Theta}{d\theta^2} = \text{constant.}$$

The solution will be *single valued* provided $\Theta(\theta)$ has period 2π. The constant must therefore be zero or equal to $n^2, n = 1, 2, \ldots$, in which case

$$\Theta = A \cos(n\theta) + B \sin(n\theta), \quad R = \frac{C}{r^n} + Dr^n, \quad n \ge 1,$$

$$\Theta = A, \quad R = C + D \ln r, \quad n = 0,$$

where A, B, C, D are constants.

Thus, when a *bounded* solution is required in the interior of a circle $r < a$, say, the general solution is

$$u = \sum_{n=0}^{\infty} \Big(A_n \cos(n\theta) + B_n \sin(n\theta) \Big) r^n, \quad r < a. \tag{4.4.1}$$

The bounded solution in an exterior domain is, similarly,

$$u = \sum_{n=0}^{\infty} \Big(A_n \cos(n\theta) + B_n \sin(n\theta) \Big) \frac{1}{r^n}, \quad r > a. \tag{4.4.2}$$

However, a bounded solution in an annular region $a < r < b$ may include all terms $\propto r^n$ and r^{-n}, and also the term in $\ln r$.

Velocity potential of flow past a cylinder The velocity potential of incompressible, ideal steady flow at speed U in the x-direction past a stationary, rigid circular cylinder of radius a (with boundary $r = a$ in polar coordinates) is given by

$$\varphi = Ux + u(r, \theta) \equiv Ur\cos\theta + u(r, \theta),$$

where u satisfies Laplace's equation and tends to zero as $r \to \infty$.

u has the general representation (4.4.2). The normal component of velocity must vanish on the cylinder

$$\left(\frac{\partial \varphi}{\partial r} \right)_{r=a} = 0.$$

This gives the equation for the Fourier coefficients A_n, B_n:

$$\sum_{n=0}^{\infty} \frac{n}{a^{n+1}} \Big(A_n \cos(n\theta) + B_n \sin(n\theta) \Big) = U\cos\theta, \quad 0 \le \theta \le 2\pi,$$

which implies that all of coefficients vanish except $A_1 = Ua^2$. Hence,

$$\varphi = U\left(r + \frac{a^2}{r} \right) \cos\theta.$$

Example Solve the Neumann boundary value problem in $a < r < b$:

$$\nabla^2 u \equiv \frac{\partial^2 u}{\partial r^2} + \frac{1}{r}\frac{\partial r}{\partial r} + \frac{1}{r^2}\frac{\partial^2 u}{\partial \theta^2} = 0, \quad a < r < b, \quad \frac{\partial u}{\partial r} = \begin{cases} U(t)\cos\theta, & r = a, \\ 0, & r = b. \end{cases}$$

This determines the velocity potential produced by an inner cylinder oscillating back and forth with velocity $U(t)$. Although the velocity depends on time (produced by an external agency that may be assumed to force the motion), time does not appear explicitly in the equation for the velocity potential because the fluid is incompressible. Note that the boundary condition on the inner cylinder satisfies the compatibility condition

$$\int_0^{2\pi} \left(\frac{\partial u}{\partial r}\right)_{r=a} d\theta = 0.$$

Clearly

$$u = \left(\frac{A}{r} + Br\right)\cos\theta,$$

where the constants A and B are determined from the boundary conditions on $r = a, b$, i.e. by

$$B - \frac{A}{a^2} = U, \quad B - \frac{A}{b^2} = 0.$$

Hence,

$$u = \frac{-Ua^2}{(b^2 - a^2)}\left(\frac{b^2}{r} + r\right)\cos\theta, \quad a < r < b.$$

The radially symmetric solution of Laplace's equation

$$u = C + D\ln r$$

is unbounded both at $r = 0$ and $r = \infty$. It represents the velocity potential of a two-dimensional ('line') source at $r = 0$ (see §3.14). When there is no source we must take $D = 0$ for the solution to be bounded, unless $r = 0$ is outside the physical region where Laplace's equation is to be solved, for example, when a solution is required in the annular region $a < r < b$ (see Question 5 of Problems 4B).

Example Poisson's integral of the Dirichlet problem for the circle (see §4.2):

$$\nabla^2 u \equiv \frac{\partial^2 u}{\partial r^2} + \frac{1}{r}\frac{\partial r}{\partial r} + \frac{1}{r^2}\frac{\partial^2 u}{\partial \theta^2} = 0, \quad 0 \le r < R, \quad \text{where } u(R, \theta) = f(\theta).$$

The bounded solution in $0 \le r \le R$ can be written

$$u = A_0 + \sum_{n=1}^{\infty} \left(A_n \cos(n\theta) + B_n \sin(n\theta)\right)r^n,$$

where

$$A_0 + \sum_{n=1}^{\infty} \Big(A_n \cos(n\theta) + B_n \sin(n\theta) \Big) R^n = f(\theta), \quad 0 \le \theta \le 2\pi.$$

$$\therefore \quad \text{(see §1.10)} \quad A_0 = \frac{1}{2\pi} \int_0^{2\pi} f(\theta') d\theta',$$

$$(A_n, B_n) = \frac{1}{\pi R^n} \int_0^{2\pi} f(\theta')(\cos n\theta', \sin n\theta') d\theta', \quad n \ge 1.$$

Hence,

$$u = \frac{1}{\pi} \int_0^{2\pi} f(\theta') \left(\frac{1}{2} + \sum_{n=1}^{\infty} \left(\frac{r}{R}\right)^n [\cos n\theta \cos n\theta' + \sin n\theta \sin n\theta'] \right) d\theta'$$

$$= \frac{1}{\pi} \int_0^{2\pi} f(\theta') \left(\frac{1}{2} + \sum_{n=1}^{\infty} \left(\frac{r}{R}\right)^n \cos n(\theta - \theta') \right) d\theta'$$

$$= \text{Re} \, \frac{1}{\pi} \int_0^{2\pi} f(\theta') \left(\frac{1}{2} + \sum_{n=1}^{\infty} \left(\frac{r}{R}\right)^n e^{in(\theta - \theta')} \right) d\theta'$$

$$= \text{Re} \, \frac{1}{\pi} \int_0^{2\pi} f(\theta') \left(\frac{1}{2} + \frac{(r/R)e^{i(\theta - \theta')}}{1 - (r/R)e^{i(\theta - \theta')}} \right) d\theta'$$

$$= \text{Re} \, \frac{1}{2\pi} \int_0^{2\pi} f(\theta') \left(\frac{R + re^{i(\theta - \theta')}}{R - re^{i(\theta - \theta')}} \right) d\theta'.$$

Extraction of the real part leads to Poisson's formula

$$u(r, \theta) = \frac{1}{2\pi} \int_0^{2\pi} \frac{(R^2 - r^2)f(\theta') \, d\theta'}{R^2 - 2rR \cos(\theta - \theta') + r^2}.$$

This is identical with the solution (4.2.2) derived using Cauchy's integral formula (3.5.1).

Problems 4B

1. If $u(r, \theta)$ satisfies

$$\frac{\partial^2 u}{\partial r^2} + \frac{1}{r} \frac{\partial u}{\partial r} + \frac{1}{r^2} \frac{\partial^2 u}{\partial \theta^2} = 0, \quad 0 < r < a,$$

and $u(a, \theta) = \theta, -\pi < \theta < \pi$, show that

$$u = -2 \sum_{n=1}^{\infty} \left(-\frac{r}{a} \right)^n \frac{\sin n\theta}{n}.$$

For the *exterior* problem, where the same equation is to be solved in the infinite region $r > a$, subject to the same condition at $r = a$, show that

$$u = -2 \sum_{n=1}^{\infty} \left(-\frac{a}{r}\right)^n \frac{\sin n\theta}{n}.$$

2. Establish the solution

$$u = \sum_{n=0}^{\infty} \frac{4C}{\pi(2n+1)} \left(\frac{r}{a}\right)^{2n+1} \sin(2n+1)\theta$$

of the boundary value problem

$$\frac{\partial^2 u}{\partial r^2} + \frac{1}{r}\frac{\partial u}{\partial r} + \frac{1}{r^2}\frac{\partial^2 u}{\partial \theta^2} = 0, \quad 0 < r < a, \quad 0 < \theta < \pi,$$

where $u(a, \theta) = C$ (constant) for $0 < \theta < \pi$, and $u(r, \theta) = 0$ for $\theta = 0$ and $\theta = \pi$.

3. Show that the solution of the cylindrically symmetric diffusion problem

$$\frac{\partial^2 u}{\partial r^2} + \frac{1}{r}\frac{\partial u}{\partial r} = \frac{1}{\kappa}\frac{\partial u}{\partial t}, \quad 0 < r < a,$$

where $u(a, t) = 0, t > 0$, and $u(r, 0) = f(r), 0 < r < a$, is

$$u = \sum_{n=1}^{\infty} A_n J_0 \left(\frac{\alpha_n r}{a}\right) \exp\left(-\frac{\kappa \alpha_n^2 t}{a^2}\right),$$

where α_n is the nth positive zero of $J_0(x)$, and

$$A_n = \frac{\int_0^a r f(r) J_0 \left(\frac{\alpha_n r}{a}\right) dr}{\int_0^a r J_0^2 \left(\frac{\alpha_n r}{a}\right) dr} \equiv \frac{2 \int_0^a r f(r) J_0 \left(\frac{\alpha_n r}{a}\right) dr}{a^2 J_1^2(\alpha_n)}.$$

4. Radially symmetric oscillations of a circular membrane of unit radius are governed by the equation

$$\frac{\partial^2 u}{\partial r^2} + \frac{1}{r}\frac{\partial u}{\partial r} - \frac{\partial^2 u}{\partial t^2} = 0, \quad 0 < r < 1,$$

where $u(r, t) = 0$ at $r = 1$ for all time t. Show that, if

$$u = f(r), \quad \text{and} \quad \frac{\partial u}{\partial t} = 0 \quad \text{at} \quad t = 0, \ 0 < r < 1,$$

then

$$u = \sum_{n=1}^{\infty} \frac{2 J_0(\alpha_n r) \cos(\alpha_n t)}{J_1^2(\alpha_n)} \left(\int_0^1 r f(r) J_0(\alpha_n r) \, dr\right), \quad t > 0,$$

where α_n is the nth positive zero of $J_0(x)$.

Show further that, when $f(r) = 1 - r^2$,

$$u = \sum_{n=1}^{\infty} \frac{4J_2(\alpha_n)J_0(\alpha_n r)\cos(\alpha_n t)}{\alpha_n^2 J_1^2(\alpha_n)}, \quad t > 0.$$

5. $u(r, \theta)$ satisfies

$$\frac{\partial^2 u}{\partial r^2} + \frac{1}{r}\frac{\partial u}{\partial r} + \frac{1}{r^2}\frac{\partial^2 u}{\partial \theta^2} = 0, \quad a < r < b, \quad -\pi < \theta < \pi,$$

where $u(a, \theta) = 0$, $u(b, \theta) = 1$, $-\pi < \theta < \pi$. Deduce that

$$u = \frac{\ln(r/a)}{\ln(b/a)}, \quad -\pi < \theta < \pi.$$

Where does $u = \frac{1}{5}$? $[r = a(b/a)^{\frac{1}{5}}]$.

6. Show that the solution $u(r, x)$ of the boundary value problem

$$\frac{\partial^2 u}{\partial r^2} + \frac{1}{r}\frac{\partial u}{\partial r} + \frac{\partial^2 u}{\partial x^2} = 0, \quad r < 1, \quad 0 < x < a,$$

where $u(1, x) = 0, u(r, a) = 0, u(r, 0) = f(r)$, is given by

$$u = \sum_{n=1}^{\infty} A_n J_0(\lambda_n r)\sinh\{\lambda_n(a - x)\},$$

$$A_n = \frac{2}{J_1^2(\lambda_n)\sinh(\lambda_n a)}\int_0^1 f(r)rJ_0(\lambda_n r)dr,$$

$\lambda_n = n$th positive zero of $J_0(\lambda) = 0$.

7. Find $u(r, x)$ in $r < 1, 0 < x < a$ when

$$\frac{\partial^2 u}{\partial r^2} + \frac{1}{r}\frac{\partial u}{\partial r} + \frac{\partial^2 u}{\partial x^2} = 0, \quad r < 1, \quad 0 < x < a,$$

and $u(1, x) = 0, u(r, 0) = f(r), u(r, a) = g(r)$.

$$\left[u = \sum_{n=1}^{\infty} \left[A_n \sinh\{\lambda_n(a - x)\} + B_n \sinh \lambda_n x \right] J_0(\lambda_n r), \right.$$

$$(A_n, B_n) = \frac{2}{J_1^2(\lambda_n)\sinh(\lambda_n a)}\int_0^1 (f(r), g(r))rJ_0(\lambda_n r)dr,$$

$$\left. \lambda_n = n\text{th positive zero of } J_0(\lambda) = 0 \right]$$

8. When $f(r) = 1$ and $g(r) = 0$ in Problem 7, show that

$$u = \sum_{n=1}^{\infty} \frac{2\sinh\{\lambda_n(a - x)\}J_0(\lambda_n r)}{\lambda_n \sinh(\lambda_n a)J_1(\lambda_n a)}.$$

9. Let λ_k be the kth positive zero of $J_0(x)$. Show that if $u(r, t)$ satisfies

$$\frac{\partial^2 u}{\partial r^2} + \frac{1}{r}\frac{\partial u}{\partial r} = \frac{1}{\kappa}\frac{\partial u}{\partial t}, \quad 0 < r < 1, \quad t > 0, \quad u(r, 0) = u_0 J_0(\lambda_k r),$$

then $u(r, t) = u_0 J_0(\lambda_k r)e^{-\kappa\lambda_k^2 t}$ for $t > 0$.

10. Show that the solution of

$$\frac{\partial^2 u}{\partial r^2} + \frac{1}{r}\frac{\partial u}{\partial r} = \frac{1}{\kappa}\frac{\partial u}{\partial t}, \quad 0 < r < 1,$$

where $u(1, t) = u_0, t > 0$, and $u(r, 0) = 0, 0 \le r < 1$, is

$$u = u_0\left\{1 - 2\sum_{n=1}^{\infty}\frac{J_0(\lambda_n r)}{\lambda_n J_1(\lambda_n)}\exp(-\kappa\lambda_n^2 t)\right\}, \quad t > 0,$$

where λ_n is the nth positive zero of $J_0(x)$.

4.5 Application of Green's Second Identity: Green's Function

The Laplace operator ∇^2 in spherical polar coordinates (r, θ, ϕ) is given by (2.6.6). Spherically symmetric solutions of Laplace's equation therefore satisfy

$$\frac{1}{r^2}\frac{\partial}{\partial r}\left(r^2\frac{\partial\varphi}{\partial r}\right) = 0, \quad \text{i.e. } \varphi = A + \frac{B}{r}, \quad r > 0, \tag{4.5.1}$$

where A and B are constants.

By temporarily shifting the coordinate origin to a given, fixed point $\mathbf{x}_0$, so that $r \equiv |\mathbf{x} - \mathbf{x}_0|$, and taking $A = 0$ and $B = -1/4\pi$, we see that

$$G(\mathbf{x}, \mathbf{x}_0) = \frac{-1}{4\pi|\mathbf{x} - \mathbf{x}_0|}, \tag{4.5.2}$$

satisfies

$$\nabla^2 G = 0 \quad \text{for } |\mathbf{x} - \mathbf{x}_0| > 0. \tag{4.5.3}$$

G is called the *free space* Green's function for the Laplace equation, and can be used to write down a formal representation of the solution of the inhomogeneous Laplace equation (*Poisson's* equation)

$$\nabla^2\varphi = f(\mathbf{x}), \tag{4.5.4}$$

where the 'source' term $f(\mathbf{x})$ is non-zero only within a finite region of space, and $\varphi \to 0$ as $|\mathbf{x}| \to \infty$.

To do this, let V be the volume bounded by the surface Σ_0 of a small sphere of radius ϵ with centre at $\mathbf{x}_0$ and a large closed surface Σ enclosing Σ_0 and the region of space in which $f(\mathbf{x}) \neq 0$. Multiply (4.5.3) by $\varphi(\mathbf{x})$, (4.5.4) by $G(\mathbf{x}, \mathbf{x}_0)$, subtract the equations, and integrate over the volume V. Green's second identity (2.5.2) enables the result to be written

$$\oint_{\Sigma+\Sigma_0} \left(G(\mathbf{x}, \mathbf{x}_0) \frac{\partial \varphi(\mathbf{x})}{\partial n} - \varphi(\mathbf{x}) \frac{\partial G(\mathbf{x}, \mathbf{x}_0)}{\partial n} \right) dS(\mathbf{x})$$

$$= \int_V G(\mathbf{x}, \mathbf{x}_0) f(\mathbf{x}) \, dV(\mathbf{x}), \qquad (4.5.5)$$

where $\partial/\partial n$ denotes differentiation in the direction of the outward normal from V.

As the surface Σ recedes to infinity, $G \sim 1/|\mathbf{x}|$ and $\partial G/\partial n \sim 1/|\mathbf{x}|^2$. Let us tentatively assume that $\varphi(\mathbf{x})$ tends to zero sufficiently fast that the integrand on Σ goes to zero faster than $1/|\mathbf{x}|^2$. Then $\oint_\Sigma (G \partial \varphi/\partial n - \varphi \partial G/\partial n) dS \to 0$, and this integral over Σ may be discarded.

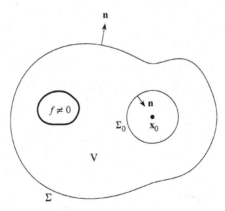

On Σ_0

$$\frac{\partial}{\partial n} = -\frac{\partial}{\partial r}, \quad r = |\mathbf{x} - \mathbf{x}_0|, \text{ so that}$$

$$G = \frac{-1}{4\pi\epsilon}, \quad \frac{\partial G}{\partial n} = \frac{-1}{4\pi\epsilon^2}, \text{ and } dS = \epsilon^2 \sin\theta d\theta d\phi,$$

where (r, θ, ϕ) are local spherical polar coordinates with origin at $\mathbf{x}_0$. Hence, (4.5.5) becomes

$$\frac{1}{4\pi\epsilon} \oint_{\Sigma_0} \frac{\partial\varphi(\mathbf{x})}{\partial r} \, dS + \frac{1}{4\pi} \oint_{\Sigma_0} \varphi(\mathbf{x}) \sin\theta d\theta d\phi = \int_V G(\mathbf{x}, \mathbf{x}_0) f(\mathbf{x}) \, dV(\mathbf{x}).$$

The first integral on the left vanishes identically by the divergence theorem (because $\nabla^2\varphi = 0$). In the second integral $\varphi(\mathbf{x})$ can be replaced by $\varphi(\mathbf{x}_0)$ as $\epsilon \to 0$, and $\oint_{\Sigma_0} \sin\theta d\theta d\phi = 4\pi$. The formula therefore reduces to

$$\varphi(\mathbf{x}_0) = \int G(\mathbf{x}, \mathbf{x}_0) f(\mathbf{x}) \, dV(\mathbf{x}) \equiv \frac{-1}{4\pi} \int \frac{f(\mathbf{x}) \, dV(\mathbf{x})}{|\mathbf{x} - \mathbf{x}_0|}, \qquad (4.5.6)$$

where the volume integral on the right is taken over the region where $f(\mathbf{x}) \neq 0$.

This formula determines φ in terms of the source distribution $f(\mathbf{x})$ at an arbitrary ('observation') point $\mathbf{x}_0$. It shows that, as $|\mathbf{x}_0| \to \infty$,

$$\varphi(\mathbf{x}_0) \sim \frac{-1}{4\pi|\mathbf{x}_0|} \int f(\mathbf{x}) dV(\mathbf{x}) - \frac{\mathbf{x}_o}{4\pi|\mathbf{x}_0|^3} \cdot \int \mathbf{x} f(\mathbf{x}) dV(\mathbf{x}) + \cdots,$$

i.e. that φ decreases at least as fast as $1/|\mathbf{x}_o|$ with increasing distance $\sim |\mathbf{x}_o|$ from the source distribution. This justifies our neglect of the surface integration over Σ.

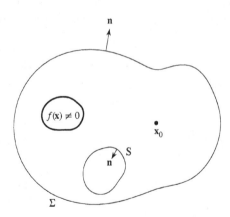

When equation (4.5.4) is to be solved in the presence of an arbitrary distribution of boundaries S, the integral on the left of (4.5.5) includes a contribution from S, and we obtain instead of (4.5.6)

$$\varphi(\mathbf{x}_0) = \int G(\mathbf{x}, \mathbf{x}_0) f(\mathbf{x}) \, dV(\mathbf{x})$$

$$+ \oint_S \left(\varphi(\mathbf{x}) \frac{\partial G(\mathbf{x}, \mathbf{x}_0)}{\partial n} - G(\mathbf{x}, \mathbf{x}_0) \frac{\partial \varphi(\mathbf{x})}{\partial n} \right) dS(\mathbf{x})$$

$$= \frac{-1}{4\pi} \int \frac{f(\mathbf{x}) \, dV(\mathbf{x})}{|\mathbf{x} - \mathbf{x}_0|}$$

$$+ \frac{1}{4\pi} \oint_S \left[\frac{1}{|\mathbf{x} - \mathbf{x}_0|} \frac{\partial \varphi(\mathbf{x})}{\partial n} - \varphi(\mathbf{x}) \frac{\partial}{\partial n} \left(\frac{1}{|\mathbf{x} - \mathbf{x}_0|} \right) \right] dS(\mathbf{x}).$$

$$(4.5.7)$$

Note   In a well-posed boundary value problem for Laplace's equation when, for example, the value of φ is specified on the boundaries, the solution actually determines the value of $\partial\varphi/\partial n$ everywhere on S. This means that φ and $\partial\varphi/\partial n$ cannot both be prescribed independently in the surface integral of (4.5.7).

4.6   The Dirac Delta Function in Three Dimensions

The integral (4.5.6), which determines the 'free space' solution of Poisson's equation $\nabla^2 \varphi = f(\mathbf{x})$, converges for all well-behaved 'source' distributions $f(\mathbf{x})$ that decrease sufficiently fast as $|\mathbf{x}| \rightarrow \infty$. The integrand becomes infinitely large as $\mathbf{x} \rightarrow \mathbf{x}_0$, but the infinity is cancelled because the volume element $dV(\mathbf{x})$ enclosing $\mathbf{x}_0$ is proportional to $|\mathbf{x} - \mathbf{x}_0|^3$. However, it is often convenient to avoid even 'harmless' infinities of this kind by observing that

$$\varphi(\mathbf{x}_0) = \frac{-1}{4\pi} \int \frac{f(\mathbf{x}) \, dV(\mathbf{x})}{|\mathbf{x} - \mathbf{x}_0|} = \lim_{\epsilon \rightarrow +0} \frac{-1}{4\pi} \int \frac{f(\mathbf{x}) \, dV(\mathbf{x})}{\sqrt{|\mathbf{x} - \mathbf{x}_0|^2 + \epsilon^2}}.$$

When $\epsilon \neq 0$ the integrand remains finite everywhere, and the result is obviously equivalent to replacing Green's function (4.5.2) by

$$G_\epsilon(\mathbf{x}, \mathbf{x}_0) = \frac{-1}{4\pi \sqrt{|\mathbf{x} - \mathbf{x}_0|^2 + \epsilon^2}}. \qquad (4.6.1)$$

Although $G_\epsilon(\mathbf{x}, \mathbf{x}_0) \to G(\mathbf{x}, \mathbf{x}_0)$ as $\epsilon \to 0$, this modified Green's function $G_\epsilon(\mathbf{x}, \mathbf{x}_0)$ does not satisfy Laplace's equation. In fact, a simple calculation shows that

$$\nabla^2 G_\epsilon = \delta_\epsilon(\mathbf{x} - \mathbf{x}_0), \qquad (4.6.2)$$

where

$$\delta_\epsilon(\mathbf{x} - \mathbf{x}_0) = \frac{3\epsilon^2}{4\pi(|\mathbf{x} - \mathbf{x}_0|^2 + \epsilon^2)^{\frac{5}{2}}}. \qquad (4.6.3)$$

The calculation described in §4.5 (using Green's second identity) may be repeated using equations (4.6.2) and (4.5.4) instead of (4.5.2) and (4.5.4). In this case G_ϵ is finite everywhere, so that the small spherical surface Σ_0 is not required; the integration over the surface Σ again gives no contribution as Σ recedes to infinity. Instead of (4.5.5) we now find

$$0 = \int G_\epsilon(\mathbf{x}, \mathbf{x}_0) f(\mathbf{x}) \, dV(\mathbf{x}) - \int \delta_\epsilon(\mathbf{x} - \mathbf{x}_0) \varphi(\mathbf{x}) \, dV(\mathbf{x}). \qquad (4.6.4)$$

As $\epsilon \to 0$ the integrand involving $\delta_\epsilon(\mathbf{x} - \mathbf{x}_0)$ vanishes except at $\mathbf{x} = \mathbf{x}_0$, where it becomes infinite like $3/4\pi\epsilon^3$, i.e. for small ϵ

$$\int \delta_\epsilon(\mathbf{x} - \mathbf{x}_0) \varphi(\mathbf{x}) \, dV(\mathbf{x}) \approx \varphi(\mathbf{x}_0) \int \delta_\epsilon(\mathbf{x} - \mathbf{x}_0) \, dV(\mathbf{x})$$

$$= \varphi(\mathbf{x}_0) \times 3\epsilon^2 \int_0^\infty \frac{r^2 dr}{(r^2 + \epsilon^2)^{\frac{5}{2}}} \equiv \varphi(\mathbf{x}_0).$$

Hence, we recover the solution (4.5.6):

$$\varphi(\mathbf{x}_0) = \lim_{\epsilon \to 0} \int G_\epsilon(\mathbf{x}, \mathbf{x}_0) f(\mathbf{x}) \, dV(\mathbf{x}) = \frac{-1}{4\pi} \int \frac{f(\mathbf{x}) \, dV(\mathbf{x})}{|\mathbf{x} - \mathbf{x}_0|}.$$

Generalised functions in three-dimensions The ϵ-sequence

$$\delta_\epsilon(\mathbf{x} - \mathbf{y}) = \frac{3\epsilon^2}{4\pi(|\mathbf{x} - \mathbf{y}|^2 + \epsilon^2)^{\frac{5}{2}}} \qquad (4.6.5)$$

defines the δ-function in space of three-dimensions (cf. §1.11). This is the *three-dimensional* generalised function, denoted by $\delta(\mathbf{x} - \mathbf{y})$, with the fundamental properties

$$\int_V \delta(\mathbf{x} - \mathbf{y})dV(\mathbf{y}) = \begin{cases} 1, & \text{when } \mathbf{x} \text{ is in V,} \\ 0, & \text{when } \mathbf{x} \text{ is not in V,} \end{cases} \qquad (4.6.6)$$

$$\int_V f(\mathbf{y})\delta(\mathbf{x} - \mathbf{y})dV(\mathbf{y}) = f(\mathbf{x}), \quad \text{when } \mathbf{x} \text{ is in V.} \qquad (4.6.7)$$

It follows from this and (4.6.2) that we can formally define Green's function for the Laplace equation as the solution of

$$\nabla^2 G(\mathbf{x}, \mathbf{x}_0) = \delta(\mathbf{x} - \mathbf{x}_0) \qquad (4.6.8)$$

that decays like $1/|\mathbf{x} - \mathbf{x}_0|$ as $|\mathbf{x} - \mathbf{x}_0| \to \infty$.

There are an infinite number of different ϵ-sequences of the type on the right of (4.6.5) that can be used to define the three-dimensional delta function. Any of the following would suffice, for example,

$$\delta_\epsilon(\mathbf{x} - \mathbf{y}) = \frac{\epsilon}{2\pi|\mathbf{x} - \mathbf{y}|(|\mathbf{x} - \mathbf{y}| + \epsilon)^3},$$

$$\frac{\exp\left[-(\mathbf{x} - \mathbf{y})^2/\epsilon^2\right]}{\pi^{\frac{3}{2}}\epsilon^3}, \quad \frac{3H(\epsilon - |\mathbf{x} - \mathbf{y}|)}{4\pi\epsilon^3},$$

where $H(x)$ is the *Heaviside* step function (1.11.4).

Example   If $\mathbf{x} = (x_1, x_2, x_3)$, $\mathbf{y} = (y_1, y_2, y_3)$, then

$$\delta(\mathbf{x} - \mathbf{y}) = \delta(x_1 - y_1)\delta(x_2 - y_2)\delta(x_3 - y_3). \qquad (4.6.9)$$

4.7   The Method of Images

Let us consider again the method based on Green's second identity for solving Poisson's equation $\nabla^2 \varphi = f(\mathbf{x})$ in a three-dimensional region V with boundaries S. Green's function is now taken to be a particular solution for $f(\mathbf{x}) = \delta(\mathbf{x} - \mathbf{x}_0)$, so that

$$\nabla^2 \varphi = f(\mathbf{x}) \qquad (4.7.1)$$

$$\nabla^2 G(\mathbf{x}, \mathbf{x}_0) = \delta(\mathbf{x} - \mathbf{x}_0). \qquad (4.7.2)$$

Multiply (4.7.1) by $G(\mathbf{x}, \mathbf{x}_0)$, (4.7.2) by $\varphi(\mathbf{x})$, subtract and integrate over V. Use Green's second identity (2.5.2) to obtain (as in §4.5)

$$\varphi(\mathbf{x}_0) = \int_V G(\mathbf{x}, \mathbf{x}_0) f(\mathbf{x}) \, dV(\mathbf{x})$$

$$+ \oint_S \left(\varphi(\mathbf{x}) \frac{\partial G(\mathbf{x}, \mathbf{x}_0)}{\partial n} - G(\mathbf{x}, \mathbf{x}_0) \frac{\partial \varphi(\mathbf{x})}{\partial n} \right) dS(\mathbf{x}). \quad (4.7.3)$$

It will be assumed that in the particular case in which the region V extends to infinity, the functions $\varphi(\mathbf{x})$ and $G(\mathbf{x}, \mathbf{x}_0)$ decay sufficiently fast that there are no contributions from the integration over the 'surface at infinity' (Σ in §4.5).

The representation (4.7.3) is valid for *any* $G(\mathbf{x}, \mathbf{x}_0)$ satisfying (4.7.2), and not just the free space Green's function (4.5.2). Put

$$G(\mathbf{x}, \mathbf{x}_0) = \frac{-1}{4\pi |\mathbf{x} - \mathbf{x}_0|} + v(\mathbf{x}, \mathbf{x}_0). \quad (4.7.4)$$

By substituting this expression into (4.7.2), and interpreting $-1/4\pi|\mathbf{x} - \mathbf{x}_0|$ as the generalised function defined by the ϵ-sequence (4.6.1), we see that $v(\mathbf{x}, \mathbf{x}_0)$ is a solution of Laplace's equation

$$\nabla^2 v(\mathbf{x}, \mathbf{x}_0) = 0.$$

Any well-behaved ('regular') solution of this equation (that decays at least as fast as $1/|\mathbf{x}|$ at infinity) can be used in our definition (4.7.4) of Green's function. Because boundary value problems involving elliptic equations of the Laplace type are 'well posed' when the value of either φ or $\partial \varphi / \partial n$ (but not both) is prescribed on the boundaries S, it is often useful to choose $v(\mathbf{x}, \mathbf{x}_0)$ such that in these respective situations $G = 0$ or $\partial G / \partial n = 0$ on S. For example, when $\partial G / \partial n = 0$ on S, equation (4.7.3) becomes

$$\varphi(\mathbf{x}_0) = \int_V G(\mathbf{x}, \mathbf{x}_0) f(\mathbf{x}) \, dV(\mathbf{x}) - \oint_S G(\mathbf{x}, \mathbf{x}_0) \frac{\partial \varphi(\mathbf{x})}{\partial n} \, dS(\mathbf{x}). \quad (4.7.5)$$

If the boundary condition on S is $\partial \varphi / \partial n = 0$, for example, in cases where φ is the velocity potential of a fluid and S is a fixed *rigid* surface, then

$$\varphi(\mathbf{x}_0) = \int G(\mathbf{x}, \mathbf{x}_0) f(\mathbf{x})\, dV(\mathbf{x}), \quad \text{provided} \quad \frac{\partial G}{\partial n} = 0 \quad \text{on S.}$$

The point source   Integrate equation (4.7.2) over the volume V' of a sphere whose centre is at $\mathbf{x}_0$:

$$\int_{V'} \nabla^2 G(\mathbf{x}, \mathbf{x}_0) dV(\mathbf{x}) = \int_{V'} \delta(\mathbf{x} - \mathbf{x}_0) dV(\mathbf{x}) \equiv 1.$$

Applying the divergence theorem (§2.3) to the integral on the left:

$$\oint_{S'} \nabla G \cdot \mathbf{n} dS = 1,$$

where S' is the surface of the sphere and $\mathbf{n}$ is the outward normal.

Thus the 'flux' through S' is independent of the radius of the sphere; indeed the same result is obtained for the flux through *any* closed surface enclosing $\mathbf{x}_0$. In the case in which G represents the velocity potential of an incompressible fluid, the result states that the net rate at which fluid is flowing out of V' is 1, i.e. that the δ-function $\delta(\mathbf{x} - \mathbf{x}_0)$ is equivalent to a *point source* of fluid of unit strength located at $\mathbf{x}_0$. This simple interpretation is the basis of the *method of images* for determining the function $v(\mathbf{x}, \mathbf{x}_0)$ for Green's function (4.7.4).

Green's function for a half-space   Let us determine $G(\mathbf{x}, \mathbf{x}_0)$ for Laplace's equation when V is the half-space $z > 0$, subject to the condition $\partial G/\partial n \equiv -\partial G/\partial z = 0$ on the boundary $z = 0$. The motion produced by a point source at $\mathbf{x}_0 = (x_0, y_0, z_0)$ is entirely radial

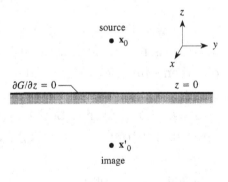

in the absence of boundaries. Evidently, the component of motion normal to the boundary will vanish at $z = 0$ if the source field is augmented by that produced by an equal 'image' source at $\mathbf{x}_0' = (x_0, y_0, -z_0)$, i.e. by taking

$$v(\mathbf{x}, \mathbf{x}_0) = \frac{-1}{4\pi |\mathbf{x} - \mathbf{x}_0'|}.$$

This function is singular at $\mathbf{x} = \mathbf{x}_0'$, but this lies in $z < 0$, so that

$$G(\mathbf{x}, \mathbf{x}_0) = -\frac{1}{4\pi} \left(\frac{1}{|\mathbf{x} - \mathbf{x}_0|} + \frac{1}{|\mathbf{x} - \mathbf{x}_0'|} \right), \tag{4.7.6}$$

is regular in V apart from the point source singularity at $\mathbf{x}_0$, and can therefore be used to solve the *Neumann* problem for Laplace's equation in V.

Similarly, to solve the *Dirichlet* problem, in which the boundary value of φ is prescribed, Green's function must be chosen to satisfy $G(\mathbf{x}, \mathbf{x}_0) = 0$ on $z = 0$. This is achieved by using an equal and *opposite* image source (a 'sink'), in which case

$$G(\mathbf{x}, \mathbf{x}_0) = -\frac{1}{4\pi} \left(\frac{1}{|\mathbf{x} - \mathbf{x}_0|} - \frac{1}{|\mathbf{x} - \mathbf{x}_0'|} \right). \tag{4.7.7}$$

Problems 4C

1. Let $\nabla^2 \varphi = 0$ in $z > 0$ and $\partial\varphi/\partial z = u(x, y)$ on $z = 0$. If $\varphi \to 0$ as $z \to \infty$, show that

$$\varphi(\mathbf{x}_0) = \frac{-1}{2\pi} \int_{-\infty}^{\infty} \frac{u(x, y)dxdy}{\sqrt{(x_0 - x)^2 + (y_0 - y)^2 + z_0^2}}, \quad z_0 \geq 0.$$

2. In Problem 1 let

$$u(x, y) = \begin{cases} u_0 = \text{constant for } x^2 + y^2 < R^2, \\ 0 \quad \text{for } x^2 + y^2 > R^2, \end{cases}$$

Show that

$$\varphi(0, 0, z_0) = u_0 \left[z_0 - \sqrt{R^2 + z_0^2} \right], \quad z_0 \geq 0.$$

3. Let $\nabla^2 \varphi = 0$ in $z > 0$ and $\varphi = u(x, y)$ on $z = 0$. If $\varphi \to 0$ as $z \to \infty$, show that

$$\varphi(\mathbf{x}_0) = \frac{z_0}{2\pi} \int_{-\infty}^{\infty} \frac{u(x, y)dx\,dy}{[(x_0 - x)^2 + (y_0 - y)^2 + z_0^2]^{\frac{3}{2}}}, \quad z_0 \geq 0.$$

4. In Problem 3 let

$$u(x, y) = \begin{cases} u_0 = \text{constant for } x^2 + y^2 < R^2, \\ 0 \quad \text{for } x^2 + y^2 > R^2, \end{cases}$$

Show that

$$\varphi(0, 0, z_0) = z_0 u_0 \left[\frac{1}{z_0} - \frac{1}{\sqrt{R^2 + z_0^2}} \right], \quad z_0 \geq 0.$$

4.8 Green's Function for the Wave Equation

In an unbounded medium, and in the absence of waves 'arriving from infinity', the solution of the inhomogeneous three-dimensional wave equation

$$\frac{1}{c^2} \frac{\partial^2 \varphi}{\partial t^2} - \nabla^2 \varphi = f(\mathbf{x}, t) \tag{4.8.1}$$

represents a system of 'waves' radiating away from a source region where $f(\mathbf{x}, t) \neq 0$. We define Green's function $G(\mathbf{x}, \mathbf{x}_0, t, t_0)$ for this problem to be the particular solution of

$$\frac{1}{c^2} \frac{\partial^2 G}{\partial t^2} - \nabla^2 G = \delta(\mathbf{x} - \mathbf{x}_0)\delta(t - t_0), \tag{4.8.2}$$

that permits the solution of (4.8.1) at any given observer position and time $(\mathbf{x}_0, t_0)$ to be expressed in the form

$$\varphi(\mathbf{x}_0, t_0) = \int_{-\infty}^{\infty} f(\mathbf{x}, t) G(\mathbf{x}, \mathbf{x}_0, t, t_0) \, dV(\mathbf{x}) dt, \tag{4.8.3}$$

the integrations being over the whole of space and all times t.

A unique G can be found which accords with physical intuition by imposing the following condition:

$$G(\mathbf{x}, \mathbf{x}_0, t, t_0) = 0 \quad \text{for } t > t_0 \text{ and for } \textit{all} \text{ values of } \mathbf{x}.$$

This is an expression of the *causality principle*, that waves arriving at $\mathbf{x}_0$ at time t_0 must have been generated by the sources $f(\mathbf{x}, t)$ at *earlier* times $t < t_0$.

The 'source' on the right of (4.8.2) is *impulsive*; it exists only for one instant $t = t_0$ and is concentrated at $\mathbf{x}_0$. Symmetry demands that $G(\mathbf{x}, \mathbf{x}_0, t, t_0)$ consists of an *incoming* spherically symmetric wave that converges onto the point $\mathbf{x} = \mathbf{x}_o$ and becomes evanescent at time $t = t_o$.

Its amplitude must therefore depend on $r = |\mathbf{x} - \mathbf{x}_0|$ alone, and its dependence on time must be as a function of $t - t_0 + |\mathbf{x} - \mathbf{x}_0|$.

For $r > 0$ equation (4.8.2) can be written in the spherically symmetric form

$$\frac{1}{c^2}\frac{\partial^2 G}{\partial t^2} - \frac{1}{r^2}\frac{\partial}{\partial r}\left(r^2\frac{\partial G}{\partial r}\right) = 0, \quad r = |\mathbf{x} - \mathbf{x}_0|.$$

This is transformed into the one-dimensional wave equation

$$\frac{1}{c^2}\frac{\partial^2}{\partial t^2}(rG) - \frac{\partial^2}{\partial r^2}(rG) = 0, \quad (r > 0), \tag{4.8.4}$$

by means of the identity $\frac{1}{r^2}\frac{\partial}{\partial r}\left(r^2\frac{\partial G}{\partial r}\right) \equiv \frac{1}{r}\frac{\partial^2}{\partial r^2}(rG)$, with the general solution (§4.1, (4.1.8))

$$G = \frac{\Phi(t - t_0 - |\mathbf{x} - \mathbf{x}_0|/c)}{4\pi|\mathbf{x} - \mathbf{x}_0|} + \frac{\Psi(t - t_0 + |\mathbf{x} - \mathbf{x}_0|/c)}{4\pi|\mathbf{x} - \mathbf{x}_0|},$$

where Φ and Ψ are arbitrary functions. The terms on the right represent spherical waves respectively radiating in the directions of increasing and decreasing values of $r = |\mathbf{x} - \mathbf{x}_0|$. Therefore $\Phi \equiv 0$, because only the incoming wave can vanish after collapsing onto $\mathbf{x} = \mathbf{x}_o$ at $t = t_o$.

The functional form of Ψ is determined by substitution into equation (4.8.2). Because the term on the right of (4.8.2) is a generalised function we first introduce the following ϵ-sequence for G, which is bounded for all values of $\mathbf{x}$,

$$G_\epsilon = \frac{\Psi(t - t_0 + r_\epsilon/c)}{4\pi r_\epsilon}, \quad \text{where } r_\epsilon = \sqrt{|\mathbf{x} - \mathbf{x}_0|^2 + \epsilon^2}.$$

Then,

$$\frac{1}{c^2}\frac{\partial^2 G_\epsilon}{\partial t^2} - \nabla^2 G_\epsilon \equiv \frac{3\epsilon^2\Psi}{4\pi r_\epsilon^5} - \frac{3\epsilon^2\Psi'}{4\pi c r_\epsilon^4} + \frac{\epsilon^2\Psi''}{4\pi c^2 r_\epsilon^3}, \tag{4.8.5}$$

where $\Psi' = \partial\Psi/\partial t$, etc. As $\epsilon \to 0$ the right-hand side of this identity must reduce to the right-hand side $\delta(\mathbf{x} - \mathbf{x}_0)\delta(t - t_0)$ of (4.8.2).

From §4.6, equation (4.6.5),

$$\frac{3\epsilon^2\Psi}{4\pi r_\epsilon^5} \to \delta(\mathbf{x} - \mathbf{x}_0)\Psi(t - t_0 + |\mathbf{x} - \mathbf{x}_0|/c)$$

$$\equiv \delta(\mathbf{x} - \mathbf{x}_0)\Psi(t - t_0) \quad \text{as } \epsilon \to 0.$$

The remaining ϵ-sequences on the right of (4.8.5) tend to zero as $\epsilon \to 0$. For example, for the second term and a test function $f(\mathbf{x})$

$$\frac{-3\epsilon^2}{4\pi c} \int\limits_{-\infty}^{\infty} \frac{f(\mathbf{x})}{[(\mathbf{x} - \mathbf{x}_0)^2 + \epsilon^2]^2} \Psi'\left(t - t_0 + \frac{\sqrt{(\mathbf{x} - \mathbf{x}_0)^2 + \epsilon^2}}{c}\right) dV(\mathbf{x})$$

$$\approx \frac{-3\epsilon^2 f(\mathbf{x}_0)\Psi'(t - t_0)}{c} \int\limits_{0}^{\infty} \frac{r^2 dr}{(r^2 + \epsilon^2)^2}$$

$$= \frac{-3\pi\epsilon f(\mathbf{x}_0)\Psi'(t - t_0)}{4c} \to 0 \quad \text{as } \epsilon \to 0.$$

Thus, equating the limit as $\epsilon \to 0$ of the right-hand side of (4.8.5) to the right-hand side of (4.8.2), we find

$$\Psi(t - t_0) = \delta(t - t_0),$$

and therefore that

$$G(\mathbf{x}, \mathbf{x}_0, t, t_0) = \frac{\delta\left(t - t_0 + \frac{|\mathbf{x} - \mathbf{x}_0|}{c}\right)}{4\pi|\mathbf{x} - \mathbf{x}_0|}. \tag{4.8.6}$$

The retarded potential   The explicit form of the solution (4.8.3) is now derived by a simple modification of the procedure used in §4.5 for Poisson's equation. Multiply (4.8.1) by $G(\mathbf{x}, \mathbf{x}_0, t, t_0)$, (4.8.2) by $\varphi(\mathbf{x}, t)$, subtract and integrate over all values of $\mathbf{x}$ and over $-\infty < t < \infty$ to obtain

$$\int dV(\mathbf{x}) \int\limits_{-\infty}^{\infty} \left\{ \frac{1}{c^2} \frac{\partial}{\partial t} \left[\varphi \frac{\partial G}{\partial t} - G \frac{\partial \varphi}{\partial t} \right] - \operatorname{div}\left[\varphi \nabla G - G \nabla \varphi \right] \right\} dt$$

$$= \varphi(\mathbf{x}_0, t_0) - \iint\limits_{-\infty}^{\infty} fG \, dV(\mathbf{x})dt. \tag{4.8.7}$$

In the absence of boundaries, the integral on the left is zero. The integral of the first term in the brace brackets involves $[\varphi \partial G/\partial t - G \partial \varphi/\partial t]_{t=-\infty}^{\infty}$, which vanishes because $G = 0$ at $t = +\infty$ and $\varphi = 0$ at $t = -\infty$ (before the sources start radiating waves). The divergence term can be transformed into a surface integral over a distant surface Σ enclosing the source region, on which φ and its derivatives are zero because no waves have arrived at any finite time t. Hence

$$\varphi(\mathbf{x}_0, t_0) = \int\!\!\!\int\limits_{-\infty}^{\infty} f(\mathbf{x}, t) G(\mathbf{x}, \mathbf{x}_0, t, t_0) dV(\mathbf{x}) dt$$

$$= \int\!\!\!\int\limits_{-\infty}^{\infty} \frac{f(\mathbf{x}, t)}{4\pi|\mathbf{x} - \mathbf{x}_0|} \delta\left(t - t_0 + \frac{|\mathbf{x} - \mathbf{x}_0|}{c}\right) dV(\mathbf{x}) dt.$$

Performing the integration with respect to time

$$\varphi(\mathbf{x}_0, t_0) = \int\limits_{-\infty}^{\infty} \frac{f\left(\mathbf{x}, t_0 - \frac{|\mathbf{x} - \mathbf{x}_0|}{c}\right) dV(\mathbf{x})}{4\pi|\mathbf{x} - \mathbf{x}_0|}. \tag{4.8.8}$$

This integral is known as a *retarded potential*. It is formally very similar to the solution (4.5.6) of Poisson's equation, except that now the solution $\varphi(\mathbf{x}_0, t_0)$ at position $\mathbf{x}_0$ and time t_0 is given as an integral over the source distribution $f(\mathbf{x}, t)$ evaluated at the earlier time $t = t_0 - |\mathbf{x} - \mathbf{x}_0|/c$. The delay $|\mathbf{x} - \mathbf{x}_0|/c$ is precisely the time required for a wave received at $\mathbf{x}_o$ to travel at speed c from a source at $\mathbf{x}$; $t_0 - |\mathbf{x} - \mathbf{x}_0|/c$ is called the *retarded time*.

Solution of the wave equation in a bounded medium The integral relation (4.8.7) is applicable for any Green's function. In the presence of boundaries S in the medium, the divergence term on the left is transformed by the divergence theorem to a surface integral over S, and the solution becomes (compare (4.7.3) for the Laplace equation)

$$\varphi(\mathbf{x}_0, t_0) = \int\limits_{-\infty}^{\infty} dt \int\limits_{V} G(\mathbf{x}, \mathbf{x}_0, t, t_0) f(\mathbf{x}, t) \, dV(\mathbf{x})$$

$$+ \int\limits_{-\infty}^{\infty} dt \oint\limits_{S} \left(G(\mathbf{x}, \mathbf{x}_0, t, t_0) \frac{\partial \varphi(\mathbf{x}, t)}{\partial n} - \varphi(\mathbf{x}, t) \frac{\partial G(\mathbf{x}, \mathbf{x}_0, t, t_0)}{\partial n} \right) dS(\mathbf{x}),$$

$$\tag{4.8.9}$$

where V is the region occupied by the sources, and the normal on S is directed out of V.

For specialized boundary conditions we can use a modified form of G that satisfies suitable conditions on S (e.g. $\partial G/\partial n = 0$ or $G = 0$ on S), by taking

$$G(\mathbf{x}, \mathbf{x}_0, t, t_0) = \frac{\delta\left(t - t_0 + \frac{|\mathbf{x} - \mathbf{x}_0|}{c}\right)}{4\pi|\mathbf{x} - \mathbf{x}_0|} + v(\mathbf{x}, \mathbf{x}_0, t, t_0), \qquad (4.8.10)$$

where $v(\mathbf{x}, \mathbf{x}_0, t, t_0)$ is a regular, causal solution of the homogeneous wave equation

$$\frac{1}{c^2}\frac{\partial^2 v}{\partial t^2} - \nabla^2 v = 0.$$

Problems 4D

1. Use the method of images to show that, for the half-space $z > 0$, Green's function for the wave equation is given by

$$G(\mathbf{x}, \mathbf{x}_0, t, t_0) = \frac{\delta\left(t - t_0 + \frac{|\mathbf{x} - \mathbf{x}_0|}{c}\right)}{4\pi|\mathbf{x} - \mathbf{x}_0|} \pm \frac{\delta\left(t - t_0 + \frac{|\mathbf{x} - \mathbf{x}_0'|}{c}\right)}{4\pi|\mathbf{x} - \mathbf{x}_0'|}, \qquad (4.8.11)$$

where $\mathbf{x}_0' = (x_0, y_0, -z_0)$ is the image of $\mathbf{x}_0$ in the plane $z = 0$, and the $\pm$ sign is taken according as $\partial G/\partial z = 0$ or $G = 0$ on $z = 0$.

2. A circular piston of radius R flush with the plane *rigid* wall $z = 0$ oscillates at frequency ω with uniform normal velocity $\partial\varphi/\partial z = u_0 \cos\omega t$. If $\partial\varphi/\partial z = 0$ on the rigid part of the wall and

$$\frac{1}{c^2}\frac{\partial^2\varphi}{\partial t^2} - \nabla^2\varphi = 0 \quad \text{for} \quad z > 0,$$

show that

$$\varphi(\mathbf{x}_0, t_0) = \frac{-u_0}{2\pi}\int_{\text{piston}} \frac{\cos\omega\left(t - \frac{\sqrt{(x-x_0)^2+(y-y_0)^2+z_0^2}}{c}\right)dxdy}{\sqrt{(x - x_0)^2 + (y - y_0)^2 + z_0^2}}, \quad z_0 > 0.$$

Deduce that on the axis of symmetry $x_0 = y_0 = 0$,

$$\varphi = \frac{u_0 c}{\omega}\left[\sin\omega\left(t - \frac{z_0}{c}\right) - \sin\omega\left(t - \frac{\sqrt{R^2 + z_0^2}}{c}\right)\right].$$

The method of descent Let us integrate both sides of equation (4.8.2) for the three-dimensional Green's function in an unbounded medium with respect to z over $-\infty < z < \infty$. Because

$$\left[\frac{\partial G}{\partial z}\right]_{-\infty}^{\infty} = 0 \quad \text{and} \quad \int_{-\infty}^{\infty} \delta(\mathbf{x} - \mathbf{x}_0)dz = \delta(x - x_0)\delta(y - y_0),$$

we find

$$\left[\frac{1}{c^2}\frac{\partial^2}{\partial t^2} - \left(\frac{\partial^2}{\partial x^2} + \frac{\partial^2}{\partial y^2}\right)\right] G(x, y, x_0, y_0, t, t_0)$$

$$= \delta(x - x_0)\delta(y - y_0)\delta(t - t_0), \tag{4.8.12}$$

where

$$G(x, y, x_0, y_0, t, t_0) = \int_{-\infty}^{\infty} G(\mathbf{x}, \mathbf{x}_0, t, t_0)dz, \tag{4.8.13}$$

is Green's function for the wave equation in *two* space dimensions. This procedure for deriving results for lower-dimensional spaces from one of higher dimension is called the *method of descent*.

To evaluate the integral (4.8.13) we use the formula (4.8.6) for $G(\mathbf{x}, \mathbf{x}_0, t, t_0)$ as follows:

$$G(x, y, x_0, y_0, t, t_0) = \int_{-\infty}^{\infty} \frac{\delta\left(t - t_0 + \frac{|\mathbf{x}-\mathbf{x}_0|}{c}\right)dz}{4\pi|\mathbf{x} - \mathbf{x}_0|}$$

$$\equiv \int_{-\infty}^{\infty} \frac{\delta\left(t - t_0 + \frac{\sqrt{(x-x_0)^2+(y-y_0)^2+(z-z_0)^2}}{c}\right)dz}{4\pi\sqrt{(x - x_0)^2 + (y - y_0)^2 + (z - z_0)^2}}$$

$$= \int_{-\infty}^{\infty} \frac{\delta\left(t - t_0 + \frac{\sqrt{(x-x_0)^2+(y-y_0)^2+\rho^2}}{c}\right)d\rho}{4\pi\sqrt{(x - x_0)^2 + (y - y_0)^2 + \rho^2}}.$$

As ρ varies over $-\infty < \rho < \infty$, the argument of the δ-function can vanish only if

$$t_0 - t > \frac{\sqrt{(x - x_0)^2 + (y - y_0)^2}}{c},$$

and it does so at $\rho = \rho_\pm$, where

$$\frac{\rho_\pm}{c} = \pm\sqrt{(t - t_0)^2 - \frac{(x - x_0)^2 + (y - y_0)^2}{c^2}}.$$

Hence, using (1.11.16)

$$\int_{-\infty}^{\infty} \frac{\delta\left(t - t_0 + \frac{\sqrt{(x-x_0)^2+(y-y_0)^2+\rho^2}}{c}\right) d\rho}{4\pi\sqrt{(x - x_0)^2 + (y - y_0)^2 + \rho^2}}$$

$$= \frac{1}{4\pi}H\left(t_0 - t - \frac{\sqrt{(x - x_0)^2 + (y - y_0)^2}}{c}\right)\left[\frac{1}{|\rho_+/c|} + \frac{1}{|\rho_-/c|}\right].$$

Green's function for the wave equation in two space dimensions is therefore given by

$$G(x, y, x_0, y_0, t, t_0) = \frac{H\left(t_0 - t - \frac{\sqrt{(x-x_0)^2+(y-y_0)^2}}{c}\right)}{2\pi\sqrt{(t_0 - t)^2 - \frac{(x-x_0)^2+(y-y_0)^2}{c^2}}}. \qquad (4.8.14)$$

Problems 4E

1. Show that the solution with outgoing wave behavior of

$$\left[\frac{1}{c^2}\frac{\partial^2}{\partial t^2} - \left(\frac{\partial^2}{\partial x^2} + \frac{\partial^2}{\partial y^2}\right)\right]\varphi = f(x, y, t)$$

   is given by

$$\varphi(x_0, y_0, t_0) = \int_{-\infty}^{\infty} dx\,dy \int_{-\infty}^{t_0 - \frac{r}{c}} \frac{f(x, y, t)dt}{2\pi\sqrt{(t_0 - t)^2 - \frac{r^2}{c^2}}},$$

   where $r = \sqrt{(x - x_0)^2 + (y - y_0)^2}.$

2. Green's function for the one-dimensional wave equation satisfies

$$\left[\frac{1}{c^2}\frac{\partial^2}{\partial t^2} - \frac{\partial^2}{\partial x^2}\right]G = \delta(x - x_0)\delta(t - t_0), \quad -\infty < x < \infty.$$

   Use the method of descent to show that

$$G(x, x_0, t, t_0) = \frac{c}{2}H\left[t_0 - t - \frac{|x - x_0|}{c}\right]. \qquad (4.8.15)$$

   Verify this formula by direct substitution into the equation.

3. Show that the solution with outgoing wave behaviour of

$$\left[\frac{1}{c^2}\frac{\partial^2}{\partial t^2} - \frac{\partial^2}{\partial x^2}\right]\varphi = \frac{\partial F}{\partial t}(x,t), \quad -\infty < x < \infty, \quad \text{where } F \to 0 \text{ as } |x| \to \infty,$$

is

$$\varphi(x_0, t_0) = \frac{c}{2}\int\limits_{-\infty}^{\infty} F\left(x, t_0 - \frac{|x - x_0|}{c}\right)dx.$$

4. If $F(x,t) = 0$ for $|x| > a$ and

$$\left[\frac{1}{c^2}\frac{\partial^2}{\partial t^2} - \frac{\partial^2}{\partial x^2}\right]\varphi = \frac{\partial F}{\partial x}(x,t), \quad -\infty < x < \infty,$$

show that the solution with outgoing wave behaviour is

$$\varphi(x_0, t_0) = \frac{1}{2}\int\limits_{-\infty}^{\infty} \operatorname{sgn}(x - x_0)F\left(x, t_0 - \frac{|x - x_0|}{c}\right)dx.$$

4.9   Fourier Transforms

Suppose $f(x)$ is defined for $-\infty < x < \infty$. The *Fourier transform* $\hat{f}(k)$ of $f(x)$ is defined by

$$\hat{f}(k) = \frac{1}{\sqrt{2\pi}}\int\limits_{-\infty}^{\infty} f(x)e^{-ikx}\,dx. \tag{4.9.1}$$

In applications we frequently know $\hat{f}(k)$ and are required to determine the corresponding function $f(x)$. This is done by means of the *inversion formula* which is derived by making use of equation (1.11.8) in the form $\frac{1}{2\pi}\int_{-\infty}^{\infty} e^{ik(x-y)}dk = \delta(x - y)$, as follows

$$f(x) = \int\limits_{-\infty}^{\infty} f(y)\delta(x - y)dy$$

$$= \frac{1}{\sqrt{2\pi}}\int\limits_{-\infty}^{\infty}\left(\frac{1}{\sqrt{2\pi}}\int\limits_{-\infty}^{\infty} f(y)e^{-iky}dy\right)e^{ikx}dk$$

$$= \frac{1}{\sqrt{2\pi}}\int\limits_{-\infty}^{\infty} \hat{f}(k)e^{ikx}dk.$$

Hence, $\hat{f}(k)$ and $f(x)$ satisfy the reciprocal relations

$$\hat{f}(k) = \frac{1}{\sqrt{2\pi}} \int_{-\infty}^{\infty} f(x)e^{-ikx}\, dx, \quad f(x) = \frac{1}{\sqrt{2\pi}} \int_{-\infty}^{\infty} \hat{f}(k)e^{ikx}\, dk,$$

$$(4.9.2)$$

where the second equation is called the *inversion formula*.

Example   Find the Fourier transform of $f(x) = e^{-|x|}$.

$$\hat{f}(k) = \frac{1}{\sqrt{2\pi}} \int_{-\infty}^{\infty} e^{-ikx-|x|}\, dx = \frac{1}{\sqrt{2\pi}} \left(\int_{-\infty}^{0} e^{-ikx+x}\, dx + \int_{0}^{\infty} e^{-ikx-x}\, dx \right)$$

$$= \frac{1}{\sqrt{2\pi}} \left(\frac{1}{1-ik} + \frac{1}{1+ik} \right)$$

$$\therefore \quad \hat{f}(k) = \sqrt{\frac{2}{\pi}} \frac{1}{1+k^2}.$$

   The integral in (4.9.2) defining $\hat{f}(k)$ converges only under the restrictive condition that $|f(x)| \to 0$ as $x \to \pm\infty$, so that powers of x and constants do not strictly possess Fourier transforms. However, in these circumstances we can always 'force' convergence by interpreting $f(x)$ as a *generalised function* (§§1.11, 4.6), provided $f(x)$ does not grow 'too rapidly' at infinity. In practice this means that $f(x)$ should not grow exponentially fast, so that it can be defined as a generalised function by means of the ϵ-sequence

$$f_\epsilon(x) = f(x)e^{-\epsilon|x|}.$$

In this case we have the reciprocal formulae

$$\hat{f}(k) = \lim_{\epsilon \to 0} \frac{1}{\sqrt{2\pi}} \int_{-\infty}^{\infty} f(x)e^{-\epsilon|x|-ikx}\, dx, \quad f(x) = \lim_{\hat{\epsilon} \to 0} \frac{1}{\sqrt{2\pi}} \int_{-\infty}^{\infty} \hat{f}(k)e^{-\hat{\epsilon}|k|+ikx}\, dk.$$

To prove the inversion formula, we write

$$\lim_{\hat{\epsilon} \to 0} \frac{1}{\sqrt{2\pi}} \int_{-\infty}^{\infty} \hat{f}(k)e^{-\hat{\epsilon}|k|+ikx}\, dk = \lim_{\hat{\epsilon} \to 0,\ \epsilon \to 0} \frac{1}{2\pi} \iint_{-\infty}^{\infty} f(y)e^{-\epsilon|y|-\hat{\epsilon}|k|+ik(x-y)}\, dy\, dk$$

$$= \lim_{\hat{\epsilon} \to 0,\ \epsilon \to 0} \int_{-\infty}^{\infty} f(y)e^{-\epsilon|y|} \frac{\hat{\epsilon}}{\pi(\hat{\epsilon}^2 + (x-y)^2)}\, dy$$

$$= \lim_{\epsilon \to 0} \int_{-\infty}^{\infty} f(y) e^{-\epsilon|y|} \, \delta(x - y) dy$$

$$= \lim_{\epsilon \to 0} f(x) e^{-\epsilon|x|} = f(x).$$

Example Consider $f(x) = C = $ constant and the corresponding ϵ-sequence $f_\epsilon(x) = C e^{-\epsilon|x|}$. Then $f(k) = \lim_{\epsilon \to 0} \hat{f}_\epsilon(k)$, where

$$\hat{f}_\epsilon(k) = \frac{1}{\sqrt{2\pi}} \int_{-\infty}^{\infty} C e^{-\epsilon|x|-ikx} \, dx = C\sqrt{2\pi} \frac{\epsilon}{\pi(\epsilon^2 + k^2)} \to C\sqrt{2\pi} \, \delta(k) \quad \text{as } \epsilon \to 0.$$

The original function $f(x) = C$ is obviously recovered when this result (involving the δ-function) is substituted into the inversion formula of (4.9.2).

Example For $f(x) = x$, we take $f_\epsilon(x) = x e^{-\epsilon|x|}$. Then

$$\hat{f}_\epsilon(k) = \frac{1}{\sqrt{2\pi}} \int_{-\infty}^{\infty} x e^{-\epsilon|x|-ikx} \, dx$$

$$= \left(i \frac{\partial}{\partial k} \right) \frac{1}{\sqrt{2\pi}} \int_{-\infty}^{\infty} e^{-\epsilon|x|-ikx} \, dx \to \sqrt{2\pi} i \, \delta'(k) \quad \text{as } \epsilon \to 0.$$

Example For $f(x) = x^n$, where n is a positive integer, $f_\epsilon(x) = x^n e^{-\epsilon|x|}$. Then

$$\hat{f}_\epsilon(k) = \frac{1}{\sqrt{2\pi}} \int_{-\infty}^{\infty} x^n e^{-\epsilon|x|-ikx} \, dx$$

$$= \left(i \frac{\partial}{\partial k} \right)^n \frac{1}{\sqrt{2\pi}} \int_{-\infty}^{\infty} e^{-\epsilon|x|-ikx} \, dx \to \sqrt{2\pi} i^n \, \delta^{(n)}(k) \quad \text{as } \epsilon \to 0.$$

Example For $f(x) = H(x)$, $f_\epsilon(x) = H(x) e^{-\epsilon x}$. Then

$$\hat{f}_\epsilon(k) = \frac{1}{\sqrt{2\pi}} \int_{0}^{\infty} e^{-\epsilon x - ikx} \, dx = \frac{-i}{\sqrt{2\pi}} \frac{1}{k - i\epsilon}$$

$$= \sqrt{\frac{\pi}{2}} \frac{\epsilon}{\pi(\epsilon^2 + k^2)} - \frac{i}{\sqrt{2\pi}} \frac{k}{\epsilon^2 + k^2} \to \sqrt{\frac{\pi}{2}} \delta(k) - \frac{i}{\sqrt{2\pi}} P\left(\frac{1}{k}\right) \quad \text{as } \epsilon \to 0.$$

Example For $f(x) = \text{sgn}(x)$, $f_\epsilon(x) = \text{sgn}(x) e^{-\epsilon|x|}$. Then

$$\hat{f}_\epsilon(k) = \frac{-2ik}{\sqrt{2\pi}(\epsilon^2 + k^2)} \to -i\sqrt{\frac{2}{\pi}} P\left(\frac{1}{k}\right) \quad \text{as } \epsilon \to 0.$$

Example For $f(x) = \cos \alpha x$,

$$\hat{f}(k) = \frac{1}{\sqrt{2\pi}} \int_{-\infty}^{\infty} \frac{1}{2} \left(e^{ix(\alpha-k)} + e^{-ix(\alpha+k)} \right) dx$$

$$= \sqrt{\frac{\pi}{2}} \left[\delta(k - \alpha) + \delta(k + \alpha) \right] \equiv \sqrt{2\pi} |\alpha| \delta(k^2 - \alpha^2).$$

Example The Fourier transform and its inversion formula may be interpreted in terms of the limiting behaviour of the Fourier series representation of a function defined in $-\ell < x < \ell$ as $\ell \to \infty$. To see this, note that De Moivre's formula

$$\cos n\theta + i \sin n\theta = e^{in\theta}$$

enables the Fourier series (1.10.4)

$$f(x) = a_0 + \sum_{n=1}^{\infty} \left(a_n \cos \frac{n\pi x}{\ell} + b_n \sin \frac{n\pi x}{\ell} \right), \quad -\ell < x < \ell,$$

for a function of period 2ℓ to be expressed in the complex form

$$f(x) = \sum_{n=-\infty}^{\infty} c_n e^{in\pi x/\ell}. \tag{4.9.3}$$

The complex eigenfunctions $e^{in\pi x/\ell}$ form an orthogonal set over any interval of length 2ℓ, and

$$\int_{-\ell}^{\ell} e^{in\pi x/\ell} \times e^{-im\pi x/\ell} \, dx = \begin{cases} 0 & m \neq n, \\ 2\ell & m = n, \end{cases} \tag{4.9.4}$$

so that

$$c_n = \frac{1}{2\ell} \int_{-\ell}^{\ell} f(x) e^{-in\pi x/\ell} \, dx. \tag{4.9.5}$$

Let us now consider the behaviour of these formulae as $\ell \to \infty$. Set

$$k_n = \frac{n\pi}{\ell} \quad \text{and} \quad \delta k = \frac{\pi}{\ell},$$

and define $\hat{f}(k_n)$ by

$$c_n = \frac{\delta k}{\sqrt{2\pi}} \hat{f}(k_n).$$

Then (4.9.3) and (4.9.5) become

$$f(x) = \frac{1}{\sqrt{2\pi}} \sum_{n=-\infty}^{\infty} \hat{f}(k_n) e^{ik_n x} \, \delta k,$$

$$\hat{f}(k_n) = \frac{1}{\sqrt{2\pi}} \int_{-\ell}^{\ell} f(x) e^{-ik_n x} \, dx.$$

As $\ell \to \infty$ the interval δk between successive values of k_n tends to zero, and the summation becomes an integral over the range $-\infty < k < \infty$. Then, $\hat{f}(k)$ and $f(x)$ are seen to satisfy the reciprocal relations (4.9.2).

Half-range transforms The Fourier transform of a function $f(x)$ that is prescribed only over the range $0 < x < \infty$ is determined by the first of equations (4.9.2) by setting $f(x) = 0$ for $x < 0$. However, it is also possible to define *half-range* transforms for such functions, in situations where the actual behaviour of f in the range $-\infty < x < 0$ is of no interest to the problem at hand. We do this by formally defining $f(x)$ for negative x by taking $f(x)$ to be an *even* or *odd* function of x.

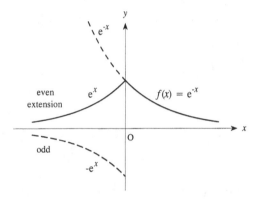

When $f(x)$ is an even function

$$\hat{f}(k) = \frac{1}{\sqrt{2\pi}} \int\limits_{-\infty}^{\infty} f(x) e^{-ikx}\, dx \equiv \sqrt{\frac{2}{\pi}} \int\limits_{0}^{\infty} f(x) \cos kx\, dx$$

is an *even* function of k, so that the inversion formula gives

$$f(x) = \frac{1}{\sqrt{2\pi}} \int\limits_{-\infty}^{\infty} \hat{f}(k) e^{ikx}\, dk \equiv \sqrt{\frac{2}{\pi}} \int\limits_{0}^{\infty} \hat{f}(k) \cos kx\, dk.$$

The Fourier transform of an even function $f(x)$ will be denoted by $\hat{f}_c(k)$, and will be called its Fourier cosine transform. For an *arbitrary* function $f(x)$ defined over $0 < x < \infty$, the Fourier cosine transform and inversion formula are given by

$$\hat{f}_c(k) = \sqrt{\frac{2}{\pi}} \int_0^\infty f(x) \cos kx \, dx, \quad f(x) = \sqrt{\frac{2}{\pi}} \int_0^\infty \hat{f}_c(k) \cos kx \, dk.$$

$$(4.9.6)$$

When $f(x)$ is an odd function, or is extended to $-\infty < x < 0$ as an odd function (so that $f(-x) = -f(x)$)

$$\hat{f}_s(k) \equiv i\hat{f}(k) = \frac{i}{\sqrt{2\pi}} \int_{-\infty}^\infty f(x) e^{-ikx} \, dx \equiv \sqrt{\frac{2}{\pi}} \int_0^\infty f(x) \sin kx \, dx$$

is an *odd* function of k, and the inversion formula of (4.9.2) yields

$$f(x) = \frac{1}{\sqrt{2\pi}} \int_{-\infty}^\infty \hat{f}(k) e^{ikx} \, dk$$

$$\equiv \sqrt{\frac{2}{\pi}} \int_0^\infty i\hat{f}(k) \sin kx \, dk \equiv \sqrt{\frac{2}{\pi}} \int_0^\infty \hat{f}_s(k) \sin kx \, dk.$$

Hence, for a function defined over $0 < x < \infty$, we can define the Fourier sine transform and inversion formula by

$$\hat{f}_s(k) = \sqrt{\frac{2}{\pi}} \int_0^\infty f(x) \sin kx \, dx, \quad f(x) = \sqrt{\frac{2}{\pi}} \int_0^\infty \hat{f}_s(k) \sin kx \, dk. \quad (4.9.7)$$

Example   Calculate the sine transform of $f(x) = x/(1 + x^2)$.

Using the residue theorem (closing the contour using a semicircular arc in the upper half-plane for $k > 0$):

$$\hat{f}_s = \sqrt{\frac{2}{\pi}} \int_0^\infty \frac{x \sin kx}{1 + x^2} \, dx = \sqrt{\frac{2}{\pi}} \mathrm{Im} \int_0^\infty \frac{x e^{ikx}}{1 + x^2} \, dx$$

$$= \sqrt{\frac{1}{2\pi}} \mathrm{Im} \int_{-\infty}^\infty \frac{x e^{ikx}}{1 + x^2} \, dx = \sqrt{\frac{1}{2\pi}} \mathrm{Im} \left(2\pi i \times \left[\frac{e^{ikx}}{2} \right]_{x=i} \right) = \sqrt{\frac{\pi}{2}} e^{-k}, \quad k > 0.$$

Fourier transform of a derivative To use the Fourier transform
to solve a partial differential equation it is necessary to express the
transform of a derivative $f^{(n)}(x)$ in terms of the transform of the original
function $f(x)$.

From the inversion formula of (4.9.2)

$$f(x) = \frac{1}{\sqrt{2\pi}} \int_{-\infty}^{\infty} \hat{f}(k) e^{ikx} \, dk$$

$$\therefore \quad f'(x) = \frac{1}{\sqrt{2\pi}} \int_{-\infty}^{\infty} ik \hat{f}(k) e^{ikx} \, dk.$$

Repeated application of this formula gives the rule:

$$\text{Fourier transform of } f^{(n)}(x) = (ik)^n \hat{f}(k). \qquad (4.9.8)$$

Provided $\hat{f}(k)$ does not grow exponentially fast as $k \to \pm\infty$ convergence of the integrals involved in these definitions can always be
assured by interpreting $\hat{f}(k)$ as a generalised function defined by the
ϵ-sequence, $\hat{f}_\epsilon(k) = \hat{f}(k) e^{-\epsilon|k|}$, say.

Problems 4F

Verify the following Fourier transform pairs:

1. $f(x) = e^{-|x|}, \quad \hat{f}(k) = \dfrac{\sqrt{2}}{\sqrt{\pi}(1 + k^2)}.$

2. $f(x) = H(x) e^{-x}, \quad \hat{f}(k) = \dfrac{1}{\sqrt{2\pi}(1 + ik)}.$

3. $f(x) = e^{-ax^2}, \quad \hat{f}(k) = \dfrac{1}{\sqrt{2a}} e^{-k^2/4a}, \quad a > 0.$

4. $f(x) = \begin{cases} x, & 0 < x < a, \\ 0, & \text{elsewhere}, \end{cases} \quad \hat{f}(k) = \dfrac{(1 + ika) e^{-ika} - 1}{\sqrt{2\pi} k^2}.$

5. $f(x) = \delta(x), \quad \hat{f}(k) = \dfrac{1}{\sqrt{2\pi}}.$

6. $f(x) = \dfrac{1}{1 + x^2}, \quad \hat{f}_c(k) = \sqrt{\dfrac{\pi}{2}} e^{-k}.$

7. $f(x) = \begin{cases} x, & 0 < x < a, \\ 0, & x > a, \end{cases}$ $\hat{f}_c(k) = \sqrt{\dfrac{2}{\pi}} \dfrac{[ka \sin ka + \cos ka - 1]}{k^2}.$

8. $f(x) = \dfrac{\cos(\pi x/2)}{1 - x^2}$ $\hat{f}_c(k) = \begin{cases} \sqrt{\frac{\pi}{2}} \cos k, & 0 < k < \dfrac{\pi}{2}, \\ 0, & k > \frac{\pi}{2}. \end{cases}$

9. $f(x) = e^{-x},$ $\hat{f}_s(k) = \sqrt{\dfrac{2}{\pi}} \dfrac{k}{1 + k^2}.$

10. $f(x) = \begin{cases} 1, & 0 < x < a, \\ 0, & x > a, \end{cases}$ $\hat{f}_s(k) = \sqrt{\dfrac{2}{\pi}} \dfrac{[1 - \cos ka]}{k}.$

11. $f(x) = x^n e^{-ax},$ $n = $ positive integer, $a > 0,$

$\qquad \hat{f}_s(k) = \sqrt{\dfrac{2}{\pi}} \dfrac{n!}{(k^2 + a^2)^{n+1}} \text{Im} \, (a + ik)^{n+1}.$

12. $f(x) = \dfrac{\cos(ax)}{x} \, (a > 0),$ $\hat{f}_s(k) = H(k - a)\sqrt{\dfrac{\pi}{2}} \equiv \begin{cases} 0, & 0 < k < a, \\ \sqrt{\frac{\pi}{2}}, & k > a. \end{cases}$

13. $f(x) = e^{-ax} \, (a > 0),$ $\hat{f}_c(k) = \sqrt{\dfrac{2}{\pi}} \dfrac{a}{a^2 + k^2}.$

14. $f(x) = xe^{-ax} \, (a > 0),$ $\hat{f}_c(k) = \sqrt{\dfrac{2}{\pi}} \dfrac{a^2 - k^2}{(a^2 + k^2)^2}.$

15. $f(x) = \dfrac{e^{-ax}}{x} \, (a > 0),$ $\hat{f}_s(k) = \sqrt{\dfrac{2}{\pi}} \tan^{-1}\left(\dfrac{k}{a}\right).$

16. $f(x) = e^{ix^2/2},$ $\hat{f}_c(k) = \dfrac{1}{\sqrt{2}}(1 + i)e^{-ik^2/2}$ (take $a = -i/2$ in Problem 3).

17. $f(x) = \cos(x^2/2),$ $\hat{f}_c(k) = \dfrac{1}{\sqrt{2}}\left(\cos(k^2/2) + \sin(k^2/2)\right).$

18. $f(x) = \sin(x^2/2),$ $\hat{f}_c(k) = \dfrac{1}{\sqrt{2}}\left(\cos(k^2/2) - \sin(k^2/2)\right).$

19. $f(x) = \dfrac{1}{x},$ $\hat{f}_s(k) = \sqrt{\dfrac{\pi}{2}}.$

20. $f(x) = \dfrac{1}{x(x^2 + a^2)},$ $\hat{f}_s(k) = \dfrac{1}{a^2}\sqrt{\dfrac{\pi}{2}}\left(1 - e^{-|ka|}\right).$

4.10 Application of Fourier Transforms to the Solution of Partial Differential Equations

Fourier transforms can be applied to solve partial differential equations (usually with constant coefficients) defined over one or more infinite or semi-infinite domains. The general procedure will be illustrated by several examples.

Example 1: The diffusion equation in $-\infty < x < \infty$

Solve the initial value problem

$$\frac{\partial^2 u}{\partial x^2} = \frac{1}{\kappa} \frac{\partial u}{\partial t}, \quad \text{for } t > 0, \quad -\infty < x < \infty, \quad \text{given that } u = f(x) \text{ at } t = 0.$$

This defines the diffusion of heat (temperature) or of a solvent in a one-dimensional infinite medium when the initial distribution $(f(x))$ is prescribed at time $t = 0$. The *diffusivity* $\kappa > 0$ is assumed to be constant.

Introduce the Fourier transform $\hat{u}(k,t) = \frac{1}{\sqrt{2\pi}} \int_{-\infty}^{\infty} u(x,t) e^{-ikx} \, dx$, and take the Fourier transform of the diffusion equation by multiplying by $\frac{1}{\sqrt{2\pi}} e^{-ikx}$ and integrating over $-\infty < x < \infty$. By making use of the rule (4.9.8), and interchanging the order of integration and differentiation in

$$\frac{1}{\sqrt{2\pi}} \int_{-\infty}^{\infty} \frac{\partial u(x,t)}{\partial t} e^{-ikx} \, dx = \frac{\partial}{\partial t} \left(\frac{1}{\sqrt{2\pi}} \int_{-\infty}^{\infty} u(x,t) e^{-ikx} \, dx \right) = \frac{\partial \hat{u}}{\partial t}(k,t),$$

we thereby derive the *ordinary* differential equation

$$-k^2 \hat{u}(k,t) = \frac{1}{\kappa} \frac{\partial \hat{u}}{\partial t}(k,t),$$

which has the general solution

$$\hat{u}(k,t) = \mathcal{A}(k) e^{-k^2 \kappa t},$$

where $\mathcal{A}(k)$ remains to be determined. The inversion formula of (4.9.2) then gives the following representation of the solution:

$$u(x,t) = \frac{1}{\sqrt{2\pi}} \int_{-\infty}^{\infty} \hat{u}(k,t) e^{ikx} \, dk \equiv \frac{1}{\sqrt{2\pi}} \int_{-\infty}^{\infty} \mathcal{A}(k) e^{-k^2 \kappa t + ikx} \, dk. \qquad (4.10.1)$$

By setting $t = 0$ and applying the initial condition $u(x,0) = f(x), -\infty < x < \infty$ we find that $\mathcal{A}(k)$ is the solution of the *integral equation*

$$f(x) = \frac{1}{\sqrt{2\pi}} \int_{-\infty}^{\infty} \mathcal{A}(k) e^{ikx} \, dk.$$

But this merely states that $A(k)$ is the Fourier transform of $f(x)$, and therefore that

$$A(k) = \frac{1}{\sqrt{2\pi}} \int_{-\infty}^{\infty} f(x)e^{-ikx}\, dx. \tag{4.10.2}$$

Equations (4.10.1) and (4.10.2) constitute the formal solution of the initial value problem. To illustrate the behaviour of the solution take the special case in which the initial distribution of heat or solvent is concentrated with infinite density at $x = 0$, i.e.

$$f(x) = u_o\delta(x).$$

Then, $A(k) = u_0/\sqrt{2\pi}$, and

$$u(x,t) = \frac{u_0}{2\pi} \int_{-\infty}^{\infty} e^{-k^2\kappa t+ikx}\, dk = \frac{u_0 e^{-x^2/4\kappa t}}{2\sqrt{\pi\kappa t}}, \quad t > 0. \tag{4.10.3}$$

This is the fundamental solution of the diffusion equation. Thermodynamic equilibrium is established in the infinite medium as $t \to +\infty$, i.e. $u(x,t) \to 0$, as indicated in the figure.

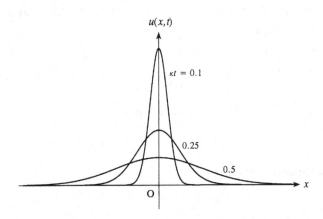

The solution for a general initial distribution $u(x,0) = f(x)$ can be expressed as a *convolution integral* involving the fundamental solution by substituting for $A(k)$ in (4.10.1) from (4.10.2) and performing the integration with respect to k:

$$u(x,t) = \frac{1}{2\pi} \int_{-\infty}^{\infty} f(\xi)d\xi \int_{-\infty}^{\infty} e^{-k^2\kappa t+ik(x-\xi)}\, dk$$

$$= \frac{1}{2\sqrt{\pi\kappa t}} \int_{-\infty}^{\infty} f(\xi)e^{-(x-\xi)^2/4\kappa t}d\xi.$$

Example 2:   Green's function for the one-dimensional wave equation

Find $G(x, x_0, t, t_0)$ satisfying

$$\left[\frac{1}{c^2}\frac{\partial^2}{\partial t^2} - \frac{\partial^2}{\partial x^2}\right] G = \delta(x - x_0)\delta(t - t_0), \quad -\infty < x < \infty, \; G = 0 \text{ for } t > t_0.$$

Take the Fourier transform with respect to x:

$$\frac{1}{c^2}\frac{\partial^2 \hat{G}}{\partial t^2} + k^2\hat{G} = \delta(t - t_0)\frac{e^{-ikx_0}}{\sqrt{2\pi}}, \qquad (4.10.4)$$

where $\hat{G} \equiv \hat{G}(k, t) = 0$ for $t > t_0$. Thus, we can write

$$\hat{G} = H(t_0 - t)\left[Ae^{ick(t-t_0)} + Be^{-ick(t-t_0)}\right],$$

where H is the Heaviside step function (1.11.4). Substituting into the left-hand side of (4.10.4), and recalling that $\frac{d}{dt}H(t_0 - t) = -\delta(t - t_0)$, we find

$$-\delta'(t - t_0)\frac{(A + B)}{c^2} - \delta(t - t_0)\frac{ik(A - B)}{c} = \delta(t - t_0)\frac{e^{-ikx_0}}{\sqrt{2\pi}},$$

which is satisfied provided that

$$A + B = 0, \quad A - B = \frac{-c}{\sqrt{2\pi}}\frac{e^{-ickx_0}}{ik}.$$

Hence $A = -B = -ce^{-ickx_0}/(2\sqrt{2\pi}ik)$ and $\hat{G} = cH(t_0 - t)\sin[ck(t_0 - t)]e^{-ikx_0}/(\sqrt{2\pi}k)$.

$$\therefore \quad G(x, x_0, t, t_0) = \frac{c}{2\pi}H(t_0 - t) \int_{-\infty}^{\infty} \frac{\sin[ck(t_0 - t)]e^{ik(x-x_0)}\, dk}{k}.$$

This integral vanishes identically when $|x - x_0| > c(t_0 - t)$, because the contour can then be displaced to $k = \text{sgn}(x - x_0)i\infty$ without encountering any singularities. When $c(t_0 - t) > |x - x_0|$ the integration contour is first shifted to a line running just below the real k-axis (without crossing any singularities). Then

$$G(x, x_0, t, t_0) = \frac{c}{2\pi}H(t_0 - t - |x - x_0|/c)$$

$$\times \int_{-\infty-i\epsilon}^{\infty-i\epsilon} \left[e^{ik[c(t_0-t)+(x-x_o)]} - e^{-ik[c(t_0-t)-(x-x_o)]}\right]\frac{dk}{2ik}.$$

The second term in the square brackets is zero for $c(t_0 - t) > |x - x_0|$. The first has a contribution from the pole at $k = 0$, so that finally

$$G(x, x_0, t, t_0) = \frac{c}{2}H\left[t_0 - t - \frac{|x - x_0|}{c}\right],$$

in agreement with (4.8.15) obtained by the method of descent.

Example 3:   Laplace's equation in $y > 0$

Find $u(x, y)$ in $y > 0$ when

$$\frac{\partial^2 u}{\partial x^2} + \frac{\partial^2 u}{\partial y^2} = 0, \quad y > 0, \quad -\infty < x < \infty, \quad \text{and} \quad u = \frac{1}{1 + x^2} \quad \text{on } y = 0.$$

$$(4.10.5)$$

Take the Fourier transform with respect to x to find that $\hat{u}(k, y)$ satisfies the ordinary differential equation

$$\frac{d^2 \hat{u}}{dy^2} - k^2 \hat{u} = 0,$$

whose general solution can be written

$$\hat{u} = \mathcal{A}(k) e^{-|k|y} + \mathcal{B}(k) e^{|k|y}, \quad y > 0.$$

We must set $\mathcal{B}(k) = 0$ to ensure that the solution is *bounded* as $y \to +\infty$. The inversion formula (4.9.2) then gives

$$u(x, y) = \frac{1}{\sqrt{2\pi}} \int_{-\infty}^{\infty} \mathcal{A}(k) e^{ikx - |k|y} dk.$$

The condition at $y = 0$ is satisfied provided

$$\frac{1}{\sqrt{2\pi}} \int_{-\infty}^{\infty} \mathcal{A}(k) e^{ikx} dk = \frac{1}{1 + x^2}.$$

Therefore $\mathcal{A}(k)$ is the Fourier transform of $1/(1 + x^2)$, so that

$$\mathcal{A}(k) = \frac{1}{\sqrt{2\pi}} \int_{-\infty}^{\infty} \frac{e^{-ikx} dx}{1 + x^2} = \sqrt{\frac{\pi}{2}} e^{-|k|},$$

and $\quad u(x, y) = \frac{1}{2} \int_{-\infty}^{\infty} e^{ikx - |k|(1+y)} dk = \frac{1 + y}{(1 + y)^2 + x^2}, \quad y > 0, \quad -\infty < x < \infty.$

The boundary value problem (4.10.5) determines, for example, the steady temperature distribution in the half-space $y > 0$ when the boundary $y = 0$ is maintained at temperature $1/(1 + x^2)$. Heat flows into/out of the half-space at those points on $y = 0$ where

$$\frac{\partial u}{\partial y} = \frac{x^2 - 1}{(1 + x^2)^2} \lessgtr 0.$$

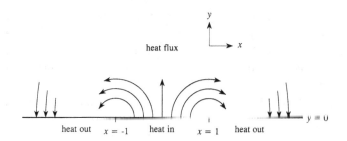

When the boundary condition on $y = 0$ is given more generally as $u(x, 0) = f(x)$, $-\infty < x < \infty$, the above procedure yields

$$A(k) = \frac{1}{\sqrt{2\pi}} \int\limits_{-\infty}^{\infty} f(\xi)e^{-ik\xi}\,d\xi,$$

and the solution becomes

$$u(x, y) = \frac{1}{\sqrt{2\pi}} \int\limits_{-\infty}^{\infty} A(k)e^{ikx - |k|y}\,dk = \frac{1}{2\pi} \iint\limits_{-\infty}^{\infty} f(\xi)e^{-|k|y + ik(x - \xi)}\,d\xi\,dk.$$

But,

$$\int\limits_{-\infty}^{\infty} e^{-|k|y + ik(x - \xi)}\,dk = \frac{2y}{y^2 + (x - \xi)^2},$$

and therefore the general solution can be expressed as the *convolution integral*

$$u(x, y) = \frac{1}{\pi} \int\limits_{-\infty}^{\infty} \frac{yf(\xi)d\xi}{y^2 + (x - \xi)^2}, \quad -\infty < x < \infty, \ y > 0.$$

This satisfies the boundary condition $u(x, y) \to f(x)$ as $y \to 0$, because

$$\frac{y}{\pi(y^2 + (x - \xi)^2)} \to \delta(x - \xi) \quad \text{as } y \to 0.$$

Example 4:   The diffusion equation

Find the steady state solution of

$$\frac{\partial^2 u}{\partial x^2} = \frac{1}{\kappa} \frac{\partial u}{\partial t}, \quad 0 < x < \infty,$$

given $\begin{cases} u = u_o e^{-i\Omega t}, & x = 0, \text{ where } u_o = \text{ constant and } \Omega > 0, \\ u \to 0 & \text{as } x \to \infty. \end{cases}$

The solution represents the temperature fluctuations in a semi-infinite slab when the temperature of the exposed face of the slab at $x = 0$ fluctuates at radian frequency Ω. In the 'steady state' the temperature everywhere in the medium varies at the same frequency, i.e. all 'transients' associated with the initial introduction of the surface heat source have decayed to zero.

Let $u(x,t) = U(x)e^{-i\Omega t}$, then $U(x)$ satisfies the ordinary differential equation

$$\frac{d^2U}{dx^2} + \frac{i\Omega}{\kappa}U = 0, \quad 0 < x < \infty.$$

Therefore

$$U = Ae^{i(i\Omega/\kappa)^{\frac{1}{2}}x} + Be^{-i(i\Omega/\kappa)^{\frac{1}{2}}x}, \quad \text{where } A \text{ and } B \text{ are constants}$$

and $(i\Omega/\kappa)^{\frac{1}{2}} = (1+i)\left(\dfrac{\Omega}{2\kappa}\right)^{\frac{1}{2}}$

$$\therefore \quad B = 0 \text{ to ensure that } u \to 0 \text{ as } x \to \infty,$$

$$\therefore \quad u(x,t) = Ae^{-i\Omega(t-x/\sqrt{2\kappa\Omega})-x\sqrt{\Omega/2\kappa}}.$$

The condition $u = u_o e^{-i\Omega t}$ at $x = 0$ implies that $A = u_o$,

$$\therefore \quad u(x,t) = u_o e^{-i\Omega(t-x/\sqrt{2\kappa\Omega})-x\sqrt{\Omega/2\kappa}}, \quad x > 0. \qquad (4.10.6)$$

Thus the temperature fluctuations within the slab decay exponentially fast with distance x from the surface, and become insensible when x exceeds the (frequency dependent) thermal penetration depth $\sqrt{2\kappa/\Omega}$. At any point x within the slab the time delay $x/\sqrt{2\kappa\Omega}$ produces a 'lag' between the phase of the temperature at x and the surface temperature.

Application of Fourier sine and cosine transforms   Consider the Fourier sine transform of $\partial^2 u(x,t)/\partial x^2$. By integration by parts, and by noting that u and $\partial u/\partial x = 0$ at $x = \infty$, we find:

$$\sqrt{\frac{2}{\pi}} \int_0^\infty \frac{\partial^2 u}{\partial x^2} \sin kx \, dx$$

$$= \sqrt{\frac{2}{\pi}} \left\{ \left[\frac{\partial u}{\partial x} \sin kx \right]_0^\infty - k \int_0^\infty \frac{\partial u}{\partial x} \cos kx \, dx \right\}$$

$$= \sqrt{\frac{2}{\pi}} \left\{ \left[\frac{\partial u}{\partial x} \sin kx \right]_0^\infty - [ku \cos kx]_0^\infty \right\} - k^2 \hat{u}_s(k,t)$$

$$= \sqrt{\frac{2}{\pi}} ku(0,t) - k^2 \hat{u}_s(k,t). \qquad (4.10.7)$$

Similarly, taking the Fourier cosine transform of $\partial^2 u(x,t)/\partial x^2$:

$$\sqrt{\frac{2}{\pi}} \int_0^\infty \frac{\partial^2 u}{\partial x^2} \cos kx \, dx$$

$$= \sqrt{\frac{2}{\pi}} \left\{ \left[\frac{\partial u}{\partial x} \cos kx \right]_0^\infty + k \int_0^\infty \frac{\partial u}{\partial x} \sin kx \, dx \right\}$$

$$= \sqrt{\frac{2}{\pi}} \left\{ \left[\frac{\partial u}{\partial x} \cos kx \right]_0^\infty + [ku \sin kx]_0^\infty \right\} - k^2 \hat{u}_c(k,t)$$

$$= -\sqrt{\frac{2}{\pi}} \left(\frac{\partial u}{\partial x} \right)_{x=0} - k^2 \hat{u}_c(k,t). \tag{4.10.8}$$

These results dictate the choice of half-range transform to be used in solving certain second-order partial differential equations defined over the semi-infinite range $0 < x < \infty$, because they reveal the information required at $x = 0$ in order to express the transform of $\partial^2 u/\partial x^2$ in terms of the corresponding transform of $u(x,t)$.

Example 5:   The diffusion equation in $0 < x < \infty$

Derive a Fourier integral representation of the solution of the initial value problem

$$\frac{\partial^2 u}{\partial x^2} = \frac{1}{\kappa} \frac{\partial u}{\partial t}, \quad \text{for } t > 0, \quad 0 < x < \infty, \quad \text{given} \begin{cases} u = 0, & x = 0, \text{ for } t > 0, \\ u = f(x), & 0 < x < \infty, \text{ at } t = 0. \end{cases}$$

Obtain explicit formulae for the special case $f(x) = T_0 = \text{constant}$.

This is a one-dimensional heat diffusion problem for a semi-infinite conductor whose temperature distribution at time $t = 0$ is $u = f(x)$ when the end $x = 0$ is maintained at constant temperature $u = 0$ for $t > 0$. Because we are given the value of $u(x,t) (= 0)$ at $x = 0$, we take the Fourier sine transform of the equation, obtaining in the usual way from (4.10.7):

$$\sqrt{\frac{2}{\pi}} ku(0,t) - k^2 \hat{u}_s(k,t) = \frac{1}{\kappa} \frac{\partial \hat{u}_s}{\partial t}(k,t).$$

Hence,

$$\frac{1}{\kappa} \frac{\partial \hat{u}_s}{\partial t} = -k^2 \hat{u}_s, \quad \text{and} \quad \therefore \quad \hat{u}_s = \mathcal{A}(k) e^{-k^2 \kappa t},$$

and the inversion formula for the sine transform then yields the formal solution

$$u(x,t) = \sqrt{\frac{2}{\pi}} \int_0^\infty A(k) e^{-k^2 \kappa t} \sin kx \, dk.$$

The initial condition $u = f(x)$ at $t = 0$ is applied by setting $t = 0$ in this formula, to obtain

$$f(x) = \sqrt{\frac{2}{\pi}} \int_0^\infty A(k) \sin kx \, dk,$$

$$\therefore \quad A(k) = \sqrt{\frac{2}{\pi}} \int_0^\infty f(x) \sin kx \, dx.$$

Thus, the required solution of the boundary value problem is

$$u(x,t) = \frac{2}{\pi} \int_0^\infty \left[\int_0^\infty f(\xi) \sin k\xi \, d\xi \right] e^{-k^2 \kappa t} \sin kx \, dk, \quad 0 < x < \infty, \quad t > 0.$$

When $f(x) = T_0 = $ constant, the temperature of the conductor is *uniform* at $t = 0$. However, the integral in the formula

$$A(k) = T_0 \sqrt{\frac{2}{\pi}} \int_0^\infty \sin kx \, dx$$

does not converge. We therefore replace the constant temperature T_0 by the ϵ-sequence $T_0 e^{-\epsilon x}, \epsilon > 0$, and consider the limiting form of the solution as $\epsilon \to 0$. Then

$$A(k) = T_0 \sqrt{\frac{2}{\pi}} \int_0^\infty e^{-\epsilon x} \sin kx \, dx = \sqrt{\frac{2}{\pi}} \frac{T_0 k}{\epsilon^2 + k^2},$$

and

$$u(x,t) = \frac{2T_0}{\pi} \int_0^\infty \frac{k \sin kx}{\epsilon^2 + k^2} e^{-k^2 \kappa t} dk$$

$$\to \frac{2T_0}{\pi} \int_0^\infty \frac{\sin kx}{k} e^{-k^2 \kappa t} dk, \quad \text{as } \epsilon \to 0,$$

$$= T_0 \operatorname{erf}\left(\frac{x}{2\sqrt{\kappa t}}\right),$$

where $\operatorname{erf}(x)$ is the error function (see §5.4 and Question 13 of Problems 5C).

The following plot of $u(x,t)/T_0$ against the non-dimensional 'similarity' variable $x/\sqrt{\kappa t}$ (κ having the dimensions of m^2/s) indicates how the temperature changes as a function of both x and t.

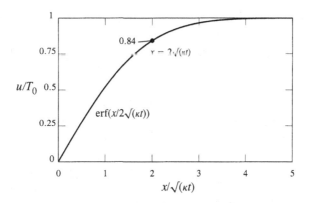

Example 6:   The diffusion equation in $0 < x < \infty$

Solve the initial value problem

$$\frac{\partial^2 u}{\partial x^2} = \frac{1}{\kappa}\frac{\partial u}{\partial t}, \quad \text{for } t > 0, \quad 0 < x < \infty,$$

$$\text{given } \begin{cases} u = u_0, & x = 0, \quad \text{for } t > 0, \\ u = 0, & 0 < x < \infty, \quad \text{at } t = 0. \end{cases}$$

The solution determines the temperature distribution in a semi-infinite rod which is initially of uniform temperature $u = 0$, when the end $x = 0$ is suddenly raised and maintained at the constant temperature $u = u_0$ for $t > 0$.

Because u is specified at $x = 0$ we take the Fourier sine transform of the equation, as in Example 5, to obtain

$$\frac{1}{\kappa}\frac{\partial \hat{u}_s}{\partial t} = \sqrt{\frac{2}{\pi}}ku_0 - k^2\hat{u}_s, \quad \therefore \quad \hat{u}_s(k,t) = \sqrt{\frac{2}{\pi}}\frac{u_0}{k} + A(k)e^{-k^2\kappa t}.$$

At $t = 0$, $u(x,t) = 0$, $0 < x < \infty$

$$\therefore \quad \hat{u}_s(k,t) = 0 \text{ at } t = 0$$

$$\therefore \quad A(k) = -\sqrt{\frac{2}{\pi}}\frac{u_0}{k} \quad \text{and} \quad \hat{u}_s(k,t) = \sqrt{\frac{2}{\pi}}\frac{u_0}{k}\left[1 - e^{-k^2\kappa t}\right].$$

Hence, using the inversion formula of (4.9.7)

$$\frac{u(x,t)}{u_0} = \frac{2}{\pi}\int_0^\infty \left(1 - e^{-k^2\kappa t}\right)\frac{\sin kx \, dk}{k}$$

$$= 1 - \text{erf}\left(\frac{x}{2\sqrt{\kappa t}}\right), \quad 0 < x < \infty, \quad t > 0.$$

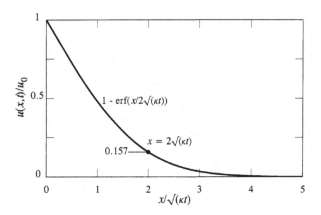

Example 7:   The diffusion equation in $0 < x < \infty$

Solve the initial value problem

$$\frac{\partial^2 u}{\partial x^2} = \frac{1}{\kappa}\frac{\partial u}{\partial t}, \quad \text{for } t > 0, \quad 0 < x < \infty,$$

$$\text{given } \begin{cases} \partial u/\partial x = -\sigma, & x = 0, \quad \text{for } t > 0, \\ u = 0, & 0 < x < \infty, \quad \text{at } t = 0. \end{cases}$$

The solution determines the temperature distribution in a semi-infinite rod which is initially of uniform temperature $u = 0$, when heat is supplied at the end $x = 0$ at a constant rate σ for $t > 0$.

Because $\partial u/\partial x$ is prescribed at $x = 0$ we use the Fourier cosine transform, and the relation (4.10.8) to deduce that

$$\hat{u}_c(k, t) = A(k)e^{-k^2\kappa t} + \sqrt{\frac{2}{\pi}}\frac{\sigma}{k^2},$$

where $A(k)$ is determined from the temperature distribution at $t = 0$ to be

$$A(k) = -\sqrt{\frac{2}{\pi}}\frac{\sigma}{k^2},$$

so that

$$\frac{u(x, t)}{\sigma} = \frac{2}{\pi}\int_0^\infty \left(1 - e^{-k^2\kappa t}\right)\frac{\cos kx \, dk}{k^2}, \quad 0 < x < \infty, \quad t > 0.$$

The solution is plotted in non-dimensional form $(u(x,t)/\sigma\sqrt{\kappa t}$ against $x/\sqrt{\kappa t}$, σ having the dimensions of u/x). The constant rate at which heat is supplied causes the temperature to rise indefinitely everywhere within the rod with increasing time, including the end $x = 0$.

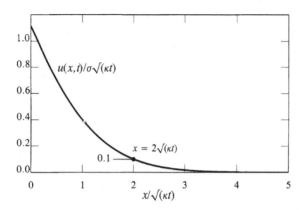

Example 8: Laplace's equation in the quadrant $x > 0, y > 0$

Calculate the steady temperature distribution $u(x, y)$ in $x > 0, y > 0$ where

$$\frac{\partial^2 u}{\partial x^2} + \frac{\partial^2 u}{\partial y^2} = 0, \quad \text{and} \quad \begin{cases} u = u_0, & x = 0, \quad 0 < y < \infty, \\ u = 0, & 0 < x < \infty, \quad y = 0. \end{cases}$$

Because $u = u_0$ at $x = 0$ we take the Fourier sine transform with respect to x (using (4.10.7)), to obtain

$$\frac{\partial^2 \hat{u}_s}{\partial y^2} - k^2 \hat{u}_s = -\sqrt{\frac{2}{\pi}} k u_0,$$

$$\therefore \quad \hat{u}_s(k, y) = \mathcal{A}(k) e^{-ky} + \mathcal{B}(k) e^{ky} + \sqrt{\frac{2}{\pi}} \frac{u_0}{k}.$$

Only positive values of k occur in the inversion integral. Thus, the solution is bounded only if $\mathcal{B}(k) \equiv 0$. Then, because $\hat{u}_s(k, y) = 0$ at $y = 0$,

$$\mathcal{A}(k) = -\sqrt{\frac{2}{\pi}} \frac{u_0}{k},$$

$$\therefore \quad \frac{u(x, y)}{u_0} = \frac{2}{\pi} \int_0^\infty \left(1 - e^{-ky}\right) \frac{\sin kx \, dk}{k} = \frac{2}{\pi} \arctan\left(\frac{y}{x}\right), \quad x > 0, \quad y > 0.$$

This is equivalent to $u(x, y)/u_o = \frac{2}{\pi} \text{Im}(\ln z) = \frac{2\theta}{\pi}, z = (x^2 + y^2)^{\frac{1}{2}} e^{i\theta}$, where $\ln z$ is regular in the first quadrant except a $z = 0$. The *isotherms*, $u(x, y)/u_0 = $ constant, are the straight lines $\theta = $ constant:

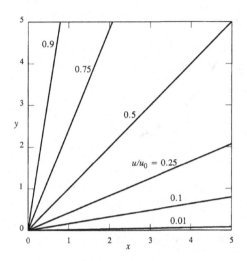

Example 9: The diffusion equation in $0 < x < \infty$

Solve the initial value problem

$$\frac{\partial^2 u}{\partial x^2} = \frac{1}{\kappa}\frac{\partial u}{\partial t}, \quad \text{for } t > 0, \quad 0 < x < \infty, \quad \text{given} \begin{cases} u = u_0, & x = 0, \quad \text{for } t > 0, \\ u = 0, & 0 < x < \infty, \quad \text{at } t = 0. \end{cases}$$

by taking the Fourier transform with respect to time.

This is the same as Example 6. The method to be described will work even when the boundary condition at $x = 0$ is 'mixed', i.e. consists of a linear relation of the form

$$\frac{\partial u}{\partial x} + \alpha u = \beta(t) \quad \alpha = \text{constant}.$$

Let

$$\hat{u}(x, \omega) = \frac{1}{\sqrt{2\pi}} \int_{-\infty}^{\infty} u(x, t) e^{-i\omega t}\, dt, \quad \text{where } u(x, t) = 0 \quad \text{for } t < 0. \qquad (4.10.9)$$

This implies that $\hat{u}(x, \omega)$ is a regular function of ω in the region $\text{Im}(\omega) < 0$. Then,

$$\frac{\partial^2 \hat{u}}{\partial x^2} = \frac{i\omega}{\kappa}\hat{u}, \quad \text{i.e.} \quad \hat{u}(x, \omega) = A(\omega) e^{-x\sqrt{i\omega/\kappa}}, \quad x > 0,$$

where a bounded solution is obtained by requiring the real part of $\sqrt{i\omega/\kappa}$ to be *positive* for $-\infty < \omega < \infty$. This can be achieved by defining the square root as the *principal value* of

$$\left(\frac{e^{\frac{i\pi}{2}}(\omega - i\epsilon)}{\kappa} \right)^{\frac{1}{2}}$$

where $\epsilon > 0$ and is subsequently allowed to tend to zero.

To determine $\mathcal{A}(\omega)$ we impose the condition $u = u_0$ at $x = 0$ for *all* $t > 0$

$$\therefore \quad \mathcal{A}(\omega) = \frac{1}{\sqrt{2\pi}} \int_0^\infty u_0 e^{-i\omega t} dt = \lim_{\epsilon \to 0} \frac{1}{\sqrt{2\pi}} \int_0^\infty u_0 e^{-i\omega t - \epsilon t} dt \equiv \lim_{\epsilon \to 0} \frac{i u_0}{\sqrt{2\pi}(\omega - i\epsilon)},$$

where to secure convergence it is necessary to replace the constant u_0 by the ϵ-sequence $u_0 e^{-\epsilon t}$, i.e. to regard u_0 as a generalised function.

Hence, using the inversion theorem for the Fourier transform

$$\frac{u(x,t)}{u_0} = \frac{-i}{2\pi} \int_{-\infty}^\infty \frac{\exp\left[-x\left(e^{\frac{i\pi}{2}}(\omega - i\epsilon)/\kappa\right)^{\frac{1}{2}}\right]}{\omega - i\epsilon} e^{i\omega t}\, d\omega$$

$$= \frac{-i}{2\pi} \int_{-\infty - i0}^{\infty - i0} \frac{e^{i\omega t - x\sqrt{i\omega/\kappa}}\, d\omega}{\omega}, \quad \text{as } \epsilon \to 0.$$

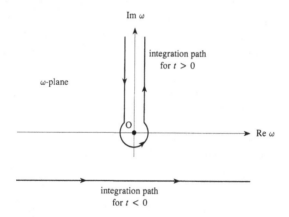

The contour in the final integral runs just below the real ω-axis. When $t < 0$ it can be displaced to $\omega = -i\infty$ without encountering any singularities, so that $u(x,t) = 0$ for $t < 0$, as required. For $t > 0$ the integration contour is deformed onto the path shown in the figure. The residue contribution from the pole at $\omega = 0$ is equal to 1. The substitution $\omega = ik^2\kappa$ in the remaining integral along both sides of the imaginary axis then transforms the result into that obtained in Example 6 using the Fourier sine transform.

Example 10:   One-dimensional wave equation

Use the Fourier *time* transform to solve the initial value problem

$$\left(\frac{1}{c^2} \frac{\partial^2}{\partial t^2} - \frac{\partial^2}{\partial x^2} \right) u = \delta(x - a) f(t), \quad -\infty < x < \infty, \quad t > 0,$$

where $a > 0$ and $u = 0$, $f(t) = 0$ for $t < 0$.

Take the Fourier time transform (defined as in (4.10.9)) of the equation:

$$\frac{\partial^2 \hat{u}}{\partial x^2} + k_o^2 \hat{u} = -\delta(x - a)\hat{f}(\omega), \quad k_o = \frac{\omega}{c}, \tag{4.10.10}$$

where, because $f(t) = 0$ for $t < 0$,

$$\hat{f}(\omega) = \frac{1}{\sqrt{2\pi}} \int_0^\infty f(t)e^{-i\omega t}\,dt$$

is *regular* in the *lower half* of the complex ω-plane, and $\hat{f}(\omega) \to 0$ as $\omega \to -i\infty$.

We solve (4.10.10) by the method of Example 2. For $x \neq a$ the general solution of (4.10.10) is

$$\hat{u}(x, \omega) = Ae^{-ik_o x} + Be^{ik_o x},$$

where A and B are arbitrary functions of ω. However, because all of the motion is generated by the 'source' at $x = a$, the solution must also satisfy a 'radiation condition'. This requires that, for $x > a$, $u(x, t)$ consists of 'waves' propagating *away* from $x = a$ towards $x = +\infty$, i.e. $u(x, t)$ must be a function of $x - ct$. This is possible only if $B = 0$ for $x > a$, because the inversion theorem gives

$$u(x, t) = \frac{1}{\sqrt{2\pi}} \int_{-\infty}^\infty \left(Ae^{i\omega(t - x/c)} + Be^{i\omega(t + x/c)} \right) d\omega.$$

Similarly, we must take $A = 0$ when $x < a$.

We therefore write the solution of (4.10.10) in the form

$$\hat{u}(x, \omega) = AH(x - a)e^{-ik_o(x - a)} + BH(a - x)e^{ik_o(x - a)}, \tag{4.10.11}$$

where H is the Heaviside step function (1.11.4). Substitute into (4.10.10):

$$\delta'(x - a)[A - B] - ik_o\delta(x - a)[A + B] = -\delta(x - a)\hat{f}(\omega),$$

$$\therefore \quad \begin{cases} A - B = 0, \\ ik_o(A + B) = \hat{f}(\omega). \end{cases}$$

Thus, $A = B$, and

$$\frac{2i\omega}{c}\hat{u}(x, \omega) = \hat{f}(\omega)e^{-ik_o|x - a|}.$$

It has been implicitly assumed that ω is *real*. In that case, when $\hat{u}(x, \omega)$ is found by dividing both sides of this equation by $2i\omega/c$, it follows from Question 3 of Problems 1J (§1.11) that

$$\hat{u}(x, \omega) = \frac{c\hat{f}(\omega)e^{-ik_o|x - a|}}{2i}\left[\mathrm{P}\left(\frac{1}{\omega}\right) + C\delta(\omega)\right],$$

where 'P' denotes the principal value, and C is arbitrary. Hence,

$$u(x,t) = \frac{c}{2i\sqrt{2\pi}} \int\limits_{-\infty}^{\infty} \left[P\left(\frac{1}{\omega}\right) + C\delta(\omega) \right] \hat{f}(\omega)e^{i\omega(t-|x-a|/c)} \, d\omega.$$

The coefficient C is determined from the condition that $u(x,t) = 0$ for $t < 0$. From Question 14 of Problems 1J (§1.11)

$$P\left(\frac{1}{\omega}\right) = -\pi i\delta(\omega) + \lim_{\epsilon \to 0} \frac{1}{\omega - i\epsilon},$$

so that

$$u(x,t) = \lim_{\epsilon \to 0} \frac{c}{2i\sqrt{2\pi}} \int\limits_{-\infty}^{\infty} \left[\frac{1}{\omega - i\epsilon} + [C - \pi i]\delta(\omega) \right] \hat{f}(\omega)e^{i\omega(t-|x-a|/c)} \, d\omega.$$

When $t < 0$ the part of the integral involving $1/(\omega - i\epsilon)$ is zero, because $\hat{f}(\omega)$ is regular in the lower half-plane and tends to zero at $\omega = -i\infty$, so that the integration contour can be displaced to $\omega = -i\infty$ (on which the integral vanishes) without crossing any singularities. The remaining part of the integral will also vanish provided $C = i\pi$. The solution is therefore given finally by

$$u(x,t) = \frac{c}{2i\sqrt{2\pi}} \int\limits_{-\infty-i0}^{\infty-i0} \frac{\hat{f}(\omega)e^{i\omega(t-|x-a|/c)} \, d\omega}{\omega}, \qquad (4.10.12)$$

where the integration contour runs from $-\infty$ to ∞ just below the real axis.
This formula implies that

$$\frac{\partial u}{\partial t}(x,t) = \frac{c}{2\sqrt{2\pi}} \int\limits_{-\infty}^{\infty} \hat{f}(\omega)e^{i\omega(t-|x-a|/c)} \, d\omega = \frac{c}{2} f(t - |x - a|/c)$$

$$\therefore \quad u(x,t) = \frac{c}{2} \int\limits_{0}^{t-|x-a|/c} f(\tau) d\tau, \quad \text{where } f(\tau) = 0 \quad \text{for } \tau < 0. \qquad (4.10.13)$$

Problems 4G

1. The steady temperature distribution $u(x,y)$ in $x > 0, y > 0$ is governed by

$$\frac{\partial^2 u}{\partial x^2} + \frac{\partial^2 u}{\partial y^2} = 0, \quad \text{and} \quad \begin{cases} u = 0, & x = 0, \quad 0 < y < \infty, \\ \partial u/\partial y = -\sigma\delta(x - a), \ (a > 0), & 0 < x < \infty, \quad y = 0. \end{cases}$$

Use the sine transform to show that

$$u = \frac{\sigma}{2\pi} \ln\left(\frac{y^2 + (a+x)^2}{y^2 + (a-x)^2} \right), \quad x > 0, \ y > 0.$$

2. By taking the Fourier transform with respect to *time*, derive the solution of the boundary value problem

$$\frac{\partial^2 u}{\partial x^2} = \frac{1}{\kappa}\frac{\partial u}{\partial t} + \delta(x)f(t), \quad -\infty < x < \infty, \quad t > t_0,$$

where $u(x,t)$, $f(t) = 0$ for $t < t_0$, in the form

$$u(x,t) = \frac{-1}{2\sqrt{2\pi}}\int_{-\infty-i0}^{\infty-i0} \frac{\hat{f}(\omega)}{\sqrt{i\omega/\kappa}} e^{i\omega t - \sqrt{i\omega/\kappa}|x|}\, d\omega,$$

$$\text{where} \quad \hat{f}(\omega) = \frac{1}{\sqrt{2\pi}}\int_{t_0}^{\infty} f(t)e^{-i\omega t}\, dt.$$

3. Show that the bounded solution of the boundary value problem

$$\frac{\partial^2 u}{\partial x^2} + \frac{\partial^2 u}{\partial y^2} = 0, \quad \text{and} \quad \begin{cases} u = 0, & x = 0, \ 0 < y < \infty, \\ u = u_0 x/(1 + x^2), & 0 < x < \infty, \ y = 0 \end{cases}$$

is

$$u(x,y) = \frac{u_0 x}{(1+y)^2 + x^2} \equiv \mathrm{Re}\left(\frac{u_0}{z+i}\right), \quad z = x + iy.$$

4. Derive the solution (4.10.13) of Example 10 by making use of Green's function (4.8.15) for the one-dimensional wave equation.

5. Solve Example 10 in the region $0 < x < \infty$ when u satisfies the condition

$$Z\frac{\partial u}{\partial x} = u \quad \text{at } x = 0, \quad Z = \text{constant}.$$

Express the solution as the sum of the infinite medium solution (4.10.12) plus a reflected 'wave' $u_R(x,t)$, where

$$u_R(x,t) = \frac{c}{2i\sqrt{2\pi}}\int_{-\infty-i0}^{\infty-i0} \frac{\mathcal{R}(\omega)\hat{f}(\omega)}{\omega} e^{i\omega(t-(x+a)/c)}\, d\omega, \quad x > 0,$$

$$\mathcal{R} = \text{reflection coefficient} = -\left(\frac{1 - ik_0 Z}{1 + ik_0 Z}\right), \quad k_0 = \frac{\omega}{c}.$$

6. Find the function $u(x,y)$ that is bounded in $y > 0$ and satisfies

$$\frac{\partial^2 u}{\partial x^2} + \frac{\partial^2 u}{\partial y^2} = 0, \quad y > 0, \quad -\infty < x < \infty \quad \text{and}$$

$$u = \delta(x - \xi) \quad \text{on } y = 0. \quad \left[u = \frac{y}{\pi[(x-\xi)^2 + y^2]} \right].$$

7. If $u(x, y)$ is bounded in $y > 0$ and satisfies

$$\frac{\partial^2 u}{\partial x^2} + \frac{\partial^2 u}{\partial y^2} = 0, \quad y > 0, \quad -\infty < x < \infty, \quad \text{and} \quad u = f(x) \quad \text{on } y = 0,$$

show that

$$u = \frac{y}{\pi} \int_{-\infty}^{\infty} \frac{f(\xi) \, d\xi}{[(x - \xi)^2 + y^2]}.$$

8. If $\nabla u(x, y)$ is bounded in $y > 0$ and

$$\frac{\partial^2 u}{\partial x^2} + \frac{\partial^2 u}{\partial y^2} = 0, \quad y > 0, \quad -\infty < x < \infty, \quad \text{and} \quad \frac{\partial u}{\partial y} = \delta(x - \xi) \quad \text{on } y = 0,$$

show that

$$\nabla u = \frac{(x - \xi, \, y)}{\pi[(x - \xi)^2 + y^2]}, \quad u = \frac{1}{2\pi} \ln[(x - \xi)^2 + y^2] + \text{constant}, \quad y > 0.$$

9. If $\nabla u(x, y)$ is bounded in $y > 0$ and

$$\frac{\partial^2 u}{\partial x^2} + \frac{\partial^2 u}{\partial y^2} = 0, \quad y > 0, \quad -\infty < x < \infty, \quad \text{and} \quad \frac{\partial u}{\partial y} = f(x) \quad \text{on } y = 0,$$

show that

$$u = \frac{1}{2\pi} \int_{-\infty}^{\infty} f(\xi) \ln[(x - \xi)^2 + y^2] \, d\xi + \text{constant}, \quad y > 0.$$

10. Use the cosine transform to show that, if $u(x, y)$ is bounded in $y > 0$ and satisfies

$$\frac{\partial^2 u}{\partial x^2} + \frac{\partial^2 u}{\partial y^2} = 0, \quad y > 0, \quad -\infty < x < \infty, \quad \text{and}$$

$$u = H(a - |x|), \quad a > 0, \quad \text{on } y = 0,$$

then

$$u = \frac{1}{\pi} \left\{ \tan^{-1} \left(\frac{a + x}{y} \right) + \tan^{-1} \left(\frac{a - x}{y} \right) \right\}, \quad y > 0.$$

11. Show that the solution $u(x, y)$ of

$$\frac{\partial^2 u}{\partial x^2} + \frac{\partial^2 u}{\partial y^2} = 0, \quad x > 0, \quad 0 < y < a,$$

$$u(x, 0) = f(x), \quad 0 < x < \infty,$$

$$u(x, a) = 0, \quad 0 < x < \infty,$$

$$u(0, y) = 0, \quad 0 < y < a,$$

is given by

$$u = \frac{2}{\pi} \int_0^{\infty} f(\xi) d\xi \int_0^{\infty} \frac{\sinh[k(a - y)]}{\sinh ka} \sin kx \sin k\xi \, dk.$$

12. If $u(x, t)$ is bounded and satisfies

$$\frac{\partial^2 u}{\partial x^2} = \frac{\partial u}{\partial t}, \quad x > 0, \quad t > 0,$$

$$u(x, 0) = xe^{-x^2/4}, \quad u(0, t) = 0,$$

$$\text{show that} \quad u = \frac{x}{(1+t)^{\frac{3}{2}}} e^{-\frac{x^2}{4}(1+t)}.$$

5

SPECIAL FUNCTIONS

5.1   The Gamma Function $\Gamma(x)$

The gamma function is defined for positive values of x by

$$\Gamma(x) = \int_0^\infty t^{x-1} \mathrm{e}^{-t}\, dt, \quad x > 0. \qquad (5.1.1)$$

It is positive for $x > 0$, and $\Gamma(1) = \int_0^\infty \mathrm{e}^{-t}\, dt = 1$.

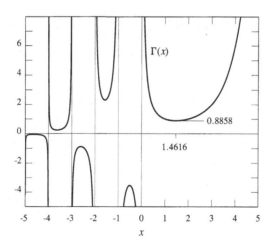

Integration by parts shows that

$$\Gamma(1+x) = \int_0^\infty t^x e^{-t}\, dt$$

$$= \left[-t^x e^{-t}\right]_0^\infty + x \int_0^\infty t^{x-1} e^{-t}\, dt \equiv x \int_0^\infty t^{x-1} e^{-t}\, dt,$$

so that $\Gamma(x)$ satisfies the *recurrence relation*

$$\Gamma(1+x) = x\Gamma(x). \tag{5.1.2}$$

In particular, $\Gamma(2) = 1 \times \Gamma(1) = 1$. Similarly, repeated application of the recurrence relation for $x = n =$ a positive integer yields $\Gamma(1+n) = n(n-1)(n-2)\ldots 2.1$, i.e.

$$n! = \Gamma(1+n).$$

The recurrence relation (5.1.2) is used to extend the definition of $\Gamma(x)$ to negative values of x. For example, the formula

$$\Gamma(x) = \frac{\Gamma(1+x)}{x}$$

permits $\Gamma(x)$ to be calculated in the range $-1 < x < 0$ in terms of the known values of $\Gamma(1+x)$ in the interval $0 < 1+x < 1$. It also shows that

$$\Gamma(x) \approx \frac{1}{x}, \quad \text{when } x \text{ is small,}$$

because $\Gamma(1+x) \to \Gamma(1) = 1$ as $x \to 0$. In the same way, $\Gamma(x)$ can be calculated for negative values of x in the range $-N < x < 0$ from

$$\Gamma(x) = \frac{\Gamma(N+x)}{x(x+1)(x+2)\ldots(x+N-1)}, \quad N = \text{positive integer.}$$

$$\tag{5.1.3}$$

This implies that $|\Gamma(x)|$ is infinite at $x = 0, -1, -2, \ldots$ (i.e. $n! = \infty$ for $n = -1, -2, -3, \ldots$) and that $\Gamma(x)$ alternates in sign for $x < 0$ (see preceding graph). The integral (5.1.1) converges also when x is replaced by $z = x+iy$, provided $x > 0$, where it defines $\Gamma(z)$ as a regular function of z. We shall see below that $\Gamma(z)$ is regular for all complex values of z except for *simple poles* at $z = 0, -1, -2, \ldots$, on the negative real axis.

Example $\Gamma(\frac{1}{2}) = \sqrt{\pi}.$ $\left[\int_0^\infty t^{\frac{1}{2}-1} e^{-t} \, dt = \int_0^\infty \frac{1}{\sqrt{t}} e^{-t} \, dt = 2 \int_0^\infty e^{-\mu^2} \, d\mu = \sqrt{\pi} \right]$

Example $\Gamma'(\frac{3}{2}) = \frac{1}{2}\Gamma(\frac{1}{2}) = \frac{1}{2}\sqrt{\pi}.$

Example $\Gamma(\frac{-5}{2}) = \frac{\Gamma(-\frac{3}{2})}{-\frac{5}{2}} = \frac{\Gamma(-\frac{1}{2})}{(-\frac{5}{2})(-\frac{3}{2})} = \frac{\Gamma(\frac{1}{2})}{(-\frac{5}{2})(-\frac{3}{2})(-\frac{1}{2})} = -\frac{8}{15}\sqrt{\pi}.$

Example Evaluate $\int_0^\infty x^6 e^{-x^3} \, dx.$ $\left[\text{Set } t = x^3, \text{ then } \int_0^\infty x^6 e^{-x^3} \, dx = \frac{1}{3}\int_0^\infty t^{\frac{7}{3}-1} e^{-t} \, dt = \frac{1}{3}\Gamma\left(\frac{7}{3}\right) \right]$

Example Set $t = \tau^2$ in (5.1.1), then

$$\Gamma(x) = 2 \int_0^\infty \tau^{2x-1} e^{-\tau^2} \, d\tau, \quad x > 0. \tag{5.1.4}$$

Stirling's formula

$$\Gamma(1+x) = \int_0^\infty t^x e^{-t} \, dt \approx \sqrt{2\pi x}\, x^x e^{-x}, \quad x \to +\infty. \tag{5.1.5}$$

This *asymptotic* formula can be derived by noting that for a fixed and large positive value of x, the function $y = t^x e^{-t} \equiv e^{(x \ln t - t)}$ has a large maximum at $t = x$, as illustrated in the figure for $x = 10$. The value of the integral as $x \to \infty$ is therefore dominated by contributions from values of t near $t = x$.

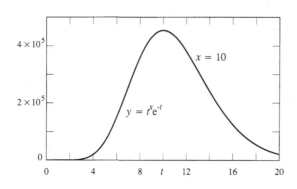

Let $\xi = t - x$, then

$$x \ln t - t = x \ln(x + \xi) - (x + \xi)$$

$$= x \ln x + x \ln\left(1 + \frac{\xi}{x}\right) - (x + \xi)$$

$$= x \ln x + x \left\{\frac{\xi}{x} - \frac{\xi^2}{2x^2} + \frac{\xi^3}{3x^3} - \frac{\xi^4}{4x^4} + \cdots\right\}$$

$$- (x + \xi), \quad x \to \infty$$

$$= x \ln x - x - \frac{\xi^2}{2x} + \frac{\xi^3}{3x^2} - \frac{\xi^4}{4x^3} + \cdots,$$

$$\therefore \quad \Gamma(1 + x) = x^x e^{-x} \int_{-x}^{\infty} e^{-\left(\frac{\xi^2}{2x} - \frac{\xi^3}{3x^2} + \frac{\xi^4}{4x^3} - \cdots\right)} d\xi$$

$$\approx x^x e^{-x} \int_{-\infty}^{\infty} e^{-\left(\frac{\xi^2}{2x} - \frac{\xi^3}{3x^2} + \frac{\xi^4}{4x^3} - \cdots\right)} d\xi.$$

By making the substitution $\mu = \xi/\sqrt{2x}$ in this integral:

$$\Gamma(1 + x) \approx \sqrt{2x} x^x e^{-x} \int_{-\infty}^{\infty} e^{-\left(\mu^2 - \frac{2\sqrt{2}\mu^3}{3\sqrt{x}} + \frac{\mu^4}{x} - \cdots\right)} d\mu$$

$$\approx \sqrt{2x} x^x e^{-x} \int_{-\infty}^{\infty} e^{-\mu^2} \left\{1 + \frac{2\sqrt{2}\mu^3}{3\sqrt{x}} + \left(\frac{4}{9} - \mu^2\right)\frac{\mu^4}{x} + \cdots\right\} d\mu.$$

$$(5.1.6)$$

This gives Stirling's formula (5.1.5) when the terms involving x in the integrand are neglected (as $x \to \infty$) and the remaining integral is evaluated using the formula $\int_{-\infty}^{\infty} e^{-\mu^2} d\mu = \sqrt{\pi}$.

A more accurate approximation, which is useful for calculating $n! \equiv \Gamma(1 + n)$ when $x = n = $ a large positive integer, is (see Problems 5A, Question 18)

$$n! \approx \sqrt{2\pi n}\, n^n e^{-n}\left(1 + \frac{1}{12n}\right), \quad n \gg 1.$$

The error is smaller than about 0.1% for n as small as one.

Example The integral definition of $\Gamma(z)$ is extended to all complex $z \neq 0, -1, -2, \ldots$, by means of the formula

$$\Gamma(z) = \frac{e^{-i\pi z}}{2i \sin \pi z} \oint_C \zeta^{z-1} e^{-\zeta} \, d\zeta, \tag{5.1.7}$$

where C is the contour shown in the figure enclosing the positive real axis.

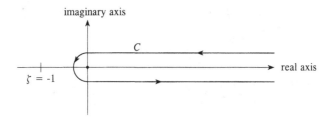

When z has a positive real part the integration contour can be collapsed down onto the real axis, where $\zeta = t$ $(0 < t < \infty)$ on the upper side of the axis and $\zeta = te^{2\pi i}$ $(0 < t < \infty)$ on the lower side. Hence, provided $z \neq 1, 2, 3, \ldots$,

$$\frac{e^{-i\pi z}}{2i \sin \pi z} \oint_C \zeta^{z-1} e^{-\zeta} \, d\zeta = \frac{e^{-i\pi z}}{2i \sin \pi z} \left\{ \int_{\infty}^{0} t^{z-1} e^{-t} \, dt + e^{2\pi i z} \int_0^{\infty} t^{z-1} e^{-t} \, dt \right\}$$

$$= \int_0^{\infty} t^{z-1} e^{-t} \, dt \equiv \Gamma(z).$$

When $z = n$ a positive integer the integral around C is zero. But also $\sin \pi z \to \sin n\pi = 0$. In this case

$$\frac{e^{-i\pi z}}{2i \sin \pi z} \oint_C \zeta^{z-1} e^{-\zeta} \, d\zeta = \frac{e^{-i\pi n}}{2i \left(\frac{d}{dz} \sin \pi z \right)_{z=n}} \left(\frac{\partial}{\partial z} \oint_C \zeta^{z-1} e^{-\zeta} \, d\zeta \right)_{z=n}$$

$$= \frac{1}{2\pi i} \oint_C \ln \zeta \, \zeta^{n-1} e^{-\zeta} \, d\zeta$$

$$= \frac{1}{2\pi i} \left(\int_{\infty}^{0} \ln t \, t^{n-1} e^{-t} dt + \int_0^{\infty} (2\pi i + \ln t) t^{n-1} e^{-t} dt \right)$$

$$= \int_0^{\infty} t^{n-1} e^{-t} dt \equiv \Gamma(n), \quad n = 1, 2, 3, \ldots.$$

5.2   The Beta Function

The Beta function $B(x, y)$ is defined by

$$B(x, y) = \int_0^1 t^{x-1}(1-t)^{y-1}\, dt, \quad x, \ y > 0 \qquad (5.2.1)$$

$$\equiv \int_0^1 (1-t)^{x-1} t^{y-1}\, dt$$

$$= B(y, x).$$

By making the change of integration variable $t = \cos^2\theta$, we also have

$$B(x, y) = 2\int_0^{\frac{\pi}{2}} \cos^{2x-1}\theta \, \sin^{2y-1}\theta \, d\theta. \qquad (5.2.2)$$

Also by setting $t = \tau/(1+\tau)$ in (5.2.1), we find

$$B(x, y) = \int_0^\infty \frac{\tau^{x-1} d\tau}{(1+\tau)^{x+y}}. \qquad (5.2.3)$$

The Beta and Gamma functions are related by

$$B(x, y) = \frac{\Gamma(x)\Gamma(y)}{\Gamma(x+y)}. \qquad (5.2.4)$$

Proof   From (5.1.4), $\Gamma(x)\Gamma(y) = 4\iint_0^\infty \xi^{2x-1}\eta^{2y-1} e^{-(\xi^2+\eta^2)} d\xi d\eta$. If we set $(\xi, \eta) = r(\cos\theta, \sin\theta)$ the integral becomes $4\int_0^\infty r^{2(x+y)-1} e^{-r^2} dr \int_0^{\frac{\pi}{2}} \cos^{2x-1}\theta \, \sin^{2y-1}\theta \, d\theta = 2\Gamma(x+y)\int_0^{\frac{\pi}{2}} \cos^{2x-1}\theta \, \sin^{2y-1}\theta \, d\theta \equiv \Gamma(x+y)B(x, y)$.

Example   Show that

$$\Gamma(x)\Gamma(1-x) = \frac{\pi}{\sin \pi x}, \quad x \neq \text{integer}.$$

Suppose first $0 < x < 1$. Then (5.2.3) and (5.2.4) (with $\Gamma(1) = 1$) and Question 10 of Problems 3F, give

$$\Gamma(x)\Gamma(1-x) = \int_0^\infty \frac{t^{x-1} dt}{1+t} = \frac{\pi}{\sin \pi x}.$$

The formula is extended to all real x by means of the recurrence relation $\Gamma(1+x) = x\Gamma(x)$.

Example Evaluate $\int_0^{\frac{\pi}{2}} \tan^{\frac{1}{3}}\theta \, d\theta$.

$$\int_0^{\frac{\pi}{2}} \tan^{\frac{1}{3}}\theta \, d\theta = \int_0^{\frac{\pi}{2}} \sin^{\frac{4}{3}-1}\theta \cos^{\frac{2}{3}-1}\theta \, d\theta$$

$$= \frac{1}{2}B\left(\frac{2}{3}, \frac{1}{3}\right) = \frac{\Gamma\left(\frac{2}{3}\right)\Gamma\left(\frac{1}{3}\right)}{2\Gamma\left(\frac{2}{3}+\frac{1}{3}\right)} = \frac{\pi}{2\sin\left(\frac{\pi}{3}\right)} = \frac{\pi}{\sqrt{3}}.$$

Problems 5A

Evaluate:

1. $\Gamma(-\frac{1}{2})$, $\Gamma(\frac{5}{2})$, $\Gamma(\frac{11}{3})/\Gamma(\frac{2}{3})$. $\quad [-2\sqrt{\pi}, \frac{3}{4}\sqrt{\pi}, \frac{80}{27}]$

2. $\int_0^\infty x^4 e^{-x^3} dx$. $\quad [\frac{2}{9}\Gamma(\frac{2}{3})]$

3. $\int_0^{\frac{\pi}{2}} \tan^{\frac{1}{2}} x \, dx$. $\quad [\frac{\pi}{\sqrt{2}}]$

4. $k^x \int_0^\infty t^{x-1} e^{-kt} dt$, $x > 0$, $k > 0$. $\quad [\Gamma(x)]$

5. $\int_{-\infty}^\infty e^{(xt-e^t)} dt$, $x > 0$. $\quad [\Gamma(x)]$

6. $\int_0^\infty \sqrt{x}\, e^{-x^3} dx$. $\quad [\frac{\sqrt{\pi}}{3}]$

7. $\int_0^\infty \sqrt{x}\, e^{-x} dx$. $\quad [\Gamma(\frac{3}{2}) = \frac{\sqrt{\pi}}{2}]$

8. $\int_0^{\frac{\pi}{2}} \cot^{\frac{1}{2}} x \, dx$. $\quad [\frac{\pi}{\sqrt{2}}]$

9. $\int_0^1 x(\ln x)^5 dx$. $\quad [-\frac{5!}{2^6}]$

10. $\int_0^{\frac{\pi}{2}} \cos^{2n+1} x \, dx$, $\int_0^{\frac{\pi}{2}} \sin^{2n+1} x \, dx$, $n =$ positive integer. [both equal to $(2^n n!)^2/(2n+1)!$]

11. $\int_0^\infty \frac{x^{\beta-1}}{1+x^\alpha} dx$, $\alpha > \beta > 0$. $\quad [\frac{\pi}{\alpha}\operatorname{cosec}\frac{\pi\beta}{\alpha}]$

12. $\int_0^1 \frac{x^4}{\sqrt{1-x^2}} dx$. $\quad [\frac{1}{2}B(\frac{5}{2}, \frac{1}{2}) = \frac{9\pi}{16}]$

13. $\int_0^\infty e^{-3x}(1-e^{-x})^5 dx$. $\quad [\Gamma(6)\Gamma(3)/\Gamma(9) = \frac{1}{168}]$

14. $\int_0^\infty x^{-n} e^{-1/x^2} dx$, $n > 1$. $\quad [\frac{1}{2}\Gamma\left(\frac{n-1}{2}\right)]$

15. If n is a positive integer, show that $\Gamma(n+\frac{1}{2}) = \frac{(2n)!}{2^{2n} n!}\sqrt{\pi}$.

16. Show that $\Gamma(x)\Gamma(-x) = -\frac{\pi}{x\sin\pi x}$.

17. Use the formula $\Gamma(x)\Gamma(x)/\Gamma(2x) = 2\int_0^{\frac{\pi}{2}} \cos^{2x-1} x \, \sin^{2x-1} x \, dx = 2^{1-2x}\int_0^{\frac{\pi}{2}} \sin^{2x-1} 2x \, dx$ to derive the *Lagrange duplication formula* $\Gamma(2x) = \frac{2^{2x-1}}{\sqrt{\pi}}\Gamma(x)\,\Gamma(x+\frac{1}{2})$.

18. Derive the following extension of Stirling's formula by retaining all of the terms involving x in the integrand of (5.1.6)

$$\Gamma(1+x) \approx \sqrt{2\pi x} x^x e^{-x} \left(1 + \frac{1}{12x}\right), \qquad x \to +\infty.$$

The error in this approximation is less than 0.008% when $x > 6$ and only about 0.1% when $x = 1$.

19. Show that the integral representation (5.1.7) of $\Gamma(z)$ satisfies the recurrence relation $\Gamma(1+z) = z\Gamma(z)$.

20. Show that if n is a positive integer, $\Gamma(z)$ has a simple pole at $z = -n$ with residue $(-1)^n/n!$.

$$\left[\begin{array}{l} \text{Set } z = -n + \xi, \text{ then using (5.1.7)} \\[2mm] \Gamma(z) \equiv \Gamma(-n+\xi) = \dfrac{(-1)^n e^{i\pi\xi}}{2i \sin(-n\pi + \xi\pi)} \displaystyle\oint_C \zeta^{\xi-(n+1)} e^{-\zeta}\, d\zeta \\[4mm] \approx \dfrac{1}{2\pi i\xi} \displaystyle\oint_C \zeta^{-(n+1)} e^{-\zeta}\, d\zeta, \quad \text{as } \xi \to 0 \\[4mm] = \dfrac{1}{2\pi i\xi} \displaystyle\oint_C \left(\dfrac{1}{\zeta^{n+1}} - \dfrac{1}{\zeta^n} + \dfrac{1}{2!\zeta^{n-1}} - \cdots + \dfrac{(-1)^n}{n!\zeta} + \cdots\right) d\zeta = \dfrac{(-1)^n}{n!\xi} \end{array}\right]$$

21. Show that $\oint_{|z|=\frac{1}{2}} \Gamma(z) \cos z\, dz = 2\pi i$, where the contour is traversed in the anticlockwise direction.

22. Show that $\frac{1}{2\pi i} \oint_{|z+n|=\frac{1}{2}} \Gamma(z) \cos \pi z\, dz = \frac{1}{n!}$, where n is a positive integer, and the contour is traversed in the anticlockwise direction.

23. Use Stirling's formula to show that

$$\lim_{n\to\infty} \frac{\Gamma(1+z+n)}{n!n^z} = 1, \qquad \text{for fixed } z \neq 0, -1, -2, \ldots.$$

Hence deduce Gauss' formula

$$\Gamma(z) = \lim_{n\to\infty} \frac{n!n^z}{z(z+1)(z+2)\cdots(z+n)}.$$

24. Show that

$$B(x,y)B(x+y,z) = B(y,z)B(y+z,x)$$

$$B(x,x) = 2^{1-2x} B\left(\frac{1}{2}, x\right)$$

$$B(x,x)B\left(x+\frac{1}{2}, x+\frac{1}{2}\right) = \frac{\pi}{2^{4x-1}x}.$$

25. Verify that the volume of the n-dimensional sphere of unit radius is

$$V = \int \int \cdots \int_{\sqrt{x_1^2 + x_2^2 + \cdots + x_n^2} < 1} dx_1 dx_2 \ldots dx_n = \frac{\pi^{\frac{n}{2}}}{\Gamma\left(\frac{n}{2} + 1\right)}.$$

Set $x_1 = \cos\varphi_1$, $x_2 = \sin\varphi_1 \cos\varphi_2$, $x_3 = \sin\varphi_1 \sin\varphi_2 \cos\varphi_3$,

$x_n = \sin\varphi_1 \sin\varphi_2 \ldots \sin\varphi_{n-1} \cos\varphi_n$, $0 < \varphi_i < \pi$, $i = 1, 2, \ldots, n$,

and show that

$$dx_1 dx_2 \ldots dx_n = \sin^n \varphi_1 \sin^{n-1} \varphi_2 \ldots \sin\varphi_n d\varphi_1 d\varphi_2 \ldots d\varphi_n \quad (\text{see } \S6.7),$$

and that $V = 2^n \displaystyle\int_0^{\frac{\pi}{2}} \sin^n \varphi_1 d\varphi_1 \int_0^{\frac{\pi}{2}} \sin^{n-1} \varphi_2 d\varphi_2 \ldots \int_0^{\frac{\pi}{2}} \sin\varphi_n d\varphi_n$, etc.

5.3 Legendre Polynomials

The Legendre polynomial of degree $n \geq 0$ is defined by the *Rodrigues formula*

$$P_n(x) = \frac{1}{2^n n!} \frac{d^n}{dx^n} (x^2 - 1)^n, \quad P_0(x) = 1. \qquad (5.3.1)$$

Thus:

$$P_1(x) = x, \quad P_2(x) = \frac{3}{2}x^2 - \frac{1}{2}, \quad P_3(x) = \frac{5}{2}x^3 - \frac{3}{2}x,$$

$$P_4(x) = \frac{35}{8}x^4 - \frac{15}{4}x^2 + \frac{3}{8}, \quad \text{etc.}$$

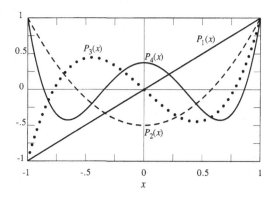

$P_n(x)$ is an even or odd function of x according as n is even or odd. Also, $(x^2 - 1)^n = (x-1)^n(x+1)^n \approx 2^n(x-1)^n$ when x is close to $x = 1$, so that

$$P_n(1) = \left(\frac{1}{2^n n!} \frac{d^n}{dx^n} \left(2^n (x-1)^n \right) \right)_{x=1} \equiv 1.$$

Similarly, $P_n(-1) = (-1)^n$.

In applications we are usually interested in the behaviour of the Legendre polynomials within the interval $-1 < x < 1$. The equation $P_n(x) = 0$ has n real roots, *all of which satisfy* $|x| < 1$. This is obviously true for the cases illustrated in the figure. The following argument shows that it is true also for larger values of n. When $|x| < 1$ the function $(x^2 - 1)^n$ has a single maximum or minimum at $x = 0$, and all of its derivatives of order up to and including the $(n-1)$st vanish at $x = \pm 1$. Therefore, successive derivatives $\frac{d^m}{dx^m}(x^2 - 1)^n$, $0 < m < n$ can only vanish in $|x| \leq 1$, and the nth derivative only in $|x| < 1$.

Legendre's equation Let $y = (x^2 - 1)^n$. Then $dy/dx = 2nx(x^2 - 1)^{n-1}$, and therefore

$$(x^2 - 1)\frac{dy}{dx} - 2nxy = 0.$$

When this is differentiated $n + 1$ times we find

$$(x^2 - 1)\frac{d^{n+2}y}{dx^{n+2}} + 2x\frac{d^{n+1}y}{dx^{n+1}} - 2(1 + 2 + 3 + \cdots + n)\frac{d^n y}{dx^n} = 0.$$

Now $1+2+3+\cdots+n = \frac{1}{2}n(n+1)$, and according to (5.3.1) $d^n y/dx^n = 2^n n! P_n(x)$. Thus the polynomial $P_n(x)$ satisfies *Legendre's equation* of order n:

$$(1 - x^2)\frac{d^2 P_n}{dx^2} - 2x\frac{dP_n}{dx} + n(n+1)P_n = 0. \qquad (5.3.2)$$

$P_n(x)$ is a bounded solution of this equation for finite values of x. A second independent solution derived by the method of §1.4 is found to be unbounded at $x = \pm 1$.

By writing Legendre's equation in the *Sturm–Liouville* form (1.9.1)

$$\frac{d}{dx}\left((1-x^2)\frac{dP_n}{dx}\right) + n(n+1)P_n = 0, \quad -1 < x < 1,$$

(where $p(x) \equiv 1 - x^2$ vanishes at $x = \pm 1$), it follows that

$$\int_{-1}^{1} P_n(x)P_m(x)dx = 0, \quad n \neq m.$$

To deal with the case $n = m$ we first use the *Rodrigues* formula (5.3.1) in the more general integral

$$\int_{-1}^{1} f(x)P_n(x)dx \equiv \frac{1}{2^n n!} \int_{-1}^{1} f(x)\frac{d^n}{dx^n}(x^2-1)^n dx.$$

Integrating by parts

$$\int_{-1}^{1} f(x)P_n(x)dx = \frac{1}{2^n n!}\left[f(x)\frac{d^{n-1}}{dx^{n-1}}(x^2-1)^n\right]_{-1}^{1}$$

$$- \frac{1}{2^n n!}\int_{-1}^{1}\frac{df}{dx}(x)\frac{d^{n-1}}{dx^{n-1}}(x^2-1)^n dx.$$

The first term on the right vanishes identically provided $f(x)$ is finite at $x = \pm 1$. Repeating this process we find

$$\int_{-1}^{1} f(x)P_n(x)dx = \frac{(-1)^n}{2^n n!}\int_{-1}^{1}(x^2-1)^n\frac{d^n f}{dx^n}(x)dx. \qquad (5.3.3)$$

If $f(x) = P_n(x)$ then

$$\frac{d^n f}{dx^n}(x) = \frac{1}{2^n n!}\frac{d^{2n}}{dx^{2n}}(x^2-1)^n \equiv \frac{(2n)!}{2^n n!},$$

so that

$$\int_{-1}^{1} \{P_n(x)\}^2 \, dx = \frac{(2n)!}{2^{2n}(n!)^2} \int_{-1}^{1} (1-x^2)^n dx$$

$$= \frac{(2n)!}{2^{2n}(n!)^2} \int_{0}^{\frac{\pi}{2}} \cos^{2n+1}\theta \, d\theta$$

$$= \frac{(2n)!}{2^{2n}(n!)^2} \frac{\Gamma\left(\frac{1}{2}\right)\Gamma(n+1)}{\Gamma\left(n+\frac{3}{2}\right)} = \frac{2}{2n+1}.$$

Hence we obtain the orthogonality relation for Legendre polynomials:

$$\int_{-1}^{1} P_n(x)P_m(x)dx = \begin{cases} \dfrac{2}{2n+1} & n = m, \\ 0 & n \neq m. \end{cases} \qquad (5.3.4)$$

Example Show that $x^4 = \frac{7}{35}P_0(x) + \frac{4}{7}P_2(x) + \frac{8}{35}P_4(x)$.

Because x^4 is an even function it can depend only on P_0, P_2, P_4. Now the definition equations

$$1 = P_0,$$

$$\frac{3}{2}x^2 - \frac{1}{2} = P_2,$$

$$\frac{35}{8}x^4 - \frac{15}{4}x^2 + \frac{3}{8} = P_4,$$

may be regarded as a system of linear equations for $1 = x^0$, x^2, x^4. The required expansion is therefore obtained by solving the equations for x^4.

Example Show that $P_{2n}(0) = \frac{(-1)^n(2n)!}{2^{2n}(n!)^2}$, $n = 0, 1, 2, \ldots$.

The coefficient of x^{2n} in the expansion of $(x^2-1)^{2n} = (1-x^2)^{2n}$ is $\frac{(-1)^n(2n)!}{(n!)^2}$.

$$\therefore \quad P_{2n}(0) = \frac{1}{2^{2n}(2n)!} \frac{d^{2n}}{dx^{2n}} \left((x^2-1)^{2n}\right)_{x=0}$$

$$= \frac{1}{2^{2n}(2n)!} \frac{d^{2n}}{dx^{2n}} \left(\frac{(-1)^n(2n)!x^{2n}}{(n!)^2}\right) = \frac{(-1)^n(2n)!}{2^{2n}(n!)^2}. \qquad (5.3.5)$$

Example   Show that $\int_{-1}^{1} x^m P_n(x)dx = 0$ if $m < n$.

Using (5.3.3): $\int_{-1}^{1} x^m P_n(x)dx = \frac{(-1)^n}{2^n n!} \int_{-1}^{1} (x^2-1)^n \frac{d^n}{dx^n} x^m \, dx \equiv 0$.

Example   Show that $\int_{-1}^{1} x^{n+2m} P_n(x)dx = \frac{(n+2m)! \Gamma(m+\frac{1}{2})}{2^n (2m)! \Gamma(n+m+\frac{3}{2})}$, $m = 0, 1, 2, \ldots$.

Using (5.3.3):

$$\int_{-1}^{1} x^{n+2m} P_n(x)dx = \frac{(-1)^n}{2^n n!} \int_{-1}^{1} (x^2-1)^n \frac{d^n}{dx^n} x^{n+2m} \, dx$$

$$= \frac{(n+2m)!}{2^{n-1} n!(2m)!} \int_{0}^{\frac{\pi}{2}} \sin^{2n+1} \theta \cos^{2m} \theta \, d\theta$$

$$= \frac{(n+2m)!}{2^n n!(2m)!} B\left(n+1, m+\frac{1}{2}\right) \equiv \frac{(n+2m)! \Gamma(m+\frac{1}{2})}{2^n (2m)! \Gamma(n+m+\frac{3}{2})}.$$

Series expansion   The orthogonality relation (5.3.4) permits an arbitrary function $f(x)$ to be expanded in the form

$$f(x) = \sum_{n=0}^{\infty} A_n P_n(x), \quad |x| < 1,$$

$$(5.3.6)$$

$$\text{where} \quad A_n = \frac{\int_{-1}^{1} f(x) P_n(x)dx}{\int_{-1}^{1} \{P_n(x)\}^2 dx} \equiv \left(n+\frac{1}{2}\right) \int_{-1}^{1} f(x) P_n(x)dx.$$

If we set $x = \cos\theta$, $0 < \theta < \pi$, these formulae become (with $f(x)$ replaced by an arbitrary function $f(\theta)$ of θ)

$$\left. \begin{array}{l} \displaystyle\int_{0}^{\pi} P_n(\cos\theta) P_m(\cos\theta) \sin\theta \, d\theta = \frac{2}{2n+1} \ (n=m), \quad 0 \ (n \neq m) \\[4mm] \displaystyle f(\theta) = \sum_{n=0}^{\infty} A_n P_n(\cos\theta), \quad 0 < \theta < \pi, \\[4mm] \displaystyle A_n = \left(n+\frac{1}{2}\right) \int_{0}^{\pi} f(\theta) P_n(\cos\theta) \sin\theta \, d\theta. \end{array} \right\}$$

$$(5.3.7)$$

Application to the axisymmetric Laplace equation When Laplace's equation is expressed in terms of spherical polar coordinates (r, θ, ϕ), as in (2.6.6), a solution $\varphi \equiv \varphi(r, \theta)$ that does not depend on the azimuthal angle ϕ is said to be axisymmetric (see the figure accompanying (2.6.6)). In this case φ is a solution of

$$\frac{\partial}{\partial r}\left(r^2 \frac{\partial \varphi}{\partial r}\right) + \frac{1}{\sin \theta} \frac{\partial}{\partial \theta}\left(\sin \theta \frac{\partial \varphi}{\partial \theta}\right) = 0, \quad 0 \leq \theta \leq \pi. \qquad (5.3.8)$$

A separable solution

$$\varphi = R(r)\Theta(\theta),$$

must satisfy

$$\frac{1}{R}\frac{d}{dr}\left(r^2 \frac{dR}{dr}\right) = -\frac{1}{\sin \theta \, \Theta}\frac{d}{d\theta}\left(\sin \theta \frac{d\Theta}{d\theta}\right) = \lambda,$$

where λ is the separation constant. Hence, R, Θ are determined by

$$r^2 \frac{d^2 R}{dr^2} + 2r \frac{dR}{dr} - \lambda R = 0,$$

$$\frac{1}{\sin \theta}\frac{d}{d\theta}\left(\sin \theta \frac{d\Theta}{d\theta}\right) + \lambda \Theta = 0, \quad 0 < \theta < \pi. \qquad (5.3.9)$$

On setting $x = \cos \theta$ in the second equation, we find

$$\frac{d}{dx}\left((1 - x^2)\frac{d\Theta}{dx}\right) + \lambda \Theta = 0, \quad -1 < x < 1.$$

We already know that this has the bounded solution $P_n(x)$ provided $\lambda = n(n + 1)$ for positive integral n. Furthermore, this is likely to be a sensible choice for λ because an arbitrary function defined in $|x| < 1$ can be expanded as an infinite series of Legendre polynomials. In this case the first of equations (5.3.9) gives $R = A_n r^n + B_n/r^{n+1}$, so that the general solution of (5.3.8) may be taken in the form

$$\varphi(r, \theta) = \sum_{n=0}^{\infty}\left\{A_n r^n + \frac{B_n}{r^{n+1}}\right\} P_n(\cos \theta), \quad 0 < \theta < \pi. \qquad (5.3.10)$$

We must take $A_n = 0$, $n > 0$ if the solution is required to be bounded as $r \to \infty$; for a bounded solution in a region including $r = 0$ we must take $B_n = 0$, $n \geq 0$.

Example Find the solution of the axisymmetric Laplace equation (5.3.8) in the region $r > a$ that vanishes as $r \to \infty$, and satisfies $\varphi = f(\theta)$ on $r = a$.

The expansion (5.3.10) $\to 0$ as $r \to \infty$ provided $A_n = 0$. Then

$$\varphi(r, \theta) = \sum_{n=0}^{\infty} \frac{B_n}{r^{n+1}} P_n(\cos\theta), \quad 0 < \theta < \pi.$$

The boundary condition on $r = a$ yields

$$\sum_{n=0}^{\infty} \frac{B_n}{a^{n+1}} P_n(\cos\theta) = f(\theta), \quad 0 < \theta < \pi,$$

from which the coefficients B_n are determined using the relations (5.3.7):

$$B_n = a^{n+1} \left(n + \frac{1}{2}\right) \int_0^\pi f(\theta) P_n(\cos\theta) \sin\theta \, d\theta.$$

Example Find the velocity potential of uniform flow at speed U in the z-direction past a rigid sphere of radius a whose centre is at the coordinate origin.

The potential satisfies the axisymmetric equation (5.3.8) and $\nabla\varphi \to (0, 0, U)$ as $r \to \infty$. Because $z = r\cos\theta \equiv rP_1(\cos\theta)$ we can put

$$\varphi = UrP_1(\cos\theta) + \sum_{n=0}^{\infty} \frac{B_n}{r^{n+1}} P_n(\cos\theta).$$

The normal component of velocity $\partial\varphi/\partial r = 0$ on the surface $r = a$ of the sphere

$$\therefore \quad \sum_{n=0}^{\infty} \frac{(n+1)B_n}{a^{n+2}} P_n(\cos\theta) = UP_1(\cos\theta), \quad 0 < \theta < \pi.$$

This shows that all of the $B_n = 0$ except for $B_1 = Ua^3/2$, i.e.

$$\varphi = Ur\cos\theta \left(1 + \frac{a^3}{2r^3}\right).$$

The generating function for Legendre polynomials The function $\varphi = 1/r$ is a solution of Laplace's equation representing a 'point source' at $r = 0$ (§4.7). When the source is placed at $(0, 0, h)$ on the

z-axis, the solution obviously becomes

$$\varphi = \frac{1}{\sqrt{x^2 + y^2 + (z - h)^2}} \equiv \frac{1}{\sqrt{r^2 - 2rh \cos\theta + h^2}}, \quad z = r \cos\theta,$$

which satisfies the axisymmetric form (5.3.8) of Laplace's equation.

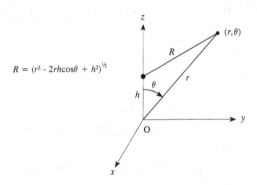

We can expand φ in ascending or descending powers of r/h depending on whether $r < h$ or $r > h$. Let the expansions be written

$$\frac{1}{\sqrt{r^2 - 2rh \cos\theta + h^2}} = \begin{cases} \dfrac{1}{h} \displaystyle\sum_{n=0}^{\infty} \left(\dfrac{r}{h}\right)^n \mathcal{P}_n(\cos\theta) \ r < h, \\[4mm] \dfrac{1}{r} \displaystyle\sum_{n=0}^{\infty} \left(\dfrac{h}{r}\right)^n \mathcal{P}_n(\cos\theta) \ h < r, \end{cases} \tag{5.3.11}$$

where symmetry requires the expansion coefficients $\mathcal{P}_n(\cos\theta)$ to be the same in both series. Substitution into equation (5.3.8) shows that each term in these series must separately satisfy Laplace's equation. Therefore, $\mathcal{P}_n(\cos\theta) = \alpha_n P_n(\cos\theta)$, where α_n is a constant.

Now, when $\theta = 0$ the expansions become

$$\frac{1}{|r - h|} = \begin{cases} \dfrac{1}{h} \displaystyle\sum_{n=0}^{\infty} \left(\dfrac{r}{h}\right)^n \mathcal{P}_n(1) \quad r < h, \\[4mm] \dfrac{1}{r} \displaystyle\sum_{n=0}^{\infty} \left(\dfrac{h}{r}\right)^n \mathcal{P}_n(1) \quad h < r, \end{cases}$$

which can only be true if $\mathcal{P}_n(1) = 1$. Thus, because $P_n(1) = 1$ we must have $\alpha_n = 1$, i.e. $\mathcal{P}_n(x) \equiv P_n(x)$.

By setting $t = r/h$ and $\mu = \cos\theta$ in the first expansion of (5.3.11), we see that we have shown that

$$\frac{1}{\sqrt{t^2 - 2t\mu + 1}} = \sum_{n=0}^{\infty} t^n P_n(\mu), \quad |t| < 1. \qquad (5.3.12)$$

The term on the left of this equation is called the *generating function* of the Legendre polynomials.

Recurrence relations The following important relations can be deduced from (5.3.12)

$$(n+1)P_{n+1}(\mu) - (2n+1)\mu P_n(\mu) + n P_{n-1}(\mu) = 0, \qquad (5.3.13)$$

$$\frac{d}{d\mu}\{P_{n+1}(\mu) - P_{n-1}(\mu)\} = (2n+1)P_n(\mu). \qquad (5.3.14)$$

Proof Take the logarithm of both sides of (5.3.12) and differentiate with respect to t:

$$\frac{\mu - t}{t^2 - 2t\mu + 1} = \frac{\sum_{n=0}^{\infty} n t^{n-1} P_n(\mu)}{\sum_{n=0}^{\infty} t^n P_n(\mu)}.$$

By cross multiplication, this becomes

$$(\mu - t)\sum_{n=0}^{\infty} t^n P_n(\mu) = (t^2 - 2t\mu + 1)\sum_{n=0}^{\infty} n t^{n-1} P_n(\mu).$$

The coefficients of each power of t must be the same on both sides, so that

$$\mu P_n(\mu) - P_{n-1}(\mu) = (n-1)P_{n-1}(\mu) - 2\mu n P_n(\mu) + (n+1)P_{n+1}(\mu),$$

which gives (5.3.13).

Next, take the logarithmic derivative with respect to μ of both sides of (5.3.12):

$$\frac{t}{t^2 - 2t\mu + 1} = \frac{\sum_{n=0}^{\infty} t^n P_n'(\mu)}{\sum_{n=0}^{\infty} t^n P_n(\mu)}, \quad \text{where} \quad P_n'(\mu) = \frac{dP_n(\mu)}{d\mu},$$

and proceed as before to obtain

$$P_n(\mu) - P_{n-1}'(\mu) + 2\mu P_n'(\mu) - P_{n+1}'(\mu) = 0.$$

The relation (5.3.14) is obtained by eliminating $P_n'(\mu)$ between this and the equation obtained by differentiating (5.3.13) with respect to μ.

Example Show that

$$\text{sgn}(x) = \sum_{n=0}^{\infty} \frac{(-1)^n (2n)!(4n+3)}{2^{2n+1} n!(n+1)!} P_{2n+1}(x), \quad |x| < 1.$$

We use the expansion (5.3.6), where

$$A_n = \left(n + \frac{1}{2}\right) \int_{-1}^{1} \text{sgn}(x) P_n(x) dx.$$

This vanishes for even values of n. When n is odd equation (5.3.14) gives:

$$A_n = (2n+1) \int_{0}^{1} P_n(x) dx$$

$$= \int_{0}^{1} \frac{d}{dx} \{P_{n+1}(x) - P_{n-1}(x)\} \, dx$$

$$= [P_{n+1}(x) - P_{n-1}(x)]_0^1 \equiv P_{n-1}(0) - P_{n+1}(0),$$

because when $n \pm 1$ is even $P_{n-1}(1) = P_{n+1}(1) = 1$. The final result is now obtained by using formula (5.3.5).

Problems 5B

Verify the following results:

1. $\int_0^{\pi} \cos\theta \sin\theta P_n(\cos\theta)d\theta = \frac{2}{3}, 0$ according as $n = 1$, $n \neq 1$.

2. $\int_0^{\pi} \cos^m\theta \sin\theta P_n(\cos\theta)d\theta = 0$ when $m < n$.

3. $\int_0^{\pi} \cos^{n+2m}\theta \sin\theta P_n(\cos\theta)d\theta = \frac{(n+2m)!\Gamma(m+\frac{1}{2})}{2^n(2m)!\Gamma(n+m+\frac{3}{2})}$, $m = 0, 1, 2, \ldots$.

4. $\int_{-1}^{1} x P_n'(x) P_n(x)dx = \frac{2n}{2n+1}$, $n = 0, 1, 2, \ldots$.

5. $\int_{-1}^{1} \frac{P_n(x)dx}{\sqrt{1-2xt+t^2}} = \frac{2t^n}{2n+1}$, $n = 0, 1, 2, \ldots$.

6. $\int_0^1 x P_4(x)dx = \frac{-1}{48}$.

7. $\int_0^1 P_{2n}(x)dx = 0$, $n = 0, 1, 2, \ldots$.

8. $x^2 = \frac{1}{3} P_0(x) + \frac{2}{3} P_2(x)$.

9. $x^3 = \frac{3}{5} P_1(x) + \frac{2}{5} P_3(x)$.

10. $x^4 = \frac{7}{35} P_0(x) + \frac{4}{7} P_2(x) + \frac{8}{35} P_4(x)$.

11. $x^{2m} = (2m)! \sum_{n=0}^{m} \frac{2^{2n}(4n+1)(m+n)! P_{2n}(x)}{(2m+2n+1)!(m-n)!}$.

12. $x^{2m+1} = (2m+1)! \sum_{n=0}^{m} \frac{2^{2n+1}(4n+3)(m+n+1)! P_{2n+1}(x)}{(2m+2n+3)!(m-n)!}$.

13. $|x| = \frac{1}{2} + \sum_{n=1}^{\infty} \frac{(-1)^{n-1}(4n+1)(2n-2)! P_{2n}(x)}{2^{2n}(n+1)!(n-1)!}$, $|x| < 1$.

14. $\int_{-1}^{1} \frac{P_n(x)\,dx}{\sqrt{1-x}} = \frac{2\sqrt{2}}{2n+1}, \quad n = 0, 1, 2, \ldots.$

If
$$\frac{\partial}{\partial r}\left(r^2 \frac{\partial \varphi}{\partial r}\right) + \frac{1}{\sin\theta} \frac{\partial}{\partial \theta}\left(\sin\theta \frac{\partial \varphi}{\partial \theta}\right) = 0, \quad 0 \le \theta \le \pi, \ r < a,$$

show that:

15. $\varphi(r, \theta) = \frac{1}{3} + \frac{2}{3}\left(\frac{r}{a}\right)^2 P_2(\cos\theta)$ when $\varphi(a, \theta) = \cos^2\theta$.

16. $\varphi(r, \theta) = -\frac{1}{3} + \frac{4}{3}\left(\frac{r}{a}\right)^2 P_2(\cos\theta)$ when $\varphi(a, \theta) = \cos 2\theta$.

17. $\varphi(r, \theta) = \frac{7}{35} + \frac{4}{7}\left(\frac{r}{a}\right)^2 P_2(\cos\theta) + \frac{8}{35}\left(\frac{r}{a}\right)^4 P_4(\cos\theta)$ when $\varphi(a, \theta) = \cos^4\theta$.

18. If
$$\frac{\partial}{\partial r}\left(r^2 \frac{\partial \varphi}{\partial r}\right) + \frac{1}{\sin\theta} \frac{\partial}{\partial \theta}\left(\sin\theta \frac{\partial \varphi}{\partial \theta}\right) = 0, \quad 0 \le \theta \le \frac{\pi}{2}, \ r < a,$$
and $\varphi(a, \theta) = \varphi_0$, $0 < \theta \le \frac{\pi}{2}$, $\varphi(r, \pi/2) = 0$, $0 \le r < a$, show that

$$\varphi = \varphi_0 \sum_{n=0}^{\infty} (-1)^n \left(\frac{4n+3}{n+1}\right) \frac{(2n)!}{2^{2n+1}(n!)^2} \left(\frac{r}{a}\right)^{2n+1} P_{2n+1}(\cos\theta).$$

19. If $\frac{\partial}{\partial r}\left(r^2 \frac{\partial \varphi}{\partial r}\right) + \frac{1}{\sin\theta} \frac{\partial}{\partial \theta}\left(\sin\theta \frac{\partial \varphi}{\partial \theta}\right) = 0, \quad 0 \le \theta \le \pi, \ r > a,$
and $\varphi \to 0$ as $r \to \infty$,

show that when $\varphi(a, \theta) = 1 - \cos^2\theta$: $\varphi = \frac{2}{3}\frac{a}{r}\left\{1 - \left(\frac{a}{r}\right)^2 P_2(\cos\theta)\right\}, \quad r > a.$

20. If $\frac{\partial}{\partial r}\left(r^2 \frac{\partial \varphi}{\partial r}\right) + \frac{1}{\sin\theta} \frac{\partial}{\partial \theta}\left(\sin\theta \frac{\partial \varphi}{\partial \theta}\right) = 0, \quad 0 \le \theta \le \pi, \ r < a,$
show that when $\varphi(a, \theta) = H(\theta - \Theta)$, where $0 < \Theta < \pi$:

$$\varphi = \frac{1}{2}\left\{1 - \cos\Theta - \sum_{n=1}^{\infty} [P_{n+1}(\cos\Theta) - P_{n-1}(\cos\Theta)]\left(\frac{r}{a}\right)^n P_n(\cos\theta)\right\}, \quad r > a.$$

21. If $\frac{\partial}{\partial r}\left(r^2 \frac{\partial \varphi}{\partial r}\right) + \frac{1}{\sin\theta} \frac{\partial}{\partial \theta}\left(\sin\theta \frac{\partial \varphi}{\partial \theta}\right) = 0, \quad 0 \le \theta \le \pi, \ r < a,$
show that when $\varphi(a, \theta) = \alpha H(\theta - \frac{\pi}{2}) + \beta H(\frac{\pi}{2} - \theta)$, where α and β are constants:

$$\varphi = \frac{1}{2}(\alpha + \beta) + \frac{3}{4}(\alpha - \beta)\left(\frac{r}{a}\right)P_1(\cos\theta) + \frac{7}{16}(\alpha - \beta)\left(\frac{r}{a}\right)^3 P_3(\cos\theta) + \cdots$$

22. Find the solution of Laplace's equation $\nabla^2\varphi = 0$ inside the unit sphere $r = 1$ when $\varphi = 1 + \cos\theta - 3\cos^2\theta$ on the surface. $[\varphi = rP_1(\cos\theta) - 2r^2 P_2(\cos\theta)]$

23. Show that when $\varphi = \sin^4\theta$ on the surface of the unit sphere and $\nabla^2\varphi = 0$ inside the sphere, then

$$\varphi = \frac{8}{15} + \frac{8}{21}r^2\left(1 - 3\cos^2\theta\right) + \frac{1}{35}r^4\left(3 - 30\cos^2\theta + 35\cos^4\theta\right), \quad r < 1.$$

5.4   The Error Function erf(x)

This is defined by

$$\text{erf}(x) = \frac{2}{\sqrt{\pi}} \int_0^x e^{-t^2} dt.$$

The principal properties of erf(x) for real values of x are evident from the figure:

$$\text{erf}(-x) = -\text{erf}(x), \quad \text{erf}(x) \to \pm 1 \text{ as } x \to \pm\infty.$$

x	erf(x)
0.5	0.5205
1.0	0.8427
1.5	0.9661
2.0	0.9953

The *complementary error function* is defined by

$$\text{erfc}(x) = 1 - \text{erf}(x) = \frac{2}{\sqrt{\pi}} \int_x^\infty e^{-t^2} dt.$$

Example   Show that $\text{erf}(x) \approx 1 - e^{-x^2}/\sqrt{\pi}x$ as $x \to +\infty$.
   By integration by parts:

$$\text{erf}(x) = 1 - \frac{2}{\sqrt{\pi}} \int_x^\infty e^{-t^2} dt$$

$$= 1 - \frac{2}{\sqrt{\pi}} \int_x^\infty \left(\frac{1}{2t}\right)\left(2te^{-t^2}\right) dt$$

$$= 1 - \frac{2}{\sqrt{\pi}} \left\{ \left[-\frac{1}{2t}e^{-t^2}\right]_x^\infty - \int_x^\infty \frac{1}{2t^2}e^{-t^2} dt \right\}$$

$$= 1 - \frac{e^{-x^2}}{\sqrt{\pi}x} + \frac{2}{\sqrt{\pi}} \int_x^\infty \left(\frac{1}{4t^3}\right)\left(2te^{-t^2}\right) dt.$$

The final integral may be neglected as $x \to +\infty$ because it gives a contribution of order e^{-x^2}/x^3.

Example Show that $\int \text{erf}(x)dx = x\text{erf}(x) + e^{-x^2}/\sqrt{\pi} + \text{constant}$.

$$\int \text{erf}(x)dx = x\text{erf}(x) - \int x\frac{d}{dx}\text{erf}(x)\,dx + C$$

$$= x\text{erf}(x) - \frac{2}{\sqrt{\pi}}\int xe^{-x^2}\,dx + C = x\text{erf}(x) + \frac{1}{\sqrt{\pi}}e^{-x^2} + C.$$

Example Find the solution $u(x,t)$ of the initial value problem

$$\frac{\partial^2 u}{\partial x^2} = \frac{1}{\kappa}\frac{\partial u}{\partial t}, \quad \text{for } t > 0, \quad -\infty < x < \infty, \quad \text{given that } u = u_0 H(1 - |x|) \text{ at } t = 0,$$

where $u_0 = \text{constant}$.

Using the general solution of §4.10, Example 1:

$$u = \frac{u_0}{2\sqrt{\pi\kappa t}}\int_{-1}^{1} e^{-(x-\xi)^2/4\kappa t}d\xi = \frac{u_0}{\sqrt{\pi}}\int_{-(1+x)/2\sqrt{\kappa t}}^{(1-x)/2\sqrt{\kappa t}} e^{-\mu^2}d\mu \quad (\mu = (\xi - x)/2\sqrt{\kappa t})$$

$$= \frac{u_0}{2}\left\{\text{erf}\left(\frac{1-x}{2\sqrt{\kappa t}}\right) + \text{erf}\left(\frac{1+x}{2\sqrt{\kappa t}}\right)\right\}, \quad t > 0.$$

Fresnel integrals These are special cases of the error function of complex argument, defined by

$$C(x) = \int_0^x \cos\left(\frac{\pi t^2}{2}\right)dt, \quad S(x) = \int_0^x \sin\left(\frac{\pi t^2}{2}\right)dt.$$

The reader can verify that $C(x)$, $S(x) \to \frac{1}{2}$ as $x \to +\infty$, and that

$$\frac{1}{\sqrt{2i}}\text{erf}\left(x\sqrt{\frac{\pi i}{2}}\right) = C(x) - iS(x).$$

$C(x), S(x)$ are odd functions whose behaviors for $x > 0$ are shown in the figure.

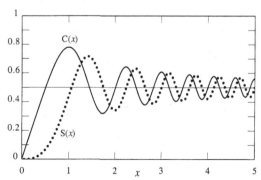

Problems 5C

1. Evaluate $\mathrm{erfc}(0)$, $\mathrm{erfc}(-\infty)$. [1, 2]

Verify the following results:

2. $\int_{-x}^{x} e^{-t^2} dt = \sqrt{\pi}\,\mathrm{erf}(x)$.

3. $\int_0^\infty e^{-(\alpha t^2 + 2\beta t + \gamma)} dt = \frac{1}{2}\sqrt{\frac{\pi}{\alpha}}e^{(\beta^2 - \alpha\gamma)/\alpha}\,\mathrm{erfc}\left(\frac{\beta}{\sqrt{\alpha}}\right)$, $\alpha > 0$.

4. $\int_0^\infty \frac{e^{-\alpha t}\,dt}{\sqrt{t+\beta}} = \sqrt{\frac{\pi}{\alpha}}e^{\alpha\beta}\,\mathrm{erfc}\left(\sqrt{\alpha\beta}\right)$, $\alpha, \beta > 0$.

5. $\int_x^\infty \frac{e^{-t^2}\,dt}{t^2} = e^{-x^2}/x - \sqrt{\pi}\,\mathrm{erfc}(x)$.

6. $\int_0^\infty e^{-\alpha t - t^2/4}\,dt = \sqrt{\pi}\,e^{\alpha^2}\,\mathrm{erfc}(\alpha)$.

7. $\int_0^\infty e^{-\alpha t}\mathrm{erf}(t)\,dt = \frac{1}{\alpha}e^{\alpha^2/4}\mathrm{erfc}\left(\frac{\alpha}{2}\right)$, $\alpha > 0$.

8. $\int_0^x \mathrm{erf}(t)\,dt = x\,\mathrm{erf}(x) + (e^{-x^2} - 1)/\sqrt{\pi}$.

9. $\int_0^x e^t\,\mathrm{erf}(t)\,dt = e^x\,\mathrm{erf}(x) - e^{\frac{1}{4}}\left\{\mathrm{erf}\left(x - \frac{1}{2}\right) + \mathrm{erf}\left(\frac{1}{2}\right)\right\}$.

10. $\int_0^\infty \cos t^2\,dt = \int_0^\infty \sin t^2\,dt = \frac{1}{2}\sqrt{\frac{\pi}{2}}$.

11. $\iint_0^\infty \cos(x^2 + y^2)\,dx\,dy = 0$.

12. $\iint_0^\infty \sin(x^2 + y^2)\,dx\,dy = \frac{\pi}{4}$.

13. $\int_0^\infty \frac{\sin kt}{t}e^{-\alpha^2 t^2}\,dt = \frac{\pi}{2}\mathrm{erf}\left(\frac{k}{2|\alpha|}\right)$.

14. Show that the solution $u(x,t)$ of the initial value problem

$$\frac{\partial^2 u}{\partial x^2} = \frac{1}{\kappa}\frac{\partial u}{\partial t}, \quad \text{for } t > 0, \quad -\infty < x < \infty, \quad u(x,0) = u_0 H(\ell - |x|),$$

where $u_0 = $ constant, can be expressed in the form

$$u = \frac{u_0}{2}\left\{\mathrm{erf}\left(\frac{\ell - x}{2\sqrt{\kappa t}}\right) + \mathrm{erf}\left(\frac{\ell + x}{2\sqrt{\kappa t}}\right)\right\}, \quad t > 0.$$

15. Using the solution of Problem 14, show that for a fixed value of $x > \ell > 0$, $u(x,t)$ attains its maximum value at $t = \ell x/\left[\kappa \ln\left(\frac{x+\ell}{x-\ell}\right)\right]$. Taking $\ell = 1$, $\kappa = 1$, sketch the graph of $u(x,t)$ as a function of t for $x = 0$ and $x = \frac{3}{2}$.

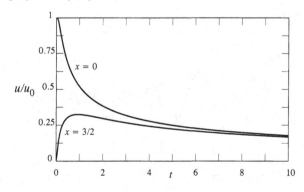

6

MATRIX ALGEBRA AND LINEAR EQUATIONS

6.1 Definitions

A matrix $\mathbf{A}$ of order $m \times n$ is a rectangular array of elements a_{ij} arranged in m rows and n columns

$$\mathbf{A} = \begin{pmatrix} a_{11} & a_{12} & \cdots & a_{1n} \\ a_{21} & a_{22} & \cdots & a_{2n} \\ \cdot & \cdot & \cdots & \cdot \\ a_{m1} & a_{m2} & \cdots & a_{mn} \end{pmatrix}.$$

We also use the notation $\mathbf{A} = [a_{ij}]$. Two matrices $\mathbf{A}$ and $\mathbf{B} = [b_{ij}]$ are equal only if they have the same order and corresponding elements are equal: $a_{ij} = b_{ij}, 1 \leq i \leq m, 1 \leq j \leq n$.

A matrix $\mathbf{x}$ of order $1 \times n$ is called a **row vector**: $\mathbf{x} = (x_1, x_2, \ldots, x_n)$.

A matrix $\mathbf{x}$ of order $m \times 1$ is called a **column vector**: $\mathbf{x} = \begin{pmatrix} x_1 \\ x_2 \\ \cdot \\ \cdot \\ x_m \end{pmatrix}$.

A **square** matrix of order n has n rows and n columns. The elements $a_{11}, a_{22}, \ldots, a_{nn}$ are called the *diagonal elements* of a square matrix.

The matrix is *symmetric* when $a_{ij} = a_{ji}$; it is *skew-symmetric* when $a_{ij} = -a_{ji}$ (in which case the diagonal elements $a_{11} = a_{22} = \cdots = a_{nn} \equiv 0$).

The **transpose** $\mathbf{A}^{\mathrm{T}}$ of the $m \times n$ matrix $\mathbf{A}$ is obtained by interchanging the rows and columns of $\mathbf{A}$, i.e. $\mathbf{A}^{\mathrm{T}}$ is the matrix of order $n \times m$ defined by

$$\mathbf{A}^{\mathrm{T}} = \left[a_{ij}^{\mathrm{T}}\right], \quad \text{where } a_{ij}^{\mathrm{T}} = a_{ji}.$$

A *symmetric* square matrix satisfies $\mathbf{A}^{\mathrm{T}} = \mathbf{A}$.

A symmetric matrix whose non-diagonal elements are all zero is called a *diagonal matrix*. A particular case is the *unit matrix*

$$\mathbf{I} = [\delta_{ij}] = \begin{pmatrix} 1 & 0 & \cdots & 0 \\ 0 & 1 & \cdots & 0 \\ . & . & \cdots & . \\ 0 & 0 & \cdots & 1 \end{pmatrix},$$

where δ_{ij} is the Kronecker delta (equal to 1 when $i = j$ and zero when $i \neq j$) introduced in §2.8.

6.2 Algebra of Matrices

Scalar multiplication $\quad \lambda \mathbf{A} \equiv \lambda[a_{ij}] = [\lambda a_{ij}]$, where λ is a real or complex number.

Matrix addition (defined only for matrices of the same order)

$$\mathbf{A} + \mathbf{B} \equiv [a_{ij}] + [b_{ij}] = [a_{ij} + b_{ij}].$$

Matrix multiplication $\quad \mathbf{AB}$ is defined only when the

number of columns of $\mathbf{A}$ = number of rows of $\mathbf{B}$,

i.e. if $\mathbf{A}$ is of order $m \times q$ then $\mathbf{B}$ must be of order $q \times n$, where m and n can be arbitrary. Then the product matrix $\mathbf{C} = \mathbf{AB}$ has order $m \times n$ and

$$\mathbf{AB} = \mathbf{C} \equiv [c_{ij}], \quad \text{where } c_{ij} = \sum_{k=1}^{q} a_{ik} b_{kj}.$$

The element c_{ij} is equal to the scalar product of the ith *row vector* of $\mathbf{A}$ and the jth *column vector* of $\mathbf{B}$. In general $\mathbf{AB} \neq \mathbf{BA}$; matrix multiplication is *not* commutative.

The multiplication rule is derived from the concept of a matrix as an *operator*. For example, the linear simultaneous equations

$$\left.\begin{array}{l} x+y+z = 5, \\ 2x-y+z = 7, \\ 3x+y-5z = 13, \end{array}\right\} \text{ are equivalent to } \begin{pmatrix} 1 & 1 & 1 \\ 2 & -1 & 1 \\ 3 & 1 & -5 \end{pmatrix} \begin{pmatrix} x \\ y \\ z \end{pmatrix} = \begin{pmatrix} 5 \\ 7 \\ 13 \end{pmatrix}.$$

We also write

$$\mathbf{Ax} = \mathbf{b}, \text{ where } \mathbf{A} = \begin{pmatrix} 1 & 1 & 1 \\ 2 & -1 & 1 \\ 3 & 1 & -5 \end{pmatrix}, \quad \mathbf{x} = \begin{pmatrix} x \\ y \\ z \end{pmatrix} \text{ and } \mathbf{b} = \begin{pmatrix} 5 \\ 7 \\ 13 \end{pmatrix}.$$

The algebra of real and complex numbers satisfies the postulates:

1. Commutative law of addition $\qquad\qquad a+b = b+a$
2. Associative law of addition $\qquad\quad (a+b)+c = a+(b+c)$
3. Commutative law of multiplication $\qquad\quad ab = ba$
4. Associative law of multiplication $\qquad\quad (ab)c = a(bc)$
5. Distributive law of multiplication $\qquad (a+b)c = ac+bc$
6. Nonfactorability of zero $\qquad\qquad ab = 0$ implies $a = 0$, or
$$b = 0, \text{ or } a = b = 0.$$

All of these laws *except* 3 and 6 are satisfied in matrix algebra.

Example   Verify that $\mathbf{AB} = 0$ but $\mathbf{BA} \neq 0$ when

$$\mathbf{A} = \begin{pmatrix} 1 & 1 & 1 \\ 2 & 2 & 2 \\ 5 & 5 & 5 \end{pmatrix}, \quad \mathbf{B} = \begin{pmatrix} 3 & 4 & 2 \\ -2 & -1 & -1 \\ -1 & -3 & -1 \end{pmatrix}. \quad \left[\mathbf{BA} = \begin{pmatrix} 21 & 21 & 21 \\ -9 & -9 & -9 \\ -12 & -12 & -12 \end{pmatrix}\right]$$

Thus we have to distinguish between $\mathbf{AB}$ and $\mathbf{BA}$. In $\mathbf{AB}$, $\mathbf{B}$ is said to be *premultiplied* by $\mathbf{A}$; in $\mathbf{BA}$, $\mathbf{B}$ is said to be *postmultiplied* by $\mathbf{A}$. For an $m \times n$ matrix $\mathbf{A}$

$$\mathbf{A} = \mathbf{IA} = \mathbf{AI},$$

where the unit matrix has order $m \times m$ in the first product, and order $n \times n$ in the second.

Transpose of a product

$$(\mathbf{AB})^{\mathrm{T}} = \sum_k (a_{ik} b_{kj})^{\mathrm{T}} = \sum_k a_{jk} b_{ki} \equiv \sum_k a_{kj}^{\mathrm{T}} b_{ik}^{\mathrm{T}} = \mathbf{B}^{\mathrm{T}} \mathbf{A}^{\mathrm{T}}.$$

Problems 6A

Verify

1. If $\mathbf{A} = \begin{pmatrix} 3 & 2 \\ -2 & -1 \\ -1 & -1 \end{pmatrix}$, then $\mathbf{A}^{\mathrm{T}} = \begin{pmatrix} 3 & -2 & -1 \\ 2 & -1 & -1 \end{pmatrix}$.

2. $\begin{pmatrix} 2 & 1 & -3 \\ -1 & 6 & 1 \\ 1 & 0 & -2 \end{pmatrix} \begin{pmatrix} 2 \\ 3 \\ -1 \end{pmatrix} = \begin{pmatrix} 10 \\ 15 \\ 4 \end{pmatrix}$.

3. $\begin{pmatrix} 1 & -2 & 3 \\ -4 & 2 & 5 \end{pmatrix} \begin{pmatrix} 1 & 3 \\ -1 & 0 \\ 2 & 4 \end{pmatrix} = \begin{pmatrix} 9 & 15 \\ 4 & 8 \end{pmatrix}$.

4. $\begin{pmatrix} 4 & 2 & -1 & 2 \\ 3 & -7 & 1 & -8 \\ 2 & 4 & -3 & 1 \end{pmatrix} \begin{pmatrix} 2 & 3 \\ -3 & 0 \\ 1 & 5 \\ 3 & 1 \end{pmatrix} = \begin{pmatrix} 7 & 9 \\ 4 & 6 \\ -8 & -8 \end{pmatrix}$.

5. $\begin{pmatrix} 3 & 4 \\ -2 & -1 \end{pmatrix} \begin{pmatrix} 1 & 2 \\ 2 & 5 \end{pmatrix} \begin{pmatrix} 2 & -1 \\ 0 & 3 \end{pmatrix} = \begin{pmatrix} 22 & 67 \\ -8 & -23 \end{pmatrix}$.

6. $\begin{pmatrix} 1 & 0 & 0 \\ 0 & -4 & 0 \\ 0 & 0 & 3 \end{pmatrix} \begin{pmatrix} 2 & 0 & 5 \\ -1 & 4 & 5 \\ 3 & 1 & 2 \end{pmatrix} = \begin{pmatrix} 2 & 0 & 5 \\ 4 & -16 & -20 \\ 9 & 3 & 6 \end{pmatrix}$.

6.3   Linear Equations

The system of m linear, *inhomogeneous* equations in the n unknowns x_n:

$$\left. \begin{array}{l} a_{11}x_1 + a_{12}x_2 + \cdots + a_{1n}x_n = b_1 \\ a_{21}x_1 + a_{22}x_2 + \cdots + a_{2n}x_n = b_2 \\ \quad\cdots \qquad\quad \cdots \qquad \cdots \qquad\qquad \cdots \\ a_{m1}x_1 + a_{m2}x_2 + \cdots + a_{mn}x_n = b_m \end{array} \right\}, \tag{6.3.1}$$

is equivalent to the matrix equation

$$\begin{pmatrix} a_{11} & a_{12} & \cdots & a_{1n} \\ a_{21} & a_{22} & \cdots & a_{2n} \\ \cdot & \cdot & \cdots & \cdot \\ a_{m1} & a_{m2} & \cdots & a_{mn} \end{pmatrix} \begin{pmatrix} x_1 \\ x_2 \\ \cdot \\ x_n \end{pmatrix} = \begin{pmatrix} b_1 \\ b_2 \\ \cdot \\ b_m \end{pmatrix}, \tag{6.3.2}$$

which in turn can be expressed in the shortened form

$$\mathbf{Ax} = \mathbf{b}, \tag{6.3.3}$$

where $\mathbf{A}$ is the $m \times n$ matrix $[a_{ij}]$, $\mathbf{x}$ is the $n \times 1$ column vector of unknowns on the left of (6.3.2) and $\mathbf{b}$ is the prescribed $m \times 1$ column vector on the right-hand side. When $\mathbf{b} - 0$ the equations (6.3.1) form a *homogenous* system.

The possibility of solving equations (6.3.1) for $\mathbf{x}$ is a well-defined problem only if the equations are consistent, i.e. are not in some way self-contradictory. When $m = n$ there are n equations in n unknowns and it appears that there is just enough information to determine $\mathbf{x}$. If the number of equations is less than n $(m < n)$ it is obvious that the equations cannot supply enough information to give a unique solution, and the system is said to be *underdetermined*. If, however, $m > n$ the system is apparently *overdetermined*, and there is a strong likelihood that the additional equations will make the system incompatible. We shall see, however, that the question of the system being over or underdetermined is distinct from the question of the compatibility of the equations. If the equations are incompatible no solution will exist, but one or many solutions are possible for $m \neq n$ provided the system is compatible.

In simple cases the question of compatibility can be resolved directly by application of a procedure known as *Gauss elimination*:

Example 1 Solve the equations:

$$\begin{aligned}
x_1 + 2x_2 + 3x_3 + 3x_4 &= 1 \\
2x_1 + 5x_2 + 10x_3 + 7x_4 &= 4 \\
2x_1 + 5x_2 + 9x_3 + 5x_4 &= 1 \\
x_1 + 2x_2 + 3x_3 + 4x_4 &= 2.
\end{aligned}$$

Start by eliminating x_1 from all equations except the first. Thus, subtract the first equation from the fourth, twice the first equation from the third, and twice the first equation from the second:

$$\begin{aligned}
x_1 + 2x_2 + 3x_3 + 3x_4 &= 1 \\
x_2 + 4x_3 + x_4 &= 2 \\
x_2 + 3x_3 - x_4 &= -1 \\
x_4 &= 1.
\end{aligned}$$

Next use the second of these equations to eliminate x_2 from the third and the fourth equation. Because x_2 has already disappeared from the fourth equation, we merely need to subtract the second equation from the third. If we then multiply the resulting third equation by -1 we obtain:

$$x_1 + 2x_2 + 3x_3 + 3x_4 = 1$$
$$x_2 + 4x_3 + x_4 = 2$$
$$x_3 + 2x_4 = 3$$
$$x_4 = 1. \tag{6.3.4}$$

The system of equations is now said to be in *row echelon* form, and x_1, x_2, x_3, x_4 are easily found by working backwards from the last equation, a procedure known as *back substitution*. From the last equation $x_4 = 1$; substitute this into the third equation, to get $x_3 = 1$; these values for x_4, x_3 are next substituted into the second equation, and so on. In this way we find the unique solution

$$x_1 = 1, \quad x_2 = -3, \quad x_3 = 1, \quad x_4 = 1.$$

Example 2 Solve the equations

$$x_1 + 2x_2 + 3x_3 + 3x_4 = 1$$
$$2x_1 + 5x_2 + 10x_3 + 7x_4 = 4$$
$$2x_1 + 5x_2 + 9x_3 + 5x_4 = 1$$
$$x_1 + 2x_2 + 3x_3 + 3x_4 = 2. \tag{6.3.5}$$

This system differs from that in Example 1 in that the coefficient of x_4 in the fourth equation has been changed from 4 to 3. This makes the left-hand side of the fourth equation identical to the left-hand side of the first. When the system is reduced to row echelon form (by the same operations used in Example 1) we find

$$x_1 + 2x_2 + 3x_3 + 3x_4 = 1$$
$$x_2 + 4x_3 + x_4 = 2$$
$$x_3 + 2x_4 = 3$$
$$0 = 1.$$

The logical contradiction contained in the last of these equations reveals that there is no solution to the original system of equations.

Example 3 Solve the equations

$$\left. \begin{array}{l} x_1 + 2x_2 + 3x_3 + 3x_4 = 1 \\ 2x_1 + 5x_2 + 10x_3 + 7x_4 = 4 \\ 2x_1 + 5x_2 + 9x_3 + 5x_4 = 1 \\ x_1 + 2x_2 + 3x_3 + 3x_4 = 1 \end{array} \right\}. \tag{6.3.6}$$

In this case the first and last equations are identical. At the outset we could discard the last equation as being redundant. We should then have three equations in four unknowns, i.e. an underdetermined system. However, in general it will not be obvious that two or more equations in a system are equivalent, so we shall therefore proceed as before by retaining the fourth equation. The row echelon form then becomes

$$\left.\begin{array}{rcl} x_1 + 2x_2 + 3x_3 + 3x_4 &=& 1 \\ x_2 + 4x_3 + x_4 &=& 2 \\ x_3 + 2x_4 &=& 3 \\ 0 &=& 0 \end{array}\right\}. \tag{6.3.7}$$

By back substitution we find

$$x_1 = 12 - 11x_4, \ x_2 = -10 + 7x_4, \ x_3 = 3 - 2x_4, \text{where } x_4 \text{ is arbitrary.}$$

This is the general solution of the system of underdetermined equations.

Note If the fourth of equations (6.3.6) had been discarded at the outset the Gauss elimination procedure would have been applied to a system of three equations in four unknowns. We should then have arrived at (6.3.7) but *without* the last equation $0 = 0$.

The augmented matrix The *row operations* applied to the simultaneous equations in Examples 1 to 3 to reduce the equations to row echelon form are equivalent to applying the same operations to the matrix of the array of coefficients of $x_1 - x_4$ *and* to the coefficients $b_1 - b_4$ on the right-hand sides, i.e. in the general case, to

$$\begin{pmatrix} a_{11} & a_{12} & \cdots & a_{1n} \\ a_{21} & a_{22} & \cdots & a_{2n} \\ . & . & \cdots & . \\ a_{m1} & a_{m2} & \cdots & a_{mn} \end{pmatrix} \quad \text{and} \quad \begin{pmatrix} b_1 \\ b_2 \\ . \\ b_m \end{pmatrix}.$$

The procedure can therefore be formalized by applying these row operations to the so-called *augmented matrix*

$$\hat{\mathbf{A}} = \begin{pmatrix} a_{11} & a_{12} & \cdots & a_{1n} & b_1 \\ a_{21} & a_{22} & \cdots & a_{2n} & b_2 \\ . & . & \cdots & . & . \\ a_{m1} & a_{m2} & \cdots & a_{mn} & b_m \end{pmatrix}.$$

Thus, in Example 1

$$\hat{\mathbf{A}} = \begin{pmatrix} 1 & 2 & 3 & 3 & 1 \\ 2 & 5 & 10 & 7 & 4 \\ 2 & 5 & 9 & 5 & 1 \\ 1 & 2 & 3 & 4 & 2 \end{pmatrix}.$$

Successive row operations reduce this as follows:

$$\begin{pmatrix} 1 & 2 & 3 & 3 & 1 \\ 0 & 1 & 4 & 1 & 2 \\ 0 & 1 & 3 & -1 & -1 \\ 0 & 0 & 0 & 1 & 1 \end{pmatrix} \rightarrow \begin{pmatrix} 1 & 2 & 3 & 3 & 1 \\ 0 & 1 & 4 & 1 & 2 \\ 0 & 0 & 1 & 2 & 3 \\ 0 & 0 & 0 & 1 & 1 \end{pmatrix}.$$

The final matrix is equivalent to equations (6.3.4), and the solution is now obtained, as in Example 1, by back substitution.

Linear dependence and independence   The vectors $c_1, c_2, \ldots, c_m$ are *linearly dependent* if there exist scalar constants $\alpha_1, \alpha_2, \ldots, \alpha_m$ *not all equal to zero* such that

$$\alpha_1 c_1 + \alpha_2 c_2 + \cdots + \alpha_m c_m = 0. \tag{6.3.8}$$

When this is satisfied only for $\alpha_1 = \alpha_2 = \cdots = \alpha_m = 0$, the vectors $c_1, c_2, \ldots, c_m$ are said to be *linearly independent*.

Let c_i be the vector formed by the ith row of the matrix of coefficients (6.3.2) of the system of equations (6.3.1). In Example 2 the matrix $\mathbf{A}$ for equations (6.3.5) is

$$\mathbf{A} = \begin{pmatrix} 1 & 2 & 3 & 3 \\ 3 & 5 & 10 & 7 \\ 3 & 5 & 9 & 5 \\ 1 & 2 & 3 & 3 \end{pmatrix}.$$

The rows are obviously linearly dependent, because the final row is identical to the first. The row echelon form of $\mathbf{A}$ is

$$\mathbf{A} = \begin{pmatrix} 1 & 2 & 3 & 3 \\ 0 & 1 & 4 & 1 \\ 0 & 0 & 1 & 2 \\ 0 & 0 & 0 & 0 \end{pmatrix}.$$

Now the linear dependence or independence of the rows cannot be altered by adding multiples of one row to another. This means the rows of $\mathbf{A}$ are linearly dependent if at least one row of the row echelon form contains nothing but zeros. When the rows containing nothing but zero are ignored, the remaining rows are linearly independent.

The *rank* of a matrix is the number of linearly independent rows. In the above example $\mathbf{A}$ is of rank 3.

Consider now the final forms of $\mathbf{A}$ and $\hat{\mathbf{A}}$ in Examples 1 to 3 above:

Example 1 : $\quad \mathbf{A} = \begin{pmatrix} 1 & 2 & 3 & 3 \\ 0 & 1 & 4 & 1 \\ 0 & 0 & 1 & 2 \\ 0 & 0 & 0 & 1 \end{pmatrix}, \quad \hat{\mathbf{A}} = \begin{pmatrix} 1 & 2 & 3 & 3 & 1 \\ 0 & 1 & 4 & 1 & 2 \\ 0 & 0 & 1 & 2 & 3 \\ 0 & 0 & 0 & 1 & 1 \end{pmatrix}.$

$$\therefore \quad \text{rank } \mathbf{A} = 4, \text{ rank } \hat{\mathbf{A}} = 4$$

$$x_1 = 1, \quad x_2 = -3, \quad x_3 = 1, \quad x_4 = 1.$$

Example 2 : $\quad \mathbf{A} = \begin{pmatrix} 1 & 2 & 3 & 3 \\ 0 & 1 & 4 & 1 \\ 0 & 0 & 1 & 2 \\ 0 & 0 & 0 & 0 \end{pmatrix}, \quad \hat{\mathbf{A}} = \begin{pmatrix} 1 & 2 & 3 & 3 & 1 \\ 0 & 1 & 4 & 1 & 2 \\ 0 & 0 & 1 & 2 & 3 \\ 0 & 0 & 0 & 0 & 1 \end{pmatrix}.$

$$\therefore \quad \text{rank } \mathbf{A} = 3, \text{ rank } \hat{\mathbf{A}} = 4$$
$$\text{No solution.}$$

Example 3 : $\quad \mathbf{A} = \begin{pmatrix} 1 & 2 & 3 & 3 \\ 0 & 1 & 4 & 1 \\ 0 & 0 & 1 & 2 \\ 0 & 0 & 0 & 0 \end{pmatrix}, \quad \hat{\mathbf{A}} = \begin{pmatrix} 1 & 2 & 3 & 3 & 1 \\ 0 & 1 & 4 & 1 & 2 \\ 0 & 0 & 1 & 2 & 3 \\ 0 & 0 & 0 & 0 & 0 \end{pmatrix}.$

$$\therefore \quad \text{rank } \mathbf{A} = 3, \text{ rank } \hat{\mathbf{A}} = 3$$

$x_1 = 12 - 11x_4, \; x_2 = -10 + 7x_4, \; x_3 = 3 - 2x_4$, where x_4 is arbitrary.

These results can be formalised into the *Fundamental theorem*.

Fundamental theorem The $m \times n$ linear system of equations

$$\mathbf{Ax} = \mathbf{b},$$

has a solution if and only if rank $\mathbf{A} = \text{rank } \hat{\mathbf{A}}$. When a solution exists, all solutions can be expressed in terms of $n - \mathcal{R}$ arbitrary parameters, where $\mathcal{R} = \text{rank } \mathbf{A}$.

Problems 6B

Solve by Gauss elimination:

1.
$$\begin{aligned} 3x_1 - x_2 + x_3 &= -4 \\ x_1 + 5x_2 + 2x_3 &= 12 \\ 2x_1 + 3x_2 + x_3 &= 0 \end{aligned} \qquad \left[\mathbf{x} = \begin{pmatrix} -4 \\ 0 \\ 8 \end{pmatrix} \right]$$

2.
$$\begin{aligned} x_1 + x_2 + x_3 &= 3 \\ -3x_1 - 17x_2 + x_3 + 2x_4 &= 1 \\ 4x_1 - 17x_2 + 8x_3 - 5x_4 &= 1 \\ -5x_2 - 2x_3 + x_4 &= 1 \end{aligned} \qquad \left[\mathbf{x} = \begin{pmatrix} 2 \\ 0 \\ 1 \\ 3 \end{pmatrix} \right]$$

3.
$$\begin{aligned} 4x_1 + 3x_2 + 2x_3 - x_4 &= 4 \\ 5x_1 + 4x_2 + 3x_3 - x_4 &= 4 \\ -2x_1 - 2x_2 - x_3 + 2x_4 &= -3 \\ 11x_1 + 6x_2 + 4x_3 + x_4 &= 11 \end{aligned} \qquad \left[\mathbf{x} = \begin{pmatrix} 1 - x_4 \\ 2 + 3x_4 \\ -3 - 2x_4 \\ x_4 \end{pmatrix}, \ \mathcal{R} = 3, x_4 = \text{arbitrary} \right]$$

4.
$$\begin{aligned} x_1 + 2x_2 - 3x_3 + x_4 &= 0 \\ 3x_1 - x_2 + 5x_3 - x_4 &= 0 \\ 2x_1 + x_2 + x_4 &= 0 \end{aligned} \qquad \left[\mathbf{x} = \mu \begin{pmatrix} -1 \\ 2 \\ 1 \\ 0 \end{pmatrix}, \ \mu = \text{arbitrary} \right]$$

5.
$$\begin{pmatrix} 1 & 3 & 5 & -2 \\ 3 & -2 & -7 & 5 \\ 2 & 1 & 0 & 1 \end{pmatrix} \begin{pmatrix} x_1 \\ x_2 \\ x_3 \\ x_4 \end{pmatrix} = \begin{pmatrix} 11 \\ 0 \\ 7 \end{pmatrix}$$

$$\left[\mathbf{x} = \begin{pmatrix} 2 + \lambda - \mu \\ 3 - 2\lambda + \mu \\ \lambda \\ \mu \end{pmatrix}, \ \mathcal{R} = 2, \ \lambda, \mu = \text{arbitrary} \right]$$

6.
$$\begin{pmatrix} 1 & 3 & 2 & 5 \\ 3 & -2 & -5 & 4 \\ 2 & 1 & -1 & 5 \end{pmatrix} \begin{pmatrix} x_1 \\ x_2 \\ x_3 \\ x_4 \end{pmatrix} = \begin{pmatrix} 10 \\ -5 \\ 5 \end{pmatrix}$$

$$\left[\text{no solution: rank } \mathbf{A} = 2, \ \text{rank } \hat{\mathbf{A}} = 3 \right]$$

7.
$$\begin{pmatrix} 1 & 1 \\ 2 & 3 \\ 3 & 2 \\ 1 & -1 \\ 3 & 5 \end{pmatrix} \begin{pmatrix} x_1 \\ x_2 \end{pmatrix} = \begin{pmatrix} 0 \\ -1 \\ 1 \\ 2 \\ -2 \end{pmatrix} \qquad \left[\mathbf{x} = \begin{pmatrix} 1 \\ -1 \end{pmatrix}, \mathcal{R} = 2 \right]$$

8. $\begin{pmatrix} 1 & -1 & 2 & 1 \\ 2 & 1 & 1 & -1 \\ 1 & 2 & -1 & -2 \\ 1 & 0 & 1 & 0 \end{pmatrix} \begin{pmatrix} x_1 \\ x_2 \\ x_3 \\ x_4 \end{pmatrix} = \begin{pmatrix} -1 \\ 4 \\ 5 \\ 1 \end{pmatrix}$

$$\left[\mathbf{x} = \begin{pmatrix} 1 - \lambda \\ 2 + \lambda + \mu \\ \lambda \\ \mu \end{pmatrix}, \mathcal{R} = 2, \lambda, \mu - \text{arbitrary} \right]$$

9. $\begin{pmatrix} 3 & -2 & 0 & -1 \\ 0 & 2 & 2 & 1 \\ 1 & -2 & -3 & -2 \\ 0 & 1 & 2 & 1 \end{pmatrix} \begin{pmatrix} x_1 \\ x_2 \\ x_3 \\ x_4 \end{pmatrix} = \begin{pmatrix} 7 \\ 5 \\ -1 \\ 6 \end{pmatrix}$ $\left[\mathbf{x} = \begin{pmatrix} -10 \\ -1 \\ 21 \\ -35 \end{pmatrix} \right]$

10. $\begin{pmatrix} 1 & 2 & 3 & 4 \\ 1 & 3 & 5 & 7 \end{pmatrix} \begin{pmatrix} x_1 \\ x_2 \\ x_3 \\ x_4 \end{pmatrix} = \begin{pmatrix} 0 \\ 0 \\ 0 \\ 0 \end{pmatrix}$

$$\left[\mathbf{x} = \begin{pmatrix} \lambda + 2\mu \\ -2\lambda - 3\mu \\ \lambda \\ \mu \end{pmatrix}, \mathcal{R} = 2, \lambda, \mu = \text{arbitrary} \right]$$

6.4 Further Discussion of Compatibility

A systematic method for determining the compatibility of the $m \times n$ system of equations

$$\mathbf{Ax} = \mathbf{b}, \tag{6.4.1}$$

is obtained by considering the solution $\mathbf{y}$ of the homogeneous *adjoint* equation

$$\mathbf{A}^{\mathrm{T}}\mathbf{y} = \mathbf{0}, \tag{6.4.2}$$

where $\mathbf{A}^{\mathrm{T}}$ is the transpose of $\mathbf{A}$.

In equation (6.4.2) $\mathbf{y}$ is an $m \times 1$ column vector. Its transpose $\mathbf{y}^{\mathrm{T}}$ is a $1 \times m$ row vector, which therefore satisfies

$$\mathbf{y}^{\mathrm{T}}\mathbf{A} = \mathbf{0}.$$

Suppose there are ν distinct *non-trivial* solutions of this equation. Because the right-hand side vanishes the solution will have the general form

$$\mathbf{y}^{\mathrm{T}} = \sum_{i=1}^{\nu} \lambda_i \mathbf{e}_i,$$

where the coefficients λ_i are arbitrary constants, and the $\mathbf{e}_i$ are constant, linearly independent row vectors. It follows from this that rank $\mathbf{A} = m - \nu$, because $\mathbf{e}_i \mathbf{A} = 0$ for $1 \leq i \leq \nu$, i.e. there are precisely ν linear relations between the rows of $\mathbf{A}$. Thus, we can conclude that the augmented matrix also has rank $m - \nu$, thereby making equations (6.4.1) compatible, if it is also true that

$$\mathbf{e}_i \cdot \mathbf{b} = 0, \quad 1 \leq i \leq \nu. \tag{6.4.3}$$

This is the required compatibility condition.

In words: the equations $\mathbf{A}\mathbf{x} = \mathbf{b}$ form a compatible system provided every solution of the homogeneous adjoint equation $\mathbf{A}^{\mathrm{T}}\mathbf{y} = 0$ is *orthogonal* to $\mathbf{b}$.

In the case where $\nu = 0$, rank $\mathbf{A} = $ rank $\hat{\mathbf{A}}$ and the system of equations (6.4.1) always has n solutions.

Example Let $\mathbf{A}$ be the matrix of Problems 6B, Question 6. Then

$$\mathbf{A}^{\mathrm{T}} = \begin{pmatrix} 1 & 3 & 2 \\ 3 & -2 & 1 \\ 2 & -5 & -1 \\ 5 & 4 & 5 \end{pmatrix} \quad \therefore \quad \left. \begin{array}{l} y_1 + 3y_2 + 2y_3 = 0 \\ 3y_1 - 2y_2 + y_3 = 0 \\ 2y_1 - 5y_2 - y_3 = 0 \\ 5y_1 + 4y_2 + 5y_3 = 0 \end{array} \right\} \rightarrow \nu = 1, \mathbf{e}_1 = (7, 5, -11).$$

Then $\mathbf{e}_1 \cdot \mathbf{b} = (7, 5, -11) \cdot (10, -5, 5) = -10 \neq 0$, so that the equations of Question 6 are incompatible.

Example Let $\mathbf{A}$ be the matrix of Problems 6B, Question 7. Then

$$\mathbf{A}^{\mathrm{T}} = \begin{pmatrix} 1 & 2 & 3 & 1 & 3 \\ 1 & 3 & 2 & -1 & 5 \end{pmatrix} \quad \therefore \quad \begin{cases} y_1 + 2y_2 + 3y_3 + y_4 + 3y_5 = 0 \\ y_1 + 3y_2 + 2y_3 - y_4 + 5y_5 = 0 \end{cases}.$$

These equations have $\nu = 3$ independent solutions

$$\mathbf{e}_1 = (-5, 1, 1, 0, 0), \quad \mathbf{e}_2 = (-5, 2, 0, 1, 0), \quad \mathbf{e}_3 = (1, -2, 0, 0, 1).$$

Hence,

$$\mathbf{e}_1 \cdot \mathbf{b} = (-5, 1, 1, 0, 0) \cdot (0, -1, 1, 2, -2) = 0$$
$$\mathbf{e}_2 \cdot \mathbf{b} = (-5, 2, 0, 1, 0) \cdot (0, -1, 1, 2, -2) = 0$$
$$\mathbf{e}_3 \cdot \mathbf{b} = (1, -2, 0, 0, 1) \cdot (0, -1, 1, 2, -2) = 0$$

which confirms that the equations of Question 7 are compatible. Further, because rank $\mathbf{A} = 5 - \nu = 2$, there is a single and unique solution.

Example Let $\mathbf{A}$ be the matrix of Problems 6B, Question 9. Then

$$\mathbf{A}^{\mathrm{T}} = \begin{pmatrix} 3 & 0 & 1 & 0 \\ -2 & 2 & -2 & 1 \\ 0 & 2 & -3 & 2 \\ -1 & 1 & -2 & 1 \end{pmatrix} \qquad \begin{cases} 3y_1 & + y_3 & = 0 \\ -2y_1 + 2y_2 - 2y_3 + y_4 = 0 \\ 2y_2 - 3y_3 + 2y_4 = 0 \\ -y_1 + y_2 - 2y_3 + y_4 = 0. \end{cases}$$

These equations are linearly independent and have only the trivial solution $\mathbf{y} = 0$. Hence the equations of Question 9 are compatible and possess a single and unique solution.

Problems 6C

1. Use the method of this section to verify that the following system of equations is compatible:

$$\begin{aligned} x_1 + 3x_2 - 2x_3 &= 11 \\ 2x_1 - 5x_2 + 7x_3 &= -11 \\ -x_1 + 2x_2 - 3x_3 &= 4 \\ x_1 + 2x_2 - x_3 &= 8 \end{aligned} \qquad \left[\nu = 2, \begin{array}{l} \mathbf{e}_1 = (1,5,11,0) \\ \mathbf{e}_2 = (9,1,0,-11) \end{array} \right]$$

2. Show that the following equations are incompatible:

$$\begin{aligned} x_1 + x_2 + x_3 &= 1 \\ 2x_1 + 2x_2 + 2x_3 &= 3 \end{aligned} \qquad \left[\nu = 1, \ \mathbf{e}_1 = (-2,1) \right]$$

3. Show that the following equations are incompatible:

$$\begin{aligned} x_1 - 2x_2 - 3x_3 + 4x_4 &= 1 \\ 4x_1 - x_2 - 5x_3 + 6x_4 &= 2 \\ 2x_1 + 3x_2 + x_3 - 2x_4 &= 2 \end{aligned} \qquad \left[\nu = 1, \ \mathbf{e}_1 = (2,-1,1) \right]$$

4. Show that the following equations are incompatible:

$$\begin{aligned} x_1 + x_2 + 3x_3 &= 6 \\ x_1 - x_2 + x_3 &= 2 \\ 5x_1 + x_2 + 11x_3 &= 5 \end{aligned} \qquad \left[\nu = 1, \ \mathbf{e}_1 = (-3,-2,1) \right]$$

6.5 Determinants

Simple second- and third-order determinants have been used without formal definition in previous chapters. We now consider their generalisations to higher orders.

A determinant is a scalar defined for a *square* array (or matrix) of numbers $\mathbf{A} = [a_{ij}], 1 \leq i, j \leq n$. For $n = 2$ we know that

$$\det \mathbf{A} \equiv \begin{vmatrix} a_{11} & a_{12} \\ a_{21} & a_{22} \end{vmatrix} = a_{11}a_{22} - a_{12}a_{21}.$$

For a 3×3 array

$$\det \mathbf{A} \equiv \begin{vmatrix} a_{11} & a_{12} & a_{13} \\ a_{21} & a_{22} & a_{23} \\ a_{31} & a_{32} & a_{33} \end{vmatrix}$$

$$= a_{11}a_{22}a_{33} - a_{11}a_{23}a_{32} + a_{12}a_{23}a_{31} - a_{12}a_{21}a_{33}$$

$$+ a_{13}a_{21}a_{32} - a_{13}a_{22}a_{31}. \tag{6.5.1}$$

Here, the determinant may be regarded as a notation for the sum of all the *signed* products of triplets of the form $a_{1i}a_{2j}a_{3k}$ where the suffixes i, j, k are a permutation of $1, 2, 3$. The meaning of 'signed' product is as follows: the natural order $1, 2, 3$ for i, j, k is taken to be positive; the signs of all other products is positive or negative according as the number of interchanges of i, j, k required to return to the natural order is respectively even or odd. Thus, $a_{13}a_{22}a_{31}$ requires one interchange of 3 and 1, so the sign of this term is negative; $a_{13}a_{21}a_{32}$ requires two interchanges, so the sign is positive, etc. Note also that the same result is obtained when the sum is regarded as consisting of the signed products of the form $a_{i1}a_{j2}a_{k3}$. Therefore

$$\det \mathbf{A} = \det \mathbf{A}^{\mathrm{T}},$$

The definition (6.5.1) can also be written

$$\det \mathbf{A} \equiv \begin{vmatrix} a_{11} & a_{12} & a_{13} \\ a_{21} & a_{22} & a_{23} \\ a_{31} & a_{32} & a_{33} \end{vmatrix}$$

$$= a_{11} \begin{vmatrix} a_{22} & a_{23} \\ a_{32} & a_{33} \end{vmatrix} - a_{12} \begin{vmatrix} a_{21} & a_{23} \\ a_{31} & a_{33} \end{vmatrix} + a_{13} \begin{vmatrix} a_{21} & a_{22} \\ a_{31} & a_{32} \end{vmatrix} \equiv \sum_{j=1}^{3} a_{1j}\mathbf{A}_{1j},$$

where $\mathbf{A}_{1j}$ is called the *cofactor* of a_{1j}, which is equal to $(-1)^{1+j}$ times the 2×2 determinant formed from $\mathbf{A}$ by deleting the first row and jth column. This is just a special case of the more general formula

$$\begin{vmatrix} a_{11} & a_{12} & a_{13} \\ a_{21} & a_{22} & a_{23} \\ a_{31} & a_{32} & a_{33} \end{vmatrix} = \sum_{j=1}^{3} a_{ij}\mathbf{A}_{ij} = \sum_{i=1}^{3} a_{ij}\mathbf{A}_{ij}, \qquad (6.5.2)$$

where the cofactor $\mathbf{A}_{ij}$ is $(-1)^{i+j}$ times the determinant formed by deleting row i and column j from $\det \mathbf{A}$. The two cases are respectively said to give the expansion of $\det \mathbf{A}$ by the ith row and by the jth column.

All of these results can be extended to the general $n \times n$ determinant

$$\begin{vmatrix} a_{11} & a_{12} & \cdots & a_{1n} \\ a_{21} & a_{22} & \cdots & a_{2n} \\ . & . & \cdots & . \\ a_{n1} & a_{n2} & \cdots & a_{nn} \end{vmatrix} = \sum \pm a_{1\alpha} a_{2\beta} \cdots a_{n\nu} = \sum \pm a_{\alpha 1} a_{\beta 2} \cdots a_{\nu n},$$

$$(6.5.3)$$

where the summations are over the $n!$ possible permutations of the n subscripts $\alpha, \beta, \ldots, \nu$ of the integers $1, 2, \ldots, n$, and the $\pm$ sign is taken according as an even or odd number of interchanges of $\alpha, \beta, \ldots, \nu$ is required to restore them to the natural order $1, 2, \ldots, n$.

Similarly, we have the expansion of $\det \mathbf{A}$ by rows or by columns, given by

$$\begin{vmatrix} a_{11} & a_{12} & \cdots & a_{1n} \\ a_{21} & a_{22} & \cdots & a_{2n} \\ . & . & \cdots & . \\ a_{n1} & a_{n2} & \cdots & a_{nn} \end{vmatrix} = \sum_{j=1}^{n} a_{ij}\mathbf{A}_{ij} = \sum_{i=1}^{n} a_{ij}\mathbf{A}_{ij}, \qquad (6.5.4)$$

where the *cofactor* $\mathbf{A}_{ij}$ of a_{ij} is $(-1)^{i+j}$ times the determinant formed by deleting row i and column j from $\mathbf{A}$.

Properties of determinants   The following general conclusions can be drawn from these definitions:

1. $\det \mathbf{A} = \det \mathbf{A}^{\mathrm{T}}$.
2. When two rows or columns are interchanged, the absolute value of $\det \mathbf{A}$ is unchanged but the sign is changed.
3. $\det \mathbf{A} = 0$ when two rows or columns are the same.
4. If every element of a given row (or column) is multiplied by the same factor, the value of $\det \mathbf{A}$ is multiplied by that factor.
5. $\det \mathbf{A}$ is unchanged when a multiple of one row (or column) is added to another row (or column).
6. $\det \mathbf{AB} = \det \mathbf{A} \cdot \det \mathbf{B}$.

Problems 6D

Evaluate the determinants

1. $\begin{vmatrix} 1 & 1 & 1 \\ 10 & 12 & 16 \\ 14 & 17 & 21 \end{vmatrix}$. $[-4]$

2. $\begin{vmatrix} 12 & 15 & 18 \\ 11 & 14 & 17 \\ 10 & 13 & 16 \end{vmatrix}$. $[0]$

3. $\begin{vmatrix} b+c & c & b \\ c & a+c & a \\ b & a & a+b \end{vmatrix}$. $[4abc]$

4. $\begin{vmatrix} 1 & 1 & 1 \\ \tan x & \tan y & \tan z \\ \sin 2x & \sin 2y & \sin 2z \end{vmatrix}$,   where $x+y+z=\pi$. $[0]$

5. If no two of a, b, c are equal and
$$\begin{vmatrix} 1 & bc+ax & a^2 \\ 1 & ca+bx & b^2 \\ 1 & ab+cx & c^2 \end{vmatrix} = 0, \quad \text{show that} \quad x = a+b+c.$$

6.6   Inverse of a Square Matrix

The effect of Gauss elimination (§6.3) on the $m \times n$ matrix of coefficients $\mathbf{A}$ of the system of m equations in n unknowns $x_1, x_2, \ldots, x_n$

$$\mathbf{Ax} = \mathbf{b},$$

is to reduce $\mathbf{A}$ to row echelon form $\mathbf{B}$, say. The matrix $\mathbf{B}$ has the properties:

1. The first $\mathcal{R}$ rows are non-zero and all remaining rows (if any) are zero, where $\mathcal{R} = \operatorname{rank} \mathbf{A}$.
2. The first non-zero element in the ith row $(1 \leq i \leq \mathcal{R})$ is unity and occurs in column c_i, where $c_1 < c_2 < \cdots < c_{\mathcal{R}}$.

Example Verify that the following matrix is in row echelon form

$$\begin{pmatrix} 1 & 1 & 2 & 3 & 4 & 5 & 6 & 7 \\ 0 & 0 & 1 & 9 & 7 & 5 & 0 & 2 \\ 0 & 0 & 0 & 1 & 3 & 8 & 9 & 1 \\ 0 & 0 & 0 & 0 & 0 & 0 & 1 & 5 \\ 0 & 0 & 0 & 0 & 0 & 0 & 0 & 0 \end{pmatrix}.$$

The rank $\mathcal{R} = 4$ and $c_1 = 1, c_2 = 3, c_3 = 4, c_4 = 7$.

A *square matrix* $\mathbf{A}$ of order $n \times n$ whose rank $\mathcal{R} = n$ is said to be *non-singular*. The main diagonal of its row echelon form consists entirely of ones, and all elements below the main diagonal are zero (i.e. $c_i = i, 1 \leq i \leq n$). $\mathbf{A}$ is reduced to row echelon form by a finite sequence of row operations. Once this is done, however, it is obvious that the application of a further sequence of row operations will eventually reduce it to the $n \times n$ unit matrix $\mathbf{I}$.

Each of these row operations can be represented by an *elementary matrix* $\mathbf{E}$, say. It is convenient to consider three basic row operations (which are not, however, independent) represented by the following elementary matrices:

$$\begin{pmatrix} 1 & 0 & 0 & \cdots & 0 \\ 0 & \alpha & 0 & \cdots & 0 \\ 0 & 0 & 1 & \cdots & 0 \\ . & . & . & \cdots & . \\ 0 & 0 & 0 & \cdots & 1 \end{pmatrix}, \quad \begin{pmatrix} 1 & 0 & \alpha & \cdots & 0 \\ 0 & 1 & 0 & \cdots & 0 \\ 0 & 0 & 1 & \cdots & 0 \\ . & . & . & \cdots & . \\ 0 & 0 & 0 & \cdots & 1 \end{pmatrix}, \quad \begin{pmatrix} 0 & 1 & 0 & \cdots & 0 \\ 1 & 0 & 0 & \cdots & 0 \\ 0 & 0 & 1 & \cdots & 0 \\ . & . & . & \cdots & . \\ 0 & 0 & 0 & \cdots & 1 \end{pmatrix}.$$

Premultiplication of $\mathbf{A}$ by the first of these matrices causes the second row of $\mathbf{A}$ to be multiplied by α; the second elementary matrix adds a multiple α of the third row to the first row of $\mathbf{A}$; and the third causes the first and second rows of $\mathbf{A}$ to be interchanged.

Now let the sequence $\mathbf{E}_1, \mathbf{E}_2, \ldots, \mathbf{E}_r$ of elementary matrices reduce $\mathbf{A}$ to the unit matrix $\mathbf{I}$, i.e.

$$\mathbf{E}_r \cdots \mathbf{E}_2 \mathbf{E}_1 \mathbf{A} = \mathbf{I}.$$

This means that $\mathbf{A}^{-1} = \mathbf{E}_r \cdots \mathbf{E}_2 \mathbf{E}_1$ is the *inverse* of $\mathbf{A}$, i.e. $\mathbf{A}^{-1}\mathbf{A} = \mathbf{I}$. Also,

$$\mathbf{A} = \mathbf{AI} = \mathbf{A}(\mathbf{A}^{-1}\mathbf{A}) = (\mathbf{AA}^{-1})\mathbf{A}$$

$$\therefore \quad \mathbf{AA}^{-1} = \mathbf{A}^{-1}\mathbf{A} = \mathbf{I}.$$

Construction of $\mathbf{A}^{-1}$ The practical determination of $\mathbf{A}^{-1}$ can be accomplished by first noting that

$$\mathbf{E}_r \cdots \mathbf{E}_2 \mathbf{E}_1 \mathbf{A} \equiv \mathbf{E}_r \ldots \mathbf{E}_2 \mathbf{E}_1 \mathbf{I} \mathbf{A} = (\mathbf{E}_r \cdots \mathbf{E}_2 \mathbf{E}_1 \mathbf{I})\mathbf{A},$$

$$\therefore \quad \mathbf{A}^{-1} = \mathbf{E}_r \cdots \mathbf{E}_2 \mathbf{E}_1 \mathbf{I}.$$

Thus $\mathbf{A}^{-1}$ can be constructed in a step-by-step manner by applying the elementary row operations to the unit matrix $\mathbf{I}$ at the same time as they are applied to $\mathbf{A}$.

Example   Find $\mathbf{A}^{-1}$ when $\mathbf{A} = \begin{pmatrix} 1 & 2 & 3 \\ 2 & 3 & 4 \\ 3 & 4 & 6 \end{pmatrix}$.

The solution is set out starting with the two arrays for $\mathbf{A}$ and $\mathbf{I}$ and applying identical row operations to each:

$$\mathbf{A} = \begin{pmatrix} 1 & 2 & 3 \\ 2 & 3 & 4 \\ 3 & 4 & 6 \end{pmatrix} \quad \begin{pmatrix} 1 & 0 & 0 \\ 0 & 1 & 0 \\ 0 & 0 & 1 \end{pmatrix} = \mathbf{I}$$

Row $2 - 2 \times$ Row 1: $\begin{pmatrix} 1 & 2 & 3 \\ 0 & -1 & -2 \\ 3 & 4 & 6 \end{pmatrix} \quad \begin{pmatrix} 1 & 0 & 0 \\ -2 & 1 & 0 \\ 0 & 0 & 1 \end{pmatrix} = \mathbf{E}_1 \mathbf{I}$

Row $3 - 3 \times$ Row 1: $\begin{pmatrix} 1 & 2 & 3 \\ 0 & -1 & -2 \\ 0 & -2 & -3 \end{pmatrix} \quad \begin{pmatrix} 1 & 0 & 0 \\ -2 & 1 & 0 \\ -3 & 0 & 1 \end{pmatrix} = \mathbf{E}_2 \mathbf{E}_1 \mathbf{I}$

Row 3 − 2 × Row 2: $\begin{pmatrix} 1 & 2 & 3 \\ 0 & -1 & -2 \\ 0 & 0 & 1 \end{pmatrix} \begin{pmatrix} 1 & 0 & 0 \\ -2 & 1 & 0 \\ 1 & -2 & 1 \end{pmatrix} = \mathbf{E}_3\mathbf{E}_2\mathbf{E}_1\mathbf{I}$

Row 2 + 2 × Row 3: $\begin{pmatrix} 1 & 2 & 3 \\ 0 & -1 & 0 \\ 0 & 0 & 1 \end{pmatrix} \begin{pmatrix} 1 & 0 & 0 \\ 0 & -3 & 2 \\ 1 & -2 & 1 \end{pmatrix} = \mathbf{E}_4\mathbf{E}_3\mathbf{E}_2\mathbf{E}_1\mathbf{I}$

$-1 \times$ Row 2: $\begin{pmatrix} 1 & 2 & 3 \\ 0 & 1 & 0 \\ 0 & 0 & 1 \end{pmatrix} \begin{pmatrix} 1 & 0 & 0 \\ 0 & 3 & -2 \\ 1 & -2 & 1 \end{pmatrix} = \mathbf{E}_5\mathbf{E}_4\mathbf{E}_3\mathbf{E}_2\mathbf{E}_1\mathbf{I}$

Row 1 − 3 × Row 3: $\begin{pmatrix} 1 & 2 & 0 \\ 0 & 1 & 0 \\ 0 & 0 & 1 \end{pmatrix} \begin{pmatrix} -2 & 6 & -3 \\ 0 & 3 & -2 \\ 1 & -2 & 1 \end{pmatrix} = \mathbf{E}_6\mathbf{E}_5\mathbf{E}_4\mathbf{E}_3\mathbf{E}_2\mathbf{E}_1\mathbf{I}$

Row 1 − 2 × Row 2: $\begin{pmatrix} 1 & 0 & 0 \\ 0 & 1 & 0 \\ 0 & 0 & 1 \end{pmatrix} \begin{pmatrix} -2 & 0 & 1 \\ 0 & 3 & -2 \\ 1 & -2 & 1 \end{pmatrix} \begin{aligned} &= \mathbf{E}_7\mathbf{E}_6\mathbf{E}_5\mathbf{E}_4\mathbf{E}_3\mathbf{E}_2\mathbf{E}_1\mathbf{I} \\ &= \mathbf{A}^{-1}. \end{aligned}$

Problems 6E

Evaluate $\mathbf{A}^{-1}$ for

1. $\mathbf{A} = \begin{pmatrix} -23 & 11 & 1 \\ 11 & -3 & -2 \\ 1 & -2 & 1 \end{pmatrix}.$ $\left[\mathbf{A}^{-1} = \begin{pmatrix} 7 & 13 & 19 \\ 13 & 24 & 35 \\ 19 & 35 & 52 \end{pmatrix} \right]$

2. $\mathbf{A} = \begin{pmatrix} 3 & 2 & 6 \\ 1 & 1 & 2 \\ 2 & 2 & 5 \end{pmatrix}.$ $\left[\mathbf{A}^{-1} = \begin{pmatrix} 1 & 2 & -2 \\ -1 & 3 & 0 \\ 0 & -2 & 1 \end{pmatrix} \right]$

3. $\mathbf{A} = \begin{pmatrix} 3 & -2 & 0 & -1 \\ 0 & 2 & 2 & 1 \\ 1 & -2 & -3 & -2 \\ 0 & 1 & 2 & 1 \end{pmatrix}.$ $\left[\mathbf{A}^{-1} = \begin{pmatrix} 1 & 1 & -2 & -4 \\ 0 & 1 & 0 & -1 \\ -1 & -1 & 3 & 6 \\ 2 & 1 & -6 & -10 \end{pmatrix} \right]$

4. Show that elementary *column* operations on a matrix $\mathbf{A}$ can be performed by *postmultiplication* by an elementary matrix $\mathbf{E}$.

5. Prove that det $(\mathbf{EA}) = \det \mathbf{E} \cdot \det \mathbf{A}$ and det $(\mathbf{AE}) = \det \mathbf{A} \cdot \det \mathbf{E}$.

6. Deduce the formula det $(\mathbf{AB}) = \det \mathbf{A} \cdot \det \mathbf{B}$ from questions 4 and 5.

Column rank   Elementary $m \times m$ matrices can be applied to an arbitrary matrix $\mathbf{A}$ of order $m \times n$ to reduce it to row echelon form. The number of zero rows in the row echelon representation is called the **nullity** of $\mathbf{A}$. Because row operations do not affect the rank $\mathcal{R}$ of $\mathbf{A}$ we know that

$$\text{rank } \mathbf{A} = m - \text{nullity}.$$

The maximum number of linearly independent columns of $\mathbf{A}$ is called the *column rank*. It is easy to verify that the column rank of a *row echelon* matrix is the same as its row rank $\mathcal{R}$. But, it can also be verified that elementary row operations do not affect the column rank of a matrix. Hence, for any matrix $\mathbf{A}$

$$\text{Column rank of } \mathbf{A} = \text{Row rank of } \mathbf{A}.$$

6.7   Cramer's Rule

A formal representation of the solution of n linearly independent equations in n unknowns

$$\begin{pmatrix} a_{11} & a_{12} & \cdots & a_{1n} \\ a_{21} & a_{22} & \cdots & a_{2n} \\ . & . & \cdots & . \\ a_{n1} & a_{n2} & \cdots & a_{nn} \end{pmatrix} \begin{pmatrix} x_1 \\ x_2 \\ . \\ x_n \end{pmatrix} = \begin{pmatrix} b_1 \\ b_2 \\ . \\ b_n \end{pmatrix}, \quad \text{i.e.} \quad \mathbf{A}\mathbf{x} = \mathbf{b}, \quad (6.7.1)$$

is obtained by premultiplying both sides of the equation by $\mathbf{A}^{-1}$, to obtain

$$\mathbf{x} = \mathbf{A}^{-1}\mathbf{b}.$$

For large systems of equations the use of this formula is frequently impracticable compared, say, to the direct computation of $\mathbf{x}$ by Gauss elimination. However, the method is in principle sufficiently simple and of sufficiently wide application to warrant detailed discussion.

According to equation (6.5.4) the determinant of the coefficients $[a_{ij}]$ of $\mathbf{A}$ can be expanded 'by columns' in the form

$$\det \mathbf{A} = \sum_{k=1}^{n} a_{ki} \mathbf{A}_{ki},$$

where $\mathbf{A}_{ki}$ is the *cofactor* of a_{ki}, which is equal to $(-1)^{i+k}$ times the determinant formed by deleting row k and column i from $\mathbf{A}$.

Now

$$\sum_{k=1}^{n} a_{ki}\mathbf{A}_{kj} \equiv 0 \quad \text{when } i \neq j,$$

because it represents a determinant in which columns i and j are identical. Hence

$$\mathbf{A}^{-1} = \frac{1}{\det \mathbf{A}}[A_{ij}]^{\mathrm{T}} = \frac{1}{\det \mathbf{A}}\begin{pmatrix} A_{11} & A_{21} & \cdots & A_{n1} \\ A_{12} & A_{22} & \cdots & A_{n2} \\ . & . & \cdots & . \\ A_{1n} & A_{2n} & \cdots & A_{nn} \end{pmatrix}, \quad (6.7.2)$$

and the solution of (6.7.1) becomes

$$x_i = \frac{1}{\det \mathbf{A}}\sum_{k=1}^{n} A_{ki}b_k, \quad 1 \leq i \leq n.$$

The sum on the right of this equation is just the determinant Δ_i, say, obtained when the ith column of $\mathbf{A}$ is replaced by $\mathbf{b}$. This observation gives Cramer's rule for solving (6.7.1)

$$x_i = \frac{\Delta_i}{\Delta}, \quad \Delta = \det \mathbf{A}, \quad 1 \leq i \leq n. \quad (6.7.3)$$

Cramer's formula for the solution of a linear system of equations is useful in practice only for $n \leq 4$. It is evident that difficulties will arise when $\mathbf{A}$ is *singular*, i.e. when $\Delta \equiv \det \mathbf{A} = 0$. Then the n linear equations are *not* linearly independent, and $\mathbf{A}^{-1}$ does not exist. On the other hand Cramer's formula shows that the solution of a system of *homogeneous* equations (for which $\mathbf{b} = \mathbf{0}$) is $\mathbf{x} = \mathbf{0}$ unless $\Delta = \det \mathbf{A} = 0$. Non-zero solutions exist only when $\Delta = 0$.

Problems 6F

Use Cramer's rule to solve:

1. $\begin{aligned} x_1 + x_2 + x_3 &= 5 \\ 2x_1 - x_2 + x_3 &= 7 \\ 3x_1 + x_2 - 5x_3 &= 13. \end{aligned}$ $[\mathbf{x} = (4, 1, 0); \Delta_1 = 88, \Delta_2 = 22, \Delta_3 = 0, \Delta = 22]$

$$4x_1 - 3x_2 + 2x_3 = -7$$
2. $6x_1 + 2x_2 - 3x_3 = 33 \quad \left[\mathbf{x} = \left(\frac{5}{2}, 3, -4\right)\right]$
$$2x_1 - 4x_2 - x_3 = -3.$$

3. Show that the three straight lines

$$\left.\begin{array}{l} a_1x + b_1y + c_1 = 0 \\ a_2x + b_2y + c_2 = 0 \\ a_3x + b_3y + c_3 = 0 \end{array}\right\} \quad \text{intersect provided} \quad \begin{vmatrix} a_1 & b_1 & c_1 \\ a_2 & b_2 & c_2 \\ a_3 & b_3 & c_3 \end{vmatrix} = 0.$$

4. Show that the three equations

$$\left.\begin{array}{l} -\lambda x + y + 2z = 0 \\ x + \lambda y + 3z = 0 \\ x + 3y + \lambda z = 0 \end{array}\right\} \quad \text{are consistent when } \lambda = 3.$$

$$\left[\begin{vmatrix} -\lambda & 1 & 2 \\ 1 & \lambda & 3 \\ 1 & 3 & \lambda \end{vmatrix} = 0 \text{ when } \lambda = 3\right]$$

Change of variable in multiple integrals The evaluation of a multiple integral of the form

$$\iint \cdots \int f(x_1, x_2, \ldots, x_n) dx_1\, dx_2 \ldots dx_n,$$

is often simplified by means of a judicious transformation in the integration variables, from x_i $(i = 1, 2, \ldots, n)$ to $\xi_i(x_1, x_2, \ldots, x_n)$.

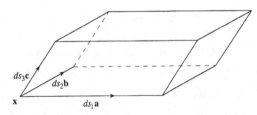

When $n = 3$ it is usual to proceed as follows. Let $\mathbf{a}, \mathbf{b}, \mathbf{c}$, be unit vectors at $\mathbf{x}$ respectively in the directions of increasing ξ_1, ξ_2, ξ_3. These vectors are not necessarily mutually perpendicular, but form the three edges (of unit length) through $\mathbf{x}$ of a parallelopiped, whose volume is just equal to $\mathbf{a} \cdot \mathbf{b} \times \mathbf{c}$ (§2.1). The volume dV of an elementary parallelopiped with edges $\mathbf{a}ds_1, \mathbf{b}ds_2, \mathbf{c}ds_3$, respectively of lengths ds_1, ds_2, ds_3, is therefore given by

$$dV = \mathbf{a} \cdot \mathbf{b} \times \mathbf{c}\, ds_1\, ds_2\, ds_3.$$

But $x_i = x_i(\xi_1, \xi_2, \xi_3)$, so that

$$
\left.
\begin{aligned}
\mathbf{a}ds_1 &= \left(\frac{\partial x_1}{\partial \xi_1}\mathbf{i} + \frac{\partial x_2}{\partial \xi_1}\mathbf{j} + \frac{\partial x_3}{\partial \xi_1}\mathbf{k} \right) d\xi_1 \\[2mm]
\mathbf{b}ds_2 &= \left(\frac{\partial x_1}{\partial \xi_2}\mathbf{i} + \frac{\partial x_2}{\partial \xi_2}\mathbf{j} + \frac{\partial x_3}{\partial \xi_2}\mathbf{k} \right) d\xi_2 \\[2mm]
\mathbf{c}ds_3 &= \left(\frac{\partial x_1}{\partial \xi_3}\mathbf{i} + \frac{\partial x_2}{\partial \xi_3}\mathbf{j} + \frac{\partial x_3}{\partial \xi_3}\mathbf{k} \right) d\xi_3.
\end{aligned}
\right\}
$$

$$
\therefore \quad dV = \begin{vmatrix} \dfrac{\partial x_1}{\partial \xi_1} & \dfrac{\partial x_2}{\partial \xi_1} & \dfrac{\partial x_3}{\partial \xi_1} \\[3mm] \dfrac{\partial x_1}{\partial \xi_2} & \dfrac{\partial x_2}{\partial \xi_2} & \dfrac{\partial x_3}{\partial \xi_2} \\[3mm] \dfrac{\partial x_1}{\partial \xi_3} & \dfrac{\partial x_2}{\partial \xi_3} & \dfrac{\partial x_3}{\partial \xi_3} \end{vmatrix} d\xi_1\, d\xi_2\, d\xi_3.
$$

Hence,

$$
\iiint f(x_1, x_2, x_3) dx_1\, dx_2\, dx_3
$$

$$
= \iiint f(x_1, x_2, x_3) \frac{\partial(x_1, x_2, x_3)}{\partial(\xi_1, \xi_2, \xi_3)} d\xi_1 d\xi_2 d\xi_3,
$$

where

$$
\frac{\partial(x_1, x_2, x_3)}{\partial(\xi_1, \xi_2, \xi_3)} = \begin{vmatrix} \dfrac{\partial x_1}{\partial \xi_1} & \dfrac{\partial x_2}{\partial \xi_1} & \dfrac{\partial x_3}{\partial \xi_1} \\[3mm] \dfrac{\partial x_1}{\partial \xi_2} & \dfrac{\partial x_2}{\partial \xi_2} & \dfrac{\partial x_3}{\partial \xi_2} \\[3mm] \dfrac{\partial x_1}{\partial \xi_3} & \dfrac{\partial x_2}{\partial \xi_3} & \dfrac{\partial x_3}{\partial \xi_3} \end{vmatrix}
$$

is the *Jacobian* of x_1, x_2, x_3 with respect to ξ_1, ξ_2, ξ_3.

We have implicitly assumed that the unit vectors $\mathbf{a}, \mathbf{b}, \mathbf{c}$ form a skew but right-handed system, so that $\mathbf{a} \cdot \mathbf{b} \times \mathbf{c} > 0$. In general this will not be true, and the formula for the change of variable when $d\xi_1, d\xi_2, d\xi_3$

are regarded as *positive* is more properly written

$$\iiint f(x_1, x_2, x_3)dx_1\, dx_2\, dx_3$$

$$= \iiint f(x_1, x_2, x_3) \left| \frac{\partial(x_1, x_2, x_3)}{\partial(\xi_1, \xi_2, \xi_3)} \right| d\xi_1\, d\xi_2\, d\xi_3. \quad (6.7.4)$$

Geometrical arguments are not easily used to extend this formula to higher dimensions $(n > 3)$. A formal analytical procedure based on Cramer's rule will now be outlined for $n = 3$. We proceed in a step-by-step manner, first eliminating the integration variable x_1 in favour of ξ_1, then x_2 in favour of ξ_2, etc.

Consider first the small changes $d\xi_i$ in ξ_i $(i = 1, 2, 3)$ produced by a small increase dx_1 in x_1 when x_2, x_3 are held fixed. These changes can be calculated from the simultaneous equations

$$\frac{\partial x_1}{\partial \xi_1}d\xi_1 + \frac{\partial x_1}{\partial \xi_2}d\xi_2 + \frac{\partial x_1}{\partial \xi_3}d\xi_3 = dx_1$$

$$\frac{\partial x_2}{\partial \xi_1}d\xi_1 + \frac{\partial x_2}{\partial \xi_2}d\xi_2 + \frac{\partial x_2}{\partial \xi_3}d\xi_3 = 0$$

$$\frac{\partial x_3}{\partial \xi_1}d\xi_1 + \frac{\partial x_3}{\partial \xi_2}d\xi_2 + \frac{\partial x_3}{\partial \xi_3}d\xi_3 = 0.$$

Solving for $d\xi_1$ by Cramer's rule (6.7.3)

$$d\xi_1 = \frac{\begin{vmatrix} \frac{\partial x_2}{\partial \xi_2} & \frac{\partial x_3}{\partial \xi_2} \\ \frac{\partial x_2}{\partial \xi_3} & \frac{\partial x_3}{\partial \xi_3} \end{vmatrix}}{\begin{vmatrix} \frac{\partial x_1}{\partial \xi_1} & \frac{\partial x_2}{\partial \xi_1} & \frac{\partial x_3}{\partial \xi_1} \\ \frac{\partial x_1}{\partial \xi_2} & \frac{\partial x_2}{\partial \xi_2} & \frac{\partial x_3}{\partial \xi_2} \\ \frac{\partial x_1}{\partial \xi_3} & \frac{\partial x_2}{\partial \xi_3} & \frac{\partial x_3}{\partial \xi_3} \end{vmatrix}} dx_1 \equiv \frac{\frac{\partial(x_2, x_3)}{\partial(\xi_2, \xi_3)}}{\frac{\partial(x_1, x_2, x_3)}{\partial(\xi_1, \xi_2, \xi_3)}} dx_1.$$

If we regard this equation as defining a change in variable from x_1 to ξ_1 then

$$\iiint f(x_1, x_2, x_3)dx_1\, dx_2\, dx_3$$

$$= \iiint f(x_1, x_2, x_3) \left[\frac{\partial(x_1, x_2, x_3)}{\partial(\xi_1, \xi_2, \xi_3)} \middle/ \frac{\partial(x_2, x_3)}{\partial(\xi_2, \xi_3)} \right] d\xi_1\, dx_2\, dx_3.$$

Next consider the changes produced when x_2 increases by dx_2 with ξ_1 and x_3 held constant. Then

$$
\left.
\begin{aligned}
\frac{\partial x_2}{\partial \xi_2}d\xi_2 + \frac{\partial x_2}{\partial \xi_3}d\xi_3 &= dx_2 \\[2mm]
\frac{\partial x_3}{\partial \xi_2}d\xi_2 + \frac{\partial x_3}{\partial \xi_3}d\xi_3 &- 0.
\end{aligned}
\right\}
\quad \therefore \quad
d\xi_2 = \frac{\frac{\partial x_3}{\partial \xi_3}}{\frac{\partial(x_2,x_3)}{\partial(\xi_1,\xi_3)}}\, dx_2,
$$

and therefore

$$
\iiint f(x_1, x_2, x_3)dx_1\, dx_2\, dx_3
$$

$$
= \iiint f(x_1, x_2, x_3)\left[\frac{\partial(x_1, x_2, x_3)}{\partial(\xi_1, \xi_2, \xi_3)}\middle/ \frac{\partial x_3}{\partial \xi_3}\right] d\xi_1 d\xi_2 dx_3.
$$

Finally, the change $d\xi_3$ produced by the variation dx_3 with ξ_1, ξ_2 held constant, is just

$$
d\xi_3 = \frac{dx_3}{\partial x_3/\partial \xi_3},
$$

$$
\therefore \quad \iiint f(x_1, x_2, x_3)dx_1\, dx_2\, dx_3
$$

$$
= \iiint f(x_1, x_2, x_3)\frac{\partial(x_1, x_2, x_3)}{\partial(\xi_1, \xi_2, \xi_3)}d\xi_1 d\xi_2 d\xi_3,
$$

as before.

In general, therefore, we arrive at the transformation formula for n-dimensional integrals

$$
\iint \cdots \int f(x_1, x_2, \ldots, x_n)dx_1 dx_2 \ldots dx_n
$$

$$
= \iint \cdots \int f(x_1, x_2, \ldots, x_n)\left|\frac{\partial(x_1, x_2, \ldots, x_n)}{\partial(\xi_1, \xi_2, \ldots, \xi_n)}\right| d\xi_1 d\xi_2 \ldots d\xi_n.
$$

$$(6.7.5)$$

Example

If
$$
\frac{\partial(x_i)}{\partial(\xi_i)} \equiv \frac{\partial(x_1, x_2, \ldots, x_n)}{\partial(\xi_1, \xi_2, \ldots, \xi_n)}, \quad \frac{\partial(\xi_i)}{\partial(\eta_i)} \equiv \frac{\partial(\xi_1, \xi_2, \ldots, \xi_n)}{\partial(\eta_1, \eta_2, \ldots, \eta_n)}
$$

show that
$$
\frac{\partial(x_i)}{\partial(\eta_i)} = \frac{\partial(x_i)}{\partial(\xi_i)}\frac{\partial(\xi_i)}{\partial(\eta_i)}, \quad \text{i.e.} \quad \frac{\partial(x_1, x_2, \ldots, x_n)}{\partial(\eta_1, \eta_2, \ldots, \eta_n)} = \frac{\partial(x_1, x_2, \ldots, x_n)}{\partial(\xi_1, \xi_2, \ldots, \xi_n)}\frac{\partial(\xi_1, \xi_2, \ldots, \xi_n)}{\partial(\eta_1, \eta_2, \ldots, \eta_n)}.
$$

$$(6.7.6)$$

This is true because

$$\frac{\partial(x_i)}{\partial(\eta_i)} = \det\left[\frac{\partial x_i}{\partial \eta_j}\right] = \det\left[\sum_{k=1}^{n}\frac{\partial x_i}{\partial \xi_k}\frac{\partial \xi_k}{\partial \eta_j}\right] \equiv \det\left[\frac{\partial x_i}{\partial \xi_j}\right]\cdot\det\left[\frac{\partial \xi_i}{\partial \eta_j}\right].$$

Problems 6G

Evaluate the Jacobians of the following transformations:

1. Polar coordinates $x = r\cos\theta, y = r\sin\theta.$ $\left[\dfrac{\partial(x,y)}{\partial(r,\theta)} = r\right]$

2. Cylindrical polar coordinates $x = r\cos\theta, y = r\sin\theta, z = z.$ $\left[\dfrac{\partial(x,y,z)}{\partial(r,\theta,z)} = r\right]$

3. Spherical polar coordinates $x = r\sin\theta\cos\phi, y = r\sin\theta\sin\phi, z = r\cos\theta.$
$\left[\dfrac{\partial(x,y,z)}{\partial(r,\theta,\phi)} = r^2\sin\theta\right]$

4. Use the substitution $\xi = x+y, \eta = y/x$ to show that

$$I = \int_0^2 dy \int_y^{4-y}\left(1+\frac{y}{x}\right)^2 e^{(x^2-y^2)/x}dx = e^4 - 5.$$

$\left[\vphantom{\int}\right.$The region of integration is the shaded area of the figure.

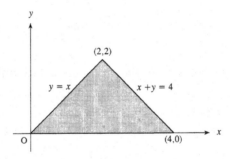

In the (ξ,η)-plane the limits of integration are $0 < \xi < 4, 0 < \eta < 1$, and

$$\frac{\partial(x,y)}{\partial(\xi,\eta)} = \frac{\xi}{(1+\eta)^2}, \quad \therefore \quad I = \int_0^4 d\xi \int_0^1 \xi e^{\xi(1-\eta)}d\eta = \int_0^4\left(e^\xi - 1\right)d\xi = e^4 - 5.\left.\vphantom{\int}\right]$$

5. By making the change of variable $\xi = xy, \eta = y - x$ show that

$$\iint_D x^2 y^2(y^2 - x^2)dx\,dy = 84,$$

where D is the region in $x > 0, y > 0$ between the hyperbolae $xy = 1$ and $xy = 4$ and the straight lines $y = x+1, y = x+3$.

6. Use the transformation $\xi = x/a, \eta = y/b, \zeta = z/c$ to show that the volume of the ellipsoid

$$\frac{x^2}{a^2} + \frac{y^2}{b^2} + \frac{z^2}{c^2} \leq 1$$

is $\frac{4}{3}\pi abc$.

6.8 Eigenvalue Problems

Set $\mathbf{b} = \mathbf{x}'$ in the $n \times n$ matrix equation (6.7.1) and write the equation in the reversed order:

$$\mathbf{x}' = \mathbf{A}\mathbf{x}. \tag{6.8.1}$$

The matrix *operator* $\mathbf{A}$ in this equation can be regarded as effecting a *linear transformation* of the n-dimensional vector $\mathbf{x}$ to a new n-dimensional vector $\mathbf{x}'$. This is a familiar procedure in two- and three-dimensional coordinate geometry. Thus, the transformation defined by

$$\mathbf{A} = \begin{pmatrix} \cos\theta & -\sin\theta \\ \sin\theta & \cos\theta \end{pmatrix} \quad \longrightarrow \quad \begin{cases} x' = x\cos\theta - y\sin\theta \\ y' = x\sin\theta + y\cos\theta. \end{cases} \tag{6.8.2}$$

rotates the point $\mathbf{x} = (x, y)$ about the origin through an angle θ onto the image point $\mathbf{x}' = (x', y')$. It can also be regarded as defining the new coordinates of $\mathbf{x}$ when the coordinate axes are rotated through angle θ in the *negative* direction; both cases are illustrated in the figure.

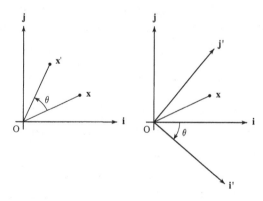

Similarly, in three dimensions, let the coordinate axes be rotated about the origin such that the new coordinate unit vectors $\mathbf{i}', \mathbf{j}', \mathbf{k}'$ are given in terms of the original coordinate unit vectors $\mathbf{i}, \mathbf{j}, \mathbf{k}$ by

$$\left.\begin{array}{l} \mathbf{i}' = l_1\mathbf{i} + m_1\mathbf{j} + n_1\mathbf{k} \\ \mathbf{j}' = l_2\mathbf{i} + m_2\mathbf{j} + n_2\mathbf{k} \\ \mathbf{k}' = l_3\mathbf{i} + m_3\mathbf{j} + n_3\mathbf{k} \end{array}\right\} \quad \text{then} \quad \mathbf{A} = \begin{pmatrix} l_1 & m_1 & n_1 \\ l_2 & m_2 & n_2 \\ l_3 & m_3 & n_3 \end{pmatrix} \tag{6.8.3}$$

and $\mathbf{x}' = \mathbf{A}\mathbf{x}$ defines the coordinates in the new reference frame in terms of the original coordinates $\mathbf{x}$.

The rotational transformations (6.8.2) and (6.8.3) are very special. A general matrix transformation (6.8.1) does not normally exhibit such a simple geometrical interpretation. It often happens that there exist one or more vectors $\mathbf{x}$ that are *parallel* to their respective image vectors $\mathbf{x}'$. Then $\mathbf{x}'$ is *proportional* to $\mathbf{x}$, and therefore

$$\mathbf{A}\mathbf{x} = \lambda\mathbf{x}, \tag{6.8.4}$$

i.e.

$$a_{11}x_1 + a_{12}x_2 + \cdots + a_{1n}x_n = \lambda x_1$$
$$a_{21}x_1 + a_{22}x_2 + \cdots + a_{2n}x_n = \lambda x_2$$
$$\cdots \qquad \cdots \qquad \cdots \qquad \cdots$$
$$a_{n1}x_1 + a_{n2}x_2 + \cdots + a_{nn}x_n = \lambda x_n,$$

where the magnification factor λ is called the *eigenvalue* for the vector $\mathbf{x}$, which is called an *eigenvector*. These n homogeneous equations would normally be written

$$\left.\begin{array}{l} (a_{11} - \lambda)x_1 + \quad a_{12}x_2 + \cdots + \quad a_{1n}x_n = 0 \\ a_{21}x_1 + (a_{22} - \lambda)x_2 + \cdots + \quad a_{2n}x_n = 0 \\ \cdots \qquad \cdots \quad \cdots \qquad \cdots \\ a_{n1}x_1 + \quad a_{n2}x_2 + \cdots + (a_{nn} - \lambda)x_n = 0 \end{array}\right\}$$

$$\text{or} \quad (\mathbf{A} - \lambda\mathbf{I})\mathbf{x} = 0. \tag{6.8.5}$$

Now the solution of a system of n homogeneous equations of this sort is $\mathbf{x} = 0$ *unless* the determinant of the coefficients vanishes, i.e. unless

$$\det(\mathbf{A} - \lambda\mathbf{I}) \equiv \begin{vmatrix} a_{11} - \lambda & a_{12} & \cdots & a_{1n} \\ a_{21} & a_{22} - \lambda & \cdots & a_{2n} \\ \cdots & \cdots & \cdots & \cdots \\ a_{n1} & a_{n2} & \cdots + a_{nn} - \lambda \end{vmatrix} = 0. \tag{6.8.6}$$

This is called the *characteristic equation* which (expanding the determinant) is equivalent to the nth order algebraic equation

$$(-1)^n \lambda^n + \left((-1)^{n-1} \sum_{j=1}^{n} a_{jj} \right) \lambda^{n-1} + \cdots + \det \mathbf{A} = 0. \qquad (6.8.7)$$

The polynomial on the left is called the *characteristic polynomial.* The eigenvalues of $\mathbf{A}$ are the roots $\lambda = \lambda_1, \lambda_2, \ldots, \lambda_n$ of the characteristic equation. Because the constant term on the left of (6.8.7) is just the product of the roots, we see that

$$\det \mathbf{A} = \lambda_1 \lambda_2 \ldots \lambda_n. \qquad (6.8.8)$$

Hence, the vanishing of an eigenvalue implies that $\mathbf{A}$ is singular. Note also that, because $\det(\mathbf{A}^T - \lambda \mathbf{I}) = \det(\mathbf{A} - \lambda \mathbf{I})$, the matrix $\mathbf{A}$ and its transpose $\mathbf{A}^T$ have the same eigenvalues.

For every root $\lambda = \lambda_i$ of the characteristic equation (6.8.6), a solution $\mathbf{x} = \mathbf{u}_i$, say, of the homogeneous equations (6.8.5) can be found. In general there are

$$n \text{ distinct eigenvalues: } \lambda_1, \lambda_2, \ldots, \lambda_n$$
$$\text{and } n \text{ corresponding eigenvectors: } \mathbf{u}_1, \mathbf{u}_2, \ldots, \mathbf{u}_n.$$

Because the eigenvectors are solutions of the homogeneous system (6.8.6) they are undetermined to within a multiplicative factor, i.e. their *orientations* are uniquely determined, but their *lengths* are arbitrary.

Example Calculate the eigenvectors and eigenvalues of

$$\mathbf{A} = \begin{pmatrix} 33 & 16 & 72 \\ -24 & -10 & -57 \\ -8 & -4 & -17 \end{pmatrix}.$$

The characteristic equation is

$$\lambda^3 - 6\lambda^2 + 11\lambda - 6 = 0$$

with roots

$$\lambda_1 = 1, \quad \lambda_2 = 2, \quad \lambda_3 = 3.$$

For $\lambda = \lambda_1 = 1$ the eigenvector equations (6.8.5) become

$$\left. \begin{array}{r} 32x_1 + 16x_2 + 72x_3 = 0 \\ -24x_1 - 11x_2 - 57x_3 = 0 \\ -8x_1 - 4x_2 - 18x_3 = 0 \end{array} \right\} \quad \therefore \quad \mathbf{u}_1 = \mu \begin{pmatrix} -15 \\ 12 \\ 4 \end{pmatrix}$$

where μ is arbitrary. (The system is of rank $\mathcal{R} = 2$, and the general solution therefore involves $3 - \mathcal{R} = 1$ arbitrary parameter; see §6.3.)

Similar calculations performed for $\lambda = \lambda_2, \lambda_3$ reveal that the eigenvectors $\mathbf{u}_2, \mathbf{u}_3$ are also one parameter vectors (in each case $\mathcal{R} = 2$). The results can be summarised as follows

$$\lambda_1 = 1, \ \mathbf{u}_1 = \begin{pmatrix} -15 \\ 12 \\ 4 \end{pmatrix}; \quad \lambda_2 = 2, \ \mathbf{u}_2 = \begin{pmatrix} -16 \\ 13 \\ 4 \end{pmatrix}; \quad \lambda_3 = 3, \ \mathbf{u}_3 = \begin{pmatrix} -4 \\ 3 \\ 1 \end{pmatrix},$$

where values of the three arbitrary parameters have been chosen to eliminate fractional components of the eigenvectors.

Example Calculate the eigenvectors and eigenvalues of

$$\mathbf{A} = \begin{pmatrix} 2 & -2 & 3 \\ -2 & -1 & 6 \\ 1 & 2 & 0 \end{pmatrix}.$$

The characteristic equation is

$$\lambda^3 - \lambda^2 - 21\lambda + 45 = 0, \quad \therefore \quad \lambda_1 = -5, \quad \lambda_2 = \lambda_3 = 3.$$

In this case we find

$$\lambda_1 = -5, \ \mathbf{u}_1 = \begin{pmatrix} 1 \\ 2 \\ -1 \end{pmatrix}; \quad \lambda_2 = \lambda_3 = 3, \ \mathbf{u}_2 = \begin{pmatrix} -2 \\ 1 \\ 0 \end{pmatrix}, \ \mathbf{u}_3 = \begin{pmatrix} 3 \\ 0 \\ 1 \end{pmatrix}.$$

In the second example the eigenvalue $\lambda = 3$ is said to have *algebraic multiplicity* 2. For this case, however, the eigenvector equation $(\mathbf{A} - 3\mathbf{I})\mathbf{x} = 0$ turns out to have rank $\mathcal{R} = 1$, so there are two linearly independent solutions, i.e. the *geometric multiplicity* of the eigenvalue $\lambda = 3$ is also 2.

When a square matrix has q *distinct* eigenvalues the corresponding eigenvectors are *linearly independent* because if, on the contrary, at most m of the eigenvectors $\mathbf{u}_1, \mathbf{u}_2, \ldots, \mathbf{u}_m$ were independent (where $m < q$) then there must exist constants $\alpha_1, \ldots, \alpha_m$ (not all zero) such that

$$\mathbf{u}_{m+1} = \alpha_1 \mathbf{u}_1 + \alpha_2 \mathbf{u}_2 + \cdots + \alpha_m \mathbf{u}_m. \tag{6.8.9}$$

However, by premultiplication by $\mathbf{A}$, this means that

$$\lambda_{m+1} \mathbf{u}_{m+1} = \alpha_1 \lambda_1 \mathbf{u}_1 + \alpha_2 \lambda_2 \mathbf{u}_2 + \cdots + \alpha_m \lambda_m \mathbf{u}_m.$$

and therefore, eliminating $\mathbf{u}_{m+1}$, that

$$\alpha_1(\lambda_{m+1} - \lambda_1)\mathbf{u}_1 + \alpha_2(\lambda_{m+1} - \lambda_2)\mathbf{u}_2 + \cdots + \alpha_m(\lambda_{m+1} - \lambda_m)\mathbf{u}_m = 0.$$

But this is only possible if, in fact, $\alpha_1 = \alpha_2 = \cdots = \alpha_m = 0$, contrary to our assumption that $\mathbf{u}_{m+1}$ can be expressed in the form (6.8.9) with at least some non-zero α_i.

Problems 6H

Find the eigenvalues and eigenvectors of:

1. $\mathbf{A} = \begin{pmatrix} 2 & 4 \\ 5 & 3 \end{pmatrix}$ $\left[\lambda_1 = -2, \; \mathbf{u}_1 = \begin{pmatrix} 1 \\ -1 \end{pmatrix}; \quad \lambda_2 = 7, \; \mathbf{u}_2 = \begin{pmatrix} 4 \\ 5 \end{pmatrix} \right]$

2. $\mathbf{A} = \begin{pmatrix} 2 & 3 & -1 \\ 0 & -4 & 2 \\ 0 & -5 & 3 \end{pmatrix}$

$\left[\lambda_1 = -2, \; \mathbf{u}_1 = \begin{pmatrix} 1 \\ -2 \\ -2 \end{pmatrix}; \quad \lambda_2 = 1, \; \mathbf{u}_2 = \begin{pmatrix} 1 \\ -2 \\ -5 \end{pmatrix}; \quad \lambda_3 = 2, \; \mathbf{u}_3 = \begin{pmatrix} 1 \\ 0 \\ 0 \end{pmatrix} \right]$

3. $\mathbf{A} = \begin{pmatrix} 2 & 1 & 2 \\ 0 & 2 & 3 \\ 0 & 0 & 5 \end{pmatrix}$

$\left[\lambda_1 = \lambda_2 = 2, \; \mathbf{u}_1 = \mathbf{u}_2 = \begin{pmatrix} 1 \\ 0 \\ 0 \end{pmatrix}; \quad \lambda_3 = 5, \; \mathbf{u}_3 = \begin{pmatrix} 1 \\ 1 \\ 1 \end{pmatrix} \right]$

4. $\mathbf{A} = \begin{pmatrix} 6 & 4 & 4 \\ 2 & 8 & 2 \\ -4 & -8 & -2 \end{pmatrix}$

$\left[\lambda_1 = 2, \; \mathbf{u}_1 = \begin{pmatrix} -1 \\ 0 \\ 1 \end{pmatrix}; \quad \lambda_2 = 4, \; \mathbf{u}_2 = \begin{pmatrix} 2 \\ -1 \\ 0 \end{pmatrix}; \quad \lambda_3 = 6, \; \mathbf{u}_3 = \begin{pmatrix} 0 \\ 1 \\ -1 \end{pmatrix} \right]$

5. $\mathbf{A} = \begin{pmatrix} 1 & 2 & -1 \\ 1 & 0 & 1 \\ 4 & -4 & 5 \end{pmatrix}$

$\left[\lambda_1 = 1, \; \mathbf{u}_1 = \begin{pmatrix} -1 \\ 1 \\ 2 \end{pmatrix}; \quad \lambda_2 = 2, \; \mathbf{u}_2 = \begin{pmatrix} -2 \\ 1 \\ 4 \end{pmatrix}; \quad \lambda_3 = 3, \; \mathbf{u}_3 = \begin{pmatrix} -1 \\ 1 \\ 4 \end{pmatrix} \right]$

6.9 Real Symmetric Matrices

If the square matrix $\mathbf{A}$ is real and symmetric we can easily show that the eigenvalues are *real* and that eigenvectors corresponding to different eigenvalues are *orthogonal*. In the first place we have (denoting complex conjugate by an asterisk)

$$\mathbf{A}\mathbf{x} = \lambda\mathbf{x}, \quad \mathbf{A}\mathbf{x}^* = \lambda^*\mathbf{x}^* \quad \therefore \quad \mathbf{x}^{*\mathrm{T}}\mathbf{A}\mathbf{x} = \lambda|\mathbf{x}|^2, \quad \mathbf{x}^{\mathrm{T}}\mathbf{A}\mathbf{x}^* = \lambda^*|\mathbf{x}|^2.$$

But $\mathbf{x}^{*\mathrm{T}}\mathbf{A}\mathbf{x} \equiv \mathbf{x}^{\mathrm{T}}\mathbf{A}\mathbf{x}^*$, so that $\lambda^* = \lambda$ and λ is therefore real.

Next, for $\lambda_i \neq \lambda_j$ and respective eigenvectors $\mathbf{u}_i, \mathbf{u}_j$

$$(\lambda_i - \lambda_j)\mathbf{u}_i \cdot \mathbf{u}_j = \mathbf{u}_j^{\mathrm{T}}\mathbf{A}\mathbf{u}_i - \mathbf{u}_i^{\mathrm{T}}\mathbf{A}\mathbf{u}_j \equiv 0,$$

$$\therefore \quad \mathbf{u}_i \cdot \mathbf{u}_j = 0 \quad \text{when } \lambda_i \neq \lambda_j.$$

The *magnitudes* of these orthogonal eigenvectors can be chosen arbitrarily, and it is frequently convenient to normalise them to have unit length, such that

$$\mathbf{u}_i \cdot \mathbf{u}_i = 1.$$

The $\mathbf{u}_i$ are then said to be *orthonormal*.

For a *non-singular* matrix $\mathbf{A}$, that has n distinct orthonormal eigenvectors $\mathbf{u}_1, \mathbf{u}_2, \ldots, \mathbf{u}_n$ corresponding to the eigenvalues $\lambda_1, \lambda_2, \ldots, \lambda_n$, we can define the $n \times n$ matrix

$$\mathbf{T} = \begin{pmatrix} \mathbf{u}_1 & \mathbf{u}_2 & \cdots & \mathbf{u}_n \end{pmatrix} \tag{6.9.1}$$

whose columns are the *orthonormal* eigenvectors of $\mathbf{A}$, in which case

$$\mathbf{T}^{\mathrm{T}}\mathbf{T} = \begin{pmatrix} 1 & 0 & \cdots & 0 \\ 0 & 1 & \cdots & 0 \\ \cdots & \cdots & \cdots & \cdots \\ 0 & 0 & \cdots & 1 \end{pmatrix}, \quad \text{i.e.} \quad \mathbf{T}^{-1} = \mathbf{T}^{\mathrm{T}}. \tag{6.9.2}$$

A matrix satisfying this condition is said to be *orthogonal*. Also, because $\det \mathbf{T} = \det \mathbf{T}^{\mathrm{T}}$, it is clear that

$$\det \mathbf{T} = \pm 1.$$

Matrices such as (6.8.2) and (6.8.3) are orthogonal; they represent *rotational* transformations (as opposed to a rotation plus a refection in the origin) and are distinguished by det $\mathbf{T} = +1$.

Next,

$$\mathbf{AT} - \begin{pmatrix} \lambda_1 \mathbf{u}_1 & \lambda_2 \mathbf{u}_2 & \cdots & \lambda_n \mathbf{u}_n \end{pmatrix},$$

and therefore $\quad \mathbf{T}^{\mathrm{T}} \mathbf{AT} = \begin{pmatrix} \lambda_1 & 0 & \cdots & 0 \\ 0 & \lambda_2 & \cdots & 0 \\ \cdots & \cdots & \cdots & \cdots \\ 0 & 0 & \cdots & \lambda_n \end{pmatrix}.$ $\qquad$ (6.9.3)

The matrix operation in (6.9.3) is said to *diagonalise* the symmetric matrix $\mathbf{A}$.

Example Let $\Lambda_1, \Lambda_2, \ldots, \Lambda_n$ be the eigenvalues of $\mathbf{T}^{\mathrm{T}}$. By (6.8.8), these are all non-zero because det $\mathbf{T} =$ det $\mathbf{T}^{\mathrm{T}} \neq 0$ (because, respectively, the columns and rows of $\mathbf{T}$ and $\mathbf{T}^{\mathrm{T}}$ are linearly independent). Let $\mathbf{U}_j$ denote an eigenvector associated with Λ_j, then

$$\mathbf{T}^{\mathrm{T}} \mathbf{U}_j = \Lambda_j \mathbf{U}_j, \quad \text{and also } \mathbf{T} \mathbf{U}_j = \frac{\mathbf{U}_j}{\Lambda_j},$$

because $\mathbf{T}\mathbf{T}^{\mathrm{T}} = \mathbf{I}$. Hence, by (6.8.8),

$$\det \mathbf{T} = \Lambda_1 \Lambda_2 \ldots \Lambda_n = \frac{1}{\Lambda_1 \Lambda_2 \ldots \Lambda_n}, \quad \text{also } (\det \mathbf{T})^2 = 1, \quad \text{i.e.} \quad \det \mathbf{T} = \pm 1.$$

Diagonalisation of quadratic forms The general Cartesian equation for a *central quadric*

$$ax^2 + by^2 + cz^2 + 2dxy + 2exz + 2fyz = 1,$$

where a, b, c, d, e, f are real constants can be written in the form

$$(x, y, z) \begin{pmatrix} a & d & e \\ d & b & f \\ e & f & c \end{pmatrix} \begin{pmatrix} x \\ y \\ z \end{pmatrix} = 1,$$

where the matrix is *symmetric*. More generally, if $\mathbf{A} = [a_{ij}]$ is a real, $n \times n$ symmetric matrix, the equation

$$\mathbf{x}^{\mathrm{T}} \mathbf{A} \mathbf{x} = 1,$$

represents the central quadric $\sum_{i,j} a_{ij} x_i x_j = 1$ in n-dimensional space.

Let $\mathbf{A}$ have n orthonormal eigenvectors $\mathbf{u}_1, \ldots, \mathbf{u}_n$ and associated eigenvalues $\lambda_1, \ldots, \lambda_n$ and define $\mathbf{T}$ as in (6.9.1), then

$$(\mathbf{x}^{\mathrm{T}}\mathbf{T})(\mathbf{T}^{\mathrm{T}}\mathbf{A}\mathbf{T})(\mathbf{T}^{\mathrm{T}}\mathbf{x}) = 1,$$

i.e. by (6.9.3), the central quadric

$$\sum_{i,j} a_{ij} x_i x_j = 1,$$

is transformed into the *principal axis* form

$$\lambda_1 y_1^2 + \lambda_2 y_2^2 + \cdots + \lambda_n y_n^2 = 1, \qquad (6.9.4)$$

under the change of variable defined by

$$\mathbf{y} = \mathbf{T}^{\mathrm{T}}\mathbf{x}, \quad \text{i.e.} \quad y_i = \sum_{j=1}^{n} u_{ij} x_j,$$

where $\mathbf{u}_i = (u_{i1}, u_{i2}, \ldots, u_{in})$.

In the particular case in which two (or more) eigenvalues of the symmetric matrix $\mathbf{A}$ become equal, the rank of the corresponding eigenvector equation is reduced such that two (or more) mutually perpendicular eigenvectors can be chosen. Thus, there will still exist, in total, n mutually perpendicular eigenvectors, but the directions of two (or more) of them can be chosen arbitrarily. Geometrically, this would correspond to an ellipsoid that is axisymmetric about one of its principal axes.

Example The transformation $\mathbf{y} = \mathbf{T}^{\mathrm{T}}\mathbf{x}$ represents a rotation of the coordinate system, or a rotation plus a reflection. This must preserve the length $|\mathbf{x}|$ of any vector and, indeed,

$$|\mathbf{y}|^2 \equiv \mathbf{y}^{\mathrm{T}}\mathbf{y} = (\mathbf{T}^{\mathrm{T}}\mathbf{x})^{\mathrm{T}}\mathbf{T}^{\mathrm{T}}\mathbf{x} = \mathbf{x}^{\mathrm{T}}(\mathbf{T}\mathbf{T}^{\mathrm{T}})\mathbf{x} = \mathbf{x}^{\mathrm{T}}\mathbf{I}\mathbf{x} = \mathbf{x}^{\mathrm{T}}\mathbf{x} = |\mathbf{x}|^2.$$

Example Find a linear transformation that reduces

$$x_1^2 + \frac{3}{2}x_2^2 + \frac{3}{2}x_3^2 + x_2 x_3 = 1,$$

to principal axis form.

In matrix notation the equation is

$$\mathbf{x}^{\mathrm{T}}\mathbf{A}\mathbf{x} \equiv (x_1, x_2, x_3) \begin{pmatrix} 1 & 0 & 0 \\ 0 & \dfrac{3}{2} & \dfrac{1}{2} \\ 0 & \dfrac{1}{2} & \dfrac{3}{2} \end{pmatrix} \begin{pmatrix} x_1 \\ x_2 \\ x_3 \end{pmatrix} = 1.$$

The eigenvalues are $\lambda = 2, 1, 1$, and a possible system of normalised eigenvectors is

$$\lambda_1 = 2, \ \mathbf{u}_1 = \begin{pmatrix} 0 \\ \dfrac{1}{\sqrt{2}} \\ \dfrac{1}{\sqrt{2}} \end{pmatrix}; \quad \lambda_2 = 1, \ \mathbf{u}_2 = \begin{pmatrix} 0 \\ \dfrac{1}{\sqrt{2}} \\ -\dfrac{1}{\sqrt{2}} \end{pmatrix}; \quad \lambda_3 = 1, \ \mathbf{u}_3' = \begin{pmatrix} \dfrac{1}{\sqrt{3}} \\ \dfrac{1}{\sqrt{3}} \\ -\dfrac{1}{\sqrt{3}} \end{pmatrix}.$$

Now $\mathbf{u}_1 \cdot \mathbf{u}_2 = 0$, $\mathbf{u}_1 \cdot \mathbf{u}_3' = 0$, but the vectors $\mathbf{u}_2$ and $\mathbf{u}_3'$ corresponding to the double eigenvalue $\lambda = 1$ are not orthogonal. However, any linear combination $\mathbf{u}_2 + \alpha \mathbf{u}_3'$ is also an eigenvector for $\lambda = 1$, and can be used instead of $\mathbf{u}_3'$. We therefore choose the value of α to make $\mathbf{u}_2 \cdot (\mathbf{u}_2 + \alpha \mathbf{u}_3') = 0$, i.e. we take $\alpha = -\sqrt{3/2}$. The new normalised eigenvector is then

$$\mathbf{u}_3 = \frac{\mathbf{u}_2 + \alpha \mathbf{u}_3'}{|\mathbf{u}_2 + \alpha \mathbf{u}_3'|} = \begin{pmatrix} -1 \\ 0 \\ 0 \end{pmatrix}.$$

Hence the required transformation is

$$\mathbf{x} = \mathbf{T}\mathbf{y}, \quad \text{where} \quad \mathbf{T} = \begin{pmatrix} 0 & 0 & -1 \\ \dfrac{1}{\sqrt{2}} & \dfrac{1}{\sqrt{2}} & 0 \\ \dfrac{1}{\sqrt{2}} & -\dfrac{1}{\sqrt{2}} & 0 \end{pmatrix},$$

and the principal axes form is

$$2y_1^2 + y_2^2 + y_3^2 = 1.$$

Note that the transformation represents a *rotation* of the axes, because the determinant of the transformation $\det \mathbf{T} = +1$.

A symmetric matrix $\mathbf{A}$ whose eigenvalues are all positive is said to be *positive definite*. In that case the principal axis form (6.9.4) can be written

$$(y_1, y_2, \ldots, y_n) \begin{pmatrix} \sqrt{\lambda_1} & 0 & \cdots & 0 \\ 0 & \sqrt{\lambda_2} & \cdots & 0 \\ \cdots & \cdots & \cdots & \cdots \\ 0 & 0 & \cdots & \sqrt{\lambda_n} \end{pmatrix}$$

$$\times \begin{pmatrix} \sqrt{\lambda_1} & 0 & \cdots & 0 \\ 0 & \sqrt{\lambda_2} & \cdots & 0 \\ \cdots & \cdots & \cdots & \cdots \\ 0 & 0 & \cdots & \sqrt{\lambda_n} \end{pmatrix} \begin{pmatrix} y_1 \\ y_2 \\ \cdot \\ y_n \end{pmatrix} = 1,$$

where all of the square roots are taken to be positive, which means that the additional transformation

$$\begin{pmatrix} z_1 \\ z_2 \\ \cdot \\ z_n \end{pmatrix} = \begin{pmatrix} \sqrt{\lambda_1} & 0 & \cdots & 0 \\ 0 & \sqrt{\lambda_2} & \cdots & 0 \\ \cdots & \cdots & \cdots & \cdots \\ 0 & 0 & \cdots & \sqrt{\lambda_n} \end{pmatrix} \begin{pmatrix} y_1 \\ y_2 \\ \cdot \\ y_n \end{pmatrix}$$

reduces the quadric to the n-dimensional *sphere*

$$z_1^2 + z_2^2 + \cdots + z_n^2 = 1.$$

Because of this, if at least one of the symmetric matrices $\mathbf{A}$, $\mathbf{B}$ defining the two central quadrics

$$\mathbf{x}^{\mathrm{T}}\mathbf{A}\mathbf{x} = 1, \quad \mathbf{x}^{\mathrm{T}}\mathbf{B}\mathbf{x} = 1$$

is positive definite, it is always possible to reduce both quadrics simultaneously to principal axis form.

We proceed as follows. Suppose $\mathbf{A}$ is positive definite with eigenvalues $\lambda_1', \lambda_2', \ldots, \lambda_n'$, and let $\mathbf{T}'$ be the $n \times n$ matrix whose columns are the corresponding orthonormal eigenvectors $\mathbf{u}_1', \mathbf{u}_2', \ldots, \mathbf{u}_n'$ of $\mathbf{A}$. Then the $n \times n$ matrix

$$\mathbf{Q} = \mathbf{T}' \begin{pmatrix} \sqrt{\lambda_1'} & 0 & \cdots & 0 \\ 0 & \sqrt{\lambda_2'} & \cdots & 0 \\ \cdots & \cdots & \cdots & \cdots \\ 0 & 0 & \cdots & \sqrt{\lambda_n'} \end{pmatrix} \equiv \left(\sqrt{\lambda_1'}\mathbf{u}_1', \sqrt{\lambda_2'}\mathbf{u}_2', \ldots, \sqrt{\lambda_n'}\mathbf{u}_n' \right)$$

converts $\mathbf{A}$ to the unit matrix, i.e.

$$\mathbf{Q}^{\mathrm{T}}\mathbf{A}\mathbf{Q} = \mathbf{I},$$

while at the same time transforming the symmetric matrix $\mathbf{B}$ to another symmetric matrix

$$\mathbf{C} = \mathbf{Q}^{\mathrm{T}}\mathbf{B}\mathbf{Q}.$$

Now let $\mathbf{C}$ have the eigenvalues and respective orthonormal eigenvectors $\lambda_1, \lambda_2, \ldots, \lambda_n$; $\mathbf{u}_1, \mathbf{u}_2, \ldots, \mathbf{u}_n$, and let

$$\mathbf{P} = (\mathbf{u}_1, \mathbf{u}_2, \ldots, \mathbf{u}_n)$$

be the $n \times n$ matrix whose columns are the eigenvectors. Then

$$\mathbf{P}^{\mathrm{T}}\mathbf{Q}^{\mathrm{T}}\mathbf{A}\mathbf{Q}\mathbf{P} = \mathbf{P}^{\mathrm{T}}\mathbf{I}\mathbf{P} \equiv \mathbf{I}, \quad \text{and} \quad \mathbf{P}^{\mathrm{T}}\mathbf{C}\mathbf{P} = \mathbf{P}^{\mathrm{T}}\mathbf{Q}^{\mathrm{T}}\mathbf{B}\mathbf{Q}\mathbf{P} = \begin{pmatrix} \lambda_1 & 0 & \cdots & 0 \\ 0 & \lambda_2 & \cdots & 0 \\ \cdots & \cdots & \cdots & \cdots \\ 0 & 0 & \cdots & \lambda_n \end{pmatrix}.$$

In other words

$$\mathbf{T}^{\mathrm{T}}\mathbf{A}\mathbf{T} = \mathbf{I}, \quad \mathbf{T}^{\mathrm{T}}\mathbf{B}\mathbf{T} = \begin{pmatrix} \lambda_1 & 0 & \cdots & 0 \\ 0 & \lambda_2 & \cdots & 0 \\ \cdots & \cdots & \cdots & \cdots \\ 0 & 0 & \cdots & \lambda_n \end{pmatrix},$$

where $\mathbf{T} = \mathbf{Q}\mathbf{P} = \left(\sqrt{\lambda_1'}\mathbf{u}_1', \sqrt{\lambda_2'}\mathbf{u}_2', \ldots, \sqrt{\lambda_n'}\mathbf{u}_n' \right) (\mathbf{u}_1, \mathbf{u}_2, \ldots, \mathbf{u}_n)$.

The eigenvalues $\lambda_1, \lambda_2, \ldots, \lambda_n$, which are the roots of the characteristic equation $\det(\mathbf{C} - \lambda\mathbf{I}) = 0$, can be calculated directly from $\mathbf{A}$ and $\mathbf{B}$ without going through a detailed evaluation of the transformations, because

$$\mathbf{C} - \lambda\mathbf{I} = \mathbf{Q}^{\mathrm{T}}\mathbf{B}\mathbf{Q} - \lambda\mathbf{Q}^{\mathrm{T}}\mathbf{A}\mathbf{Q} \equiv \mathbf{Q}^{\mathrm{T}}(\mathbf{B} - \lambda\mathbf{A})\mathbf{Q}$$

$$\therefore \quad \det(\mathbf{C} - \lambda\mathbf{I}) = \det[\mathbf{Q}^{\mathrm{T}}(\mathbf{B} - \lambda\mathbf{A})\mathbf{Q}] = (\det\mathbf{Q}^{\mathrm{T}})(\det\mathbf{Q})[\det(\mathbf{B} - \lambda\mathbf{A})] = 0.$$

Hence the eigenvalues λ_i are the n roots of

$$\det(\mathbf{B} - \lambda\mathbf{A}) = 0, \tag{6.9.5}$$

because $\det\mathbf{Q}^{\mathrm{T}} \times \det\mathbf{Q} = (\det\mathbf{Q})^2 = \lambda_1'\lambda_2'\ldots\lambda_n' \neq 0$.

Problems 6I

1. If $\mathbf{A} = \begin{pmatrix} 1 & 1 \\ 1 & 4 \end{pmatrix}$ and $\mathbf{B} = \begin{pmatrix} 0 & 3 \\ 3 & 0 \end{pmatrix}$, show that $\mathbf{A}$ is positive definite. Use (6.9.5) to reduce the matrices simultaneously respectively to $\begin{pmatrix} 1 & 0 \\ 0 & 1 \end{pmatrix}$ and $\begin{pmatrix} 1 & 0 \\ 0 & -3 \end{pmatrix}$.

2. Find a linear transformation that simultaneously puts the two central quadrics

$$4x_1^2 + 9x_2^2 = 1, \quad 2x_1x_2 = 1$$

into principal axis form.

In matrix notation we have

$$(x_1, x_2) \begin{pmatrix} 4 & 0 \\ 0 & 9 \end{pmatrix} \begin{pmatrix} x_1 \\ x_2 \end{pmatrix} = 1, \quad (x_1, x_2) \begin{pmatrix} 0 & 1 \\ 1 & 0 \end{pmatrix} \begin{pmatrix} x_1 \\ x_2 \end{pmatrix} = 1.$$

The first is already in principal axis form with positive eigenvalues $\lambda_1 = 4, \lambda_2 = 9$. The eigenvalues of the second matrix are $\lambda = \pm 1$, and it is not, therefore, positive definite.

The first quadric becomes a *circle* under the transformation

$$\begin{pmatrix} y_1 \\ y_2 \end{pmatrix} = \begin{pmatrix} \sqrt{\lambda_1} & 0 \\ 0 & \sqrt{\lambda_2} \end{pmatrix} \begin{pmatrix} x_1 \\ x_2 \end{pmatrix}$$

$$\equiv \begin{pmatrix} 2 & 0 \\ 0 & 3 \end{pmatrix} \begin{pmatrix} x_1 \\ x_2 \end{pmatrix} \quad \therefore \quad \begin{pmatrix} x_1 \\ x_2 \end{pmatrix} = \begin{pmatrix} \frac{1}{2} & 0 \\ 0 & \frac{1}{3} \end{pmatrix} \begin{pmatrix} y_1 \\ y_2 \end{pmatrix},$$

so that the quadrics are now

$$(y_1, y_2) \begin{pmatrix} 1 & 0 \\ 0 & 1 \end{pmatrix} \begin{pmatrix} y_1 \\ y_2 \end{pmatrix} = 1, \quad (y_1, y_2) \begin{pmatrix} 0 & \frac{1}{6} \\ \frac{1}{6} & 0 \end{pmatrix} \begin{pmatrix} y_1 \\ y_2 \end{pmatrix} = 1.$$

The first of these (being a circle) will be unaffected by any subsequent orthogonal transformation. The eigenvalues and eigenvectors of the second are

$$\lambda_1 = \frac{1}{6}, \ \mathbf{u}_1 = \begin{pmatrix} \dfrac{1}{\sqrt{2}} \\ \dfrac{1}{\sqrt{2}} \end{pmatrix}; \quad \lambda_2 = -\frac{1}{6}, \ \mathbf{u}_2 = \begin{pmatrix} \dfrac{1}{\sqrt{2}} \\ -\dfrac{1}{\sqrt{2}} \end{pmatrix},$$

and the appropriate orthogonal transformation is

$$\begin{pmatrix} z_1 \\ z_2 \end{pmatrix} = \begin{pmatrix} \dfrac{1}{\sqrt{2}} & \dfrac{1}{\sqrt{2}} \\ \dfrac{1}{\sqrt{2}} & -\dfrac{1}{\sqrt{2}} \end{pmatrix} \begin{pmatrix} y_1 \\ y_2 \end{pmatrix} \quad \therefore \quad \begin{pmatrix} y_1 \\ y_2 \end{pmatrix} = \begin{pmatrix} \dfrac{1}{\sqrt{2}} & \dfrac{1}{\sqrt{2}} \\ \dfrac{1}{\sqrt{2}} & -\dfrac{1}{\sqrt{2}} \end{pmatrix} \begin{pmatrix} z_1 \\ z_2 \end{pmatrix}$$

$$\therefore \quad \begin{pmatrix} x_1 \\ x_2 \end{pmatrix} = \begin{pmatrix} \dfrac{1}{2} & 0 \\ 0 & \dfrac{1}{3} \end{pmatrix} \begin{pmatrix} y_1 \\ y_2 \end{pmatrix}$$

$$= \begin{pmatrix} \dfrac{1}{2} & 0 \\ 0 & \dfrac{1}{3} \end{pmatrix} \begin{pmatrix} \dfrac{1}{\sqrt{2}} & \dfrac{1}{\sqrt{2}} \\ \dfrac{1}{\sqrt{2}} & -\dfrac{1}{\sqrt{2}} \end{pmatrix} \begin{pmatrix} z_1 \\ z_2 \end{pmatrix} = \begin{pmatrix} \dfrac{1}{2\sqrt{2}} & \dfrac{1}{2\sqrt{2}} \\ \dfrac{1}{3\sqrt{2}} & -\dfrac{1}{3\sqrt{2}} \end{pmatrix} \begin{pmatrix} z_1 \\ z_2 \end{pmatrix}.$$

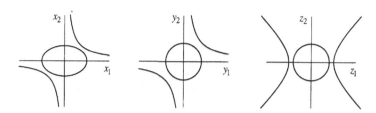

Collecting together all of these results: The transformation

$$x_1 = \frac{1}{2\sqrt{2}}z_1 + \frac{1}{2\sqrt{2}}z_2$$

$$x_2 = \frac{1}{3\sqrt{2}}z_1 - \frac{1}{3\sqrt{2}}z_2$$

reduces the quadrics to the principal axes forms

$$z_1^2 + z_2^2 = 1, \qquad \frac{z_1^2}{6} - \frac{z_2^2}{6} = 1.\,\Bigg]$$

3. Show that the central quadric

$$2x_1x_2 + 2x_2x_3 + 2x_3x_1 = 1$$

is a hyperboloid of two sheets, and is reduced to

$$2y_1^2 - y_2^2 - y_3^2 = 1$$

by means of the orthogonal transformation

$$\begin{pmatrix} x_1 \\ x_2 \\ x_3 \end{pmatrix} = \begin{pmatrix} \frac{1}{\sqrt{3}} & \frac{1}{\sqrt{2}} & \frac{1}{\sqrt{6}} \\ \frac{1}{\sqrt{3}} & -\frac{1}{\sqrt{2}} & \frac{1}{\sqrt{6}} \\ \frac{1}{\sqrt{3}} & 0 & -\frac{2}{\sqrt{6}} \end{pmatrix} \begin{pmatrix} y_1 \\ y_2 \\ y_3 \end{pmatrix}.$$

$$\left[\lambda_1 = 2, \lambda_2 = \lambda_3 = -1; \quad \mathbf{u}_1 = \begin{pmatrix} \frac{1}{\sqrt{3}} \\ \frac{1}{\sqrt{3}} \\ \frac{1}{\sqrt{3}} \end{pmatrix}, \quad \mathbf{u}_2 = \begin{pmatrix} \frac{1}{\sqrt{2}} \\ -\frac{1}{\sqrt{2}} \\ 0 \end{pmatrix}, \quad \mathbf{u}_3 = \begin{pmatrix} \frac{1}{\sqrt{6}} \\ \frac{1}{\sqrt{6}} \\ -\frac{2}{\sqrt{6}} \end{pmatrix} \right]$$

4. Show that the eigenvalue equation $\mathbf{Ax} = \lambda\mathbf{x}$ for a symmetric matrix $\mathbf{A}$ expresses the fact that the principal axes of the central quadric $\mathbf{x}^{\mathrm{T}}\mathbf{Ax} = 1$ intersect the surface at right angles.

5. Show that

$$\iint \cdots \int_{-\infty}^{\infty} e^{-\sum_{ij} a_{ij} x_i x_j} dx_1 dx_2 \cdots dx_n = \frac{\pi^{n/2}}{\sqrt{\det [a_{ij}]}}$$

where $[a_{ij}]$ is an $n \times n$ positive definite symmetric matrix.

6. An $n \times n$ matrix $\mathbf{B}$ is said to be *similar* to a second $n \times n$ matrix $\mathbf{A}$ if there exists a non-singular matrix $\mathbf{X}$ such that

$$\mathbf{B} = \mathbf{X}^{-1}\mathbf{AX}.$$

Show that $\mathbf{B}$ and $\mathbf{A}$ have the same eigenvalues, and that if $\mathbf{u}$ is an eigenvector of $\mathbf{A}$ then $\mathbf{X}^{-1}\mathbf{u}$ is an eigenvector of $\mathbf{B}$ corresponding to the same eigenvalue.

7. Let $\mathbf{A}$ be an arbitrary $n \times n$ matrix with distinct eigenvalues $\lambda_1, \lambda_2, \ldots, \lambda_n$ with corresponding (not necessarily normalised) eigenvectors $\mathbf{u}_1, \mathbf{u}_2, \ldots, \mathbf{u}_n$. Show that

$$\mathbf{X}^{-1}\mathbf{AX} = \begin{pmatrix} \lambda_1 & 0 & \cdots & 0 \\ 0 & \lambda_2 & \cdots & 0 \\ \cdots & \cdots & \cdots & \cdots \\ 0 & 0 & \cdots & \lambda_n \end{pmatrix},$$

where

$$\mathbf{X} = \begin{pmatrix} \mathbf{u}_1 & \mathbf{u}_2 & \cdots & \mathbf{u}_n \end{pmatrix}$$

is the matrix whose columns are $\mathbf{u}_1, \mathbf{u}_2, \ldots, \mathbf{u}_n$.

8. Diagonalise the matrix

$$\mathbf{A} = \begin{pmatrix} 2 & 3 & -1 \\ 0 & -4 & 2 \\ 0 & -5 & 3 \end{pmatrix}.$$

Use the method of Problem 7. Show that

$$\mathbf{X} = \begin{pmatrix} 1 & 1 & 1 \\ 0 & -2 & -2 \\ 0 & -2 & -5 \end{pmatrix}, \quad \mathbf{X}^{-1} = \begin{pmatrix} 1 & \frac{1}{2} & 0 \\ 0 & -\frac{5}{6} & \frac{1}{3} \\ 0 & \frac{1}{3} & -\frac{1}{3} \end{pmatrix}, \quad \mathbf{X}^{-1}\mathbf{AX} = \begin{pmatrix} 2 & 0 & 0 \\ 0 & -2 & 0 \\ 0 & 0 & 1 \end{pmatrix}.$$

9. Let $\mathbf{A}$ be a positive definite $n \times n$ symmetric matrix. Show that the volume bounded by the n-dimensional ellipsoid $\mathbf{x}^T\mathbf{Ax} \leq 1$ is equal to

$$\iint \cdots \int_{\mathbf{x}^T\mathbf{Ax}<1} dx_1 dx_2 \cdots dx_n \equiv \iint \cdots \int_{\sum_{j=1}^n \lambda_j y_j^2 <1} dy_1 dy_2 \cdots dy_n$$

$$= \frac{\pi^{n/2}}{\Gamma\left(\frac{n}{2}+1\right)\sqrt{\lambda_1\lambda_2\ldots\lambda_n}} \equiv \frac{\pi^{n/2}}{\Gamma\left(\frac{n}{2}+1\right)\sqrt{\det \mathbf{A}}}$$

[Use the result of Question 25 of Problems 5A.]

6.10   The Cayley–Hamilton Equation

Let the $n \times n$ matrix $\mathbf{A} = [a_{ij}]$ have the distinct eigenvalues $\lambda_1, \lambda_2, \ldots, \lambda_n$. Consider the equation

$$(\mathbf{A} - \lambda_1\mathbf{I})(\mathbf{A} - \lambda_2\mathbf{I})(\mathbf{A} - \lambda_3\mathbf{I}) \cdots (\mathbf{A} - \lambda_n\mathbf{I})\mathbf{x} = \mathbf{0} \qquad (6.10.1)$$

The matrix factors can be taken in any order, so that the equation is satisfied by any linear combination

$$\mathbf{x} = \alpha_1\mathbf{u}_1 + \alpha_2\mathbf{u}_2 + \cdots + \alpha_n\mathbf{u}_n,$$

of the eigenvectors $\mathbf{u}_1, \mathbf{u}_2, \ldots, \mathbf{u}_n$. But the eigenvectors are linearly independent, and therefore can be used to represent *any* n-dimensional vector $\mathbf{x}$. This means that (6.10.1) is satisfied by an *arbitrary* vector $\mathbf{x}$, which is possible only if $\mathbf{A}$ satisfies the *Cayley–Hamilton* equation

$$(\mathbf{A} - \lambda_1\mathbf{I})(\mathbf{A} - \lambda_2\mathbf{I})(\mathbf{A} - \lambda_3\mathbf{I}) \cdots (\mathbf{A} - \lambda_n\mathbf{I}) = \mathbf{0}. \qquad (6.10.2)$$

By comparing this with (6.8.7) and expanding the left-hand side, this can be seen to be equivalent to the characteristic equation with λ replaced by $\mathbf{A}$

$$(-1)^n\mathbf{A}^n + \left((-1)^{n-1} \sum_{j=1}^{n} a_{jj} \right) \mathbf{A}^{n-1} + \cdots + (\det \mathbf{A})\mathbf{I} = 0. \qquad (6.10.3)$$

Thus, every square matrix satisfies identically its own characteristic equation.

Note that, whereas the algebraic equation

$$(\lambda - \lambda_1)(\lambda - \lambda_2)(\lambda - \lambda_3) \cdots (\lambda - \lambda_n) = 0,$$

implies the vanishing of a root factor on the left, this is not necessarily the case in (6.10.2). The expanded form (6.10.3) actually shows that any power $\mathbf{A}^m$ of $\mathbf{A}$, where $m > n - 1$, can always be expressed as a linear combination of $\mathbf{A}, \mathbf{A}^2, \ldots, \mathbf{A}^{n-1}$. Our derivation of the Cayley–Hamilton equation depended on the eigenvalues being distinct. However, a limiting argument can be used to show that the equation remains valid when there are repeated eigenvalues.

Example　　Verify that

$$\mathbf{A} = \begin{pmatrix} 1 & 2 \\ 4 & 3 \end{pmatrix}$$

satisfies the Cayley–Hamilton equation

$$\mathbf{A}^2 - 4\mathbf{A} - 5\mathbf{I} = 0,$$

and use this formula to evaluate $\mathbf{A}^{-1}$.

First　　　　　$$\mathbf{A}^2 = \begin{pmatrix} 1 & 2 \\ 4 & 3 \end{pmatrix}\begin{pmatrix} 1 & 2 \\ 4 & 3 \end{pmatrix} = \begin{pmatrix} 9 & 8 \\ 16 & 17 \end{pmatrix},$$

$$\therefore \quad \mathbf{A}^2 - 4\mathbf{A} - 5\mathbf{I} = \begin{pmatrix} 9 & 8 \\ 16 & 17 \end{pmatrix} - 4\begin{pmatrix} 1 & 2 \\ 4 & 3 \end{pmatrix} - 5\begin{pmatrix} 1 & 0 \\ 0 & 1 \end{pmatrix} = \begin{pmatrix} 0 & 0 \\ 0 & 0 \end{pmatrix}.$$

Next,

$$\mathbf{A}^{-1}\left(\mathbf{A}^2 - 4\mathbf{A} - 5\mathbf{I}\right) = 0, \quad \therefore \quad \mathbf{A}^{-1} = \frac{1}{5}\left(\mathbf{A} - 4\mathbf{I}\right) = \begin{pmatrix} -\dfrac{3}{5} & \dfrac{2}{5} \\ \dfrac{4}{5} & -\dfrac{1}{5} \end{pmatrix}.$$

An $n \times n$ matrix with n different eigenvalues always has n linearly independent eigenvectors. In the case of a symmetric matrix with one or more equal eigenvalues, we have seen that it is still possible to find a set of n orthogonal eigenvectors, and that they are related geometrically to the principal axes of a central quadric. By contrast, for a non-symmetric matrix it is possible that the number of linearly independent eigenvectors becomes smaller than n when two or more eigenvalues become equal. When this happens the matrix is said to be *defective*. Lanzcos has illustrated the situation in terms of the following 3×3 matrices, all of which have the eigenvalue $\lambda = 1$ with multiplicity 3:

Case I　　$\mathbf{A} = \begin{pmatrix} 1 & 0 & 0 \\ 0 & 1 & 0 \\ 0 & 0 & 1 \end{pmatrix}$,　　3 eigenvectors:　$\begin{pmatrix} 1 \\ 0 \\ 0 \end{pmatrix}$, $\begin{pmatrix} 0 \\ 1 \\ 0 \end{pmatrix}$, $\begin{pmatrix} 0 \\ 0 \\ 1 \end{pmatrix}$

Case II　　$\mathbf{A} = \begin{pmatrix} 1 & 1 & 2 \\ 0 & 1 & 0 \\ 0 & 0 & 1 \end{pmatrix}$,　　2 eigenvectors:　$\begin{pmatrix} 1 \\ 0 \\ 0 \end{pmatrix}$, $\begin{pmatrix} 0 \\ -2 \\ 1 \end{pmatrix}$

Case III $\quad \mathbf{A} = \begin{pmatrix} 1 & -1 & 3 \\ 0 & 1 & 2 \\ 0 & 0 & 1 \end{pmatrix}$, $\quad$ 1 eigenvector: $\begin{pmatrix} 1 \\ 0 \\ 0 \end{pmatrix}$

All three matrices satisfy the same Cayley–Hamilton equation

$$(\mathbf{A} - \mathbf{I})^3 = \mathbf{0}.$$

In the symmetric Case I, however (which corresponds geometrically to a *sphere*) $\mathbf{A}$ actually satisfies a reduced-order equation (called the *minimal equation*)

$$\mathbf{A} - \mathbf{I} = \mathbf{0},$$

that contains the multiple root $\lambda = 1$ only *once*. This is an indication that, although the *directions* of the three eigenvectors are not unique, it is still possible to find three that are independent.

In Case II, $\mathbf{A} - \mathbf{I} \neq \mathbf{0}$, but

$$(\mathbf{A} - \mathbf{I})^2 = \begin{pmatrix} 0 & 1 & 2 \\ 0 & 0 & 0 \\ 0 & 0 & 0 \end{pmatrix} \begin{pmatrix} 0 & 1 & 2 \\ 0 & 0 & 0 \\ 0 & 0 & 0 \end{pmatrix} = \mathbf{0}.$$

The minimal equation satisfied by $\mathbf{A}$ now involves the multiple root $\lambda = 1$ *twice*. This indicates the *loss of one space dimension* in the domain spanned by the eigenvectors, and indeed there are only two independent eigenvectors.

Finally, in Case III, $\mathbf{A} - \mathbf{I} \neq \mathbf{0}$ and $(\mathbf{A} - \mathbf{I})^2 \neq \mathbf{0}$, but

$$(\mathbf{A} - \mathbf{I})^3 = \begin{pmatrix} 0 & -1 & 3 \\ 0 & 0 & 2 \\ 0 & 0 & 0 \end{pmatrix} \begin{pmatrix} 0 & -1 & 3 \\ 0 & 0 & 2 \\ 0 & 0 & 0 \end{pmatrix} \begin{pmatrix} 0 & -1 & 3 \\ 0 & 0 & 2 \\ 0 & 0 & 0 \end{pmatrix}$$

$$= \begin{pmatrix} 0 & 0 & -2 \\ 0 & 0 & 0 \\ 0 & 0 & 0 \end{pmatrix} \begin{pmatrix} 0 & -1 & 3 \\ 0 & 0 & 2 \\ 0 & 0 & 0 \end{pmatrix} = \mathbf{0}.$$

Now the multiple root $\lambda = 1$ enters the minimal equation satisfied by **A** *three* times, and *two* dimensions are lost by the space spanned by the eigenvectors, i.e. there exists only one independent eigenvector.

In general, the degree of defectiveness of a non-symmetric matrix can always be found by determining the number of times the defective eigenvalue occurs in the *minimal equation* satisfied by the matrix.

7

VARIATIONAL CALCULUS

7.1 Taylor's Theorem for Several Variables

Taylor's theorem for a function $f(x)$ of a single independent variable x provides the expansion

$$f(x+h) = f(x) + hf'(x) + \frac{h^2}{2!}f''(x) + \frac{h^3}{3!}f'''(x) + \cdots + \frac{h^n}{n!}f^{(n)}(x) + \cdots.$$

This is valid for $|h| < R$, where R is the distance from x to the 'nearest singularity' of $f(x)$. We have seen (§3.6) that this singularity may occur at a *complex* value of x. The special case in which the expansion is about $x = 0$

$$f(h) = f(0) + hf'(0) + \frac{h^2}{2!}f''(0) + \frac{h^3}{3!}f'''(0) + \cdots + \frac{h^n}{n!}f^{(n)}(0) + \cdots,$$

is called *Maclaurin's theorem*.

The Maclaurin expansion can be used to extend Taylor's theorem to two and higher dimensions. To expand $f(x+h, y+k)$ in powers of h and k we first write down the Maclaurin expansion of $f(x+ht, y+kt)$ *regarded as a function of* t. We use the formula

$$\frac{d}{dt}\Big(f(x+ht, y+kt)\Big) = h\frac{\partial f}{\partial x}(x+ht, y+kt) + k\frac{\partial f}{\partial y}(x+ht, y+kt)$$

$$= \left(h\frac{\partial}{\partial x} + k\frac{\partial}{\partial y}\right)f(x+ht, y+kt)$$

271

and, for general integer $n \geq 1$

$$\frac{d^n}{dt^n} \left(f(x + ht, y + kt) \right) = \left(h\frac{\partial}{\partial x} + k\frac{\partial}{\partial y} \right)^n f(x + ht, y + kt).$$

Then, the Maclaurin expansion of $f(x + ht, y + kt)$ with respect to t is

$$f(x + ht, y + kt) = f(x, y) + t \left(h\frac{\partial}{\partial x} + k\frac{\partial}{\partial y} \right) f(x, y)$$

$$+ \frac{t^2}{2!} \left(h\frac{\partial}{\partial x} + k\frac{\partial}{\partial y} \right)^2 f(x, y) + \cdots$$

$$+ \frac{t^n}{n!} \left(h\frac{\partial}{\partial x} + k\frac{\partial}{\partial y} \right)^n f(x, y) + \cdots$$

By setting $t = 1$ this becomes *Taylor's theorem in two dimensions*

$$f(x + h, y + k) = f(x, y) + \left(h\frac{\partial}{\partial x} + k\frac{\partial}{\partial y} \right) f(x, y)$$

$$+ \frac{1}{2!} \left(h\frac{\partial}{\partial x} + k\frac{\partial}{\partial y} \right)^2 f(x, y) + \cdots$$

$$+ \frac{1}{n!} \left(h\frac{\partial}{\partial x} + k\frac{\partial}{\partial y} \right)^n f(x, y) + \cdots \qquad (7.1.1)$$

In vector notation $\mathbf{h} = (h, k)$ this assumes the compact form

$$f(\mathbf{x} + \mathbf{h}) = f(\mathbf{x}) + (\mathbf{h} \cdot \nabla) f(\mathbf{x}) + \frac{1}{2!} (\mathbf{h} \cdot \nabla)^2 f(\mathbf{x})$$

$$+ \cdots + \frac{1}{n!} (\mathbf{h} \cdot \nabla)^n f(\mathbf{x}) + \cdots \qquad (7.1.2)$$

The extension of Taylor's theorem to an arbitrary number of m dimensions is now obvious. We can use (7.1.2) with

$$\mathbf{h} = (h_1, h_2, \ldots, h_m), \quad \nabla = \left(\frac{\partial}{\partial x_1}, \frac{\partial}{\partial x_2}, \ldots, \frac{\partial}{\partial x_m} \right),$$

or the following generalisation of (7.1.1)

$$f(x_1 + h_1, \ldots, x_m + h_m) = f(x_1, \ldots, x_m) + \left(\sum_{i=1}^{m} h_i \frac{\partial}{\partial x_i} \right) f(x_1, \ldots, x_m)$$

$$+ \frac{1}{2!} \left(\sum_{i=1}^{m} h_i \frac{\partial}{\partial x_i} \right)^2 f(x_1, \ldots, x_m) \cdots$$

$$+ \frac{1}{n!} \left(\sum_{i=1}^{m} h_i \frac{\partial}{\partial x_i} \right)^n f(x_1, \ldots, x_m) + \cdots$$

$$(7.1.3)$$

The following *second-order* approximation will be used in §7.2

$$f(x_1 + h_1, \ldots, x_m + h_m) \approx f(x_1, \ldots, x_m) + \sum_{i=1}^{m} f_i h_i + \frac{1}{2} \sum_{i,j=1}^{m} f_{ij} h_i h_j$$

$$(7.1.4)$$

where

$$f_i = \frac{\partial f}{\partial x_i}(x_1, \ldots, x_m), \quad f_{ij} = f_{ji} = \frac{\partial^2 f}{\partial x_i \partial x_j}(x_1, \ldots, x_m).$$

Example Use Taylor's theorem to expand $f(x, y) = 2(x^2 - y^2) - (x^2 + y^2)^2$ to *second order* about $(1, 0)$.

Using the suffix notation $f_x = \partial f / \partial x$, etc., to denote partial derivatives, we find

$$f_x = 4x[1 - (x^2 + y^2)], \quad f_y = -4y[1 + (x^2 + y^2)]$$

$$f_{xx} = 4(1 - 3x^2 - y^2), \quad f_{xy} = f_{yx} = -8xy, \quad f_{yy} = -4(1 + x^2 + 3y^2).$$

Then, at $x = 1$, $y = 0$: $f = 1$, $f_x = f_y = 0$, $f_{xx} = -8$, $f_{xy} = f_{yx} = 0$, $f_{yy} = -8$. Hence

$$f(1 + h, k) \approx f(1, 0) + (hf_x + kf_y) + \frac{1}{2} \left(h^2 f_{xx} + 2hk f_{xy} + k^2 f_{yy} \right)$$

$$\therefore \quad f(1 + h, k) \approx 1 - 4(h^2 + k^2).$$

The term $\frac{1}{2} \sum_{i,j=1}^{m} f_{ij} h_i h_j$ in (7.1.4) is a *quadratic form* in $\mathbf{h}$ (§6.9), and the second-order approximation may also be written

$$f(\mathbf{x} + \mathbf{h}) \approx f(\mathbf{x}) + \mathbf{h} \cdot \mathbf{f}' + \frac{1}{2} \mathbf{h}^{\mathrm{T}} \mathbf{A} \mathbf{h} \tag{7.1.5}$$

where

$$\underset{\sim}{h} = \begin{pmatrix} h_1 \\ h_2 \\ \cdot \\ h_m \end{pmatrix}, \quad \mathbf{f'} = \begin{pmatrix} f_1 \\ f_2 \\ \cdot \\ f_m \end{pmatrix}, \quad \mathbf{A} = \begin{pmatrix} f_{11} & f_{12} & \cdots & f_{1m} \\ f_{21} & f_{22} & \cdots & f_{2m} \\ \cdot & \cdot & \cdots & \cdot \\ f_{m1} & f_{m2} & \cdots & f_{mm} \end{pmatrix}.$$

The matrix $\mathbf{A} = [f_{ij}]$ is symmetric.

7.2 Maxima and Minima

A smoothly varying function $f(x)$ of a single variable attains a local maximum at $x = a$ if $f(a) > f(x)$ for all values of x close to a. Near this point

$$f(x) = f(a) + (x - a)f'(a) + \frac{1}{2}(x - a)^2 f''(a) + \cdots . \qquad (7.2.1)$$

If $f(a)$ is a maximum, $f(x)$ must start to *decrease* as x moves away from a, which is possible only if $f'(a) = 0$ (if $f'(a) \neq 0$, $f(x)$ would initially increase as x moves away from a on one side and decrease on the other). The point $x = a$ where $f'(x) = 0$ is called a *stationary point*. If this is also a *maximum* and $f''(a) \neq 0$ then we must have $f''(a) < 0$.

Similarly, $f'(a) = 0$ at a *minimum*, but if $f''(a)$ is non-zero it must now be *positive*. If $f''(a) = 0$ the local behaviour of $f(x)$ is governed by higher-order terms not shown explicitly in the Taylor series (7.2.1). Thus, the behaviour of a *smooth function* at a stationary point $x = a$ is characterised by:

$$\left. \begin{array}{l} \text{1. } f'(a) = 0, \ \ f''(a) < 0 \text{ at a maximum} \\[4pt] \text{2. } f'(a) = 0, \ \ f''(a) > 0 \text{ at a minimum} \\[4pt] \text{3. } f'(a) = 0, \ \ f''(a) = 0 \text{ further investigation required} \end{array} \right\} \qquad (7.2.2)$$

A stationary point that is also a maximum or a minimum is called an *extremal*.

Example 1 $f(x) = e^{-(x-2)^2}$ has a maximum at $x = 2$, where $f'(2) = 0$, $f''(2) = -2$.

Example 2 $f(x) = e^{-(x-2)^4}$ has a maximum at $x = 2$, where $f'(2) = 0$, $f''(2) = 0, f'''(2) = 0, f^{(4)}(2) = -24$.

Example 3 The *non-smooth* function $f(x) = e^{-|x-2|}$ has a maximum at $x = 2$, but

$$f'(x) = -\text{sgn}(x-2)e^{-|x-2|} \text{is } not \ defined \text{ at } x = a.$$

Example 3 shows that when investigating extremals attention must be given to the possible existence of values of x where $f(x)$ ceases to be analytic, where either $f'(x)$ becomes undefined or infinite. An extremal can also occur at a boundary point $x = a$ if x is restricted to lie within a certain region of the x-axis. Then $x - a$ can assume one sign only, and the usual condition $f'(a) = 0$ for a stationary point does not apply. Thus, $f = e^{-x}$ takes its maximum value at $x = 1$ where $f'(1) \neq 0$ when x is restricted to the domain $x \geq 1$.

Stationary points in two dimensions A function $z = f(x, y)$ of two independent variables has a local maximum at $x = a$, $y = b$ if at all points sufficiently close to (a, b) $f(x, y) < f(a, b)$. Similarly, in the neighbourhood of a *minimum* $f(x, y) > f(a, b)$. The figure shows that when $f(x, y)$ is a *smoothly* varying function the normal to the surface $z = f(x, y)$ (§2.2)

$$\mathbf{n} = \frac{\nabla\Big(z - f(x, y)\Big)}{\Big|\nabla\Big(z - f(x, y)\Big)\Big|} = \frac{(-f_x, -f_y, 1)}{\sqrt{1 + f_x^2 + f_y^2}}, \quad \Big(f_x = \frac{\partial f}{\partial x}, \quad f_y = \frac{\partial f}{\partial y}\Big),$$

becomes parallel to the z-axis at the maximum. The point (a, b) at which $f_x = f_y = 0$ is called a *stationary point*. At this point the rate of change of f in every possible direction from (a, b) vanishes.

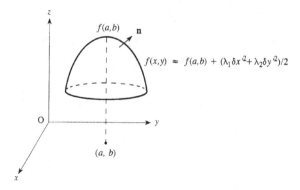

As in the case of a function of one variable, an extremal can occur on a boundary or at points where $f(x, y)$ is not *analytic*. Thus

$$z = f(x, y) = 2e^{-3\sqrt{(x-a)^2+(y-b)^2}},$$

is smoothly varying everywhere *except* at $(x, y) = (a, b)$ where it assumes its maximum value of 2, and where the partial derivatives

$$f_x = \frac{-6(x - a)}{\sqrt{(x - a)^2 + (y - b)^2}} e^{-3\sqrt{(x-a)^2+(y-b)^2}},$$

$$f_y = \frac{-6(y - b)}{\sqrt{(x - a)^2 + (y - b)^2}} e^{-3\sqrt{(x-a)^2+(y-b)^2}},$$

are undefined. Their limiting values at (a, b) depend on the path in the (x, y)-plane along which $x \to a$, $y \to b$.

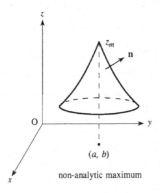

non-analytic maximum

It is necessary to undertake a special investigation of the behaviour of a function at points where it ceases to be differentiable. However, general rules can be formulated for classifying the stationary points of smooth functions. To do this we use (7.1.5) to approximate $f(x, y)$ in the neighbourhood of any point (a, b). Then for an infinitesimal displacement $\mathbf{h} = (\delta x, \delta y) \equiv (x - a, y - b)$ from (a, b)

$$f(x, y) \approx f(a, b) + f_x \delta x + f_y \delta y + \frac{1}{2}\begin{pmatrix} \delta x, & \delta y \end{pmatrix} \begin{pmatrix} f_{xx} & f_{xy} \\ f_{xy} & f_{yy} \end{pmatrix} \begin{pmatrix} \delta x \\ \delta y \end{pmatrix},$$

$$(7.2.3)$$

where the derivatives f_x, f_y, etc. are evaluated at (a, b).

The quantity

$$\delta f = f_x \delta x + f_y \delta y \equiv \nabla f \cdot \delta \mathbf{x}, \tag{7.2.4}$$

is called the *first variation* of f. An extremal can occur at (a, b) only if δf vanishes for arbitrary *orientations* of the displacement $\delta \mathbf{x}$, which then implies that $f_x = f_y = 0$. However, the vanishing of the first variation is not sufficient to ensure that the stationary point is an extremal. When $\nabla f = 0$ the local behaviour of $f(x, y)$ near (a, b) is governed by the final term on the right of (7.2.3), that is, by the quadratic form

$$\delta^2 f = \frac{1}{2} \left(f_{xx} \delta x^2 + 2 f_{xy} \delta x \delta y + f_{yy} \delta y^2 \right). \tag{7.2.5}$$

This is called the *second variation* of f. An extremal occurs at (a, b) if $\delta^2 f$ is either *positive definite* or *negative definite* for all possible directions of the displacement $\delta \mathbf{x}$; it is respectively a maximum or a minimum according as $\delta^2 f \lessgtr 0$.

To determine the sign of $\delta^2 f$ we recall that the symmetric matrix

$$\mathbf{A} = \begin{pmatrix} f_{xx} & f_{xy} \\ f_{xy} & f_{yy} \end{pmatrix},$$

can be *diagonalised* (§6.9) by means of an orthogonal transformation represented by a 2×2 matrix $\mathbf{T}$, such that

$$\begin{pmatrix} \delta x \\ \delta y \end{pmatrix} = \mathbf{T} \begin{pmatrix} \delta x' \\ \delta y' \end{pmatrix} \quad \text{and} \quad \mathbf{T}^{\mathrm{T}} \begin{pmatrix} f_{xx} & f_{xy} \\ f_{xy} & f_{yy} \end{pmatrix} \mathbf{T} = \begin{pmatrix} \lambda_1 & 0 \\ 0 & \lambda_2 \end{pmatrix}.$$

The quantities λ_1 and λ_2 are the *real valued* eigenvalues of $\mathbf{A}$, and are the solutions of the *characteristic equation*

$$\det (\mathbf{A} - \lambda \mathbf{I}) \equiv \begin{vmatrix} f_{xx} - \lambda & f_{xy} \\ f_{xy} & f_{yy} - \lambda \end{vmatrix} = 0. \tag{7.2.6}$$

The columns of $\mathbf{T}$ are the orthonormal eigenvectors associated with λ_1, λ_2. The transformation from $(\delta x, \delta y)$ to $(\delta x', \delta y')$ replaces the $(\delta x, \delta y)$ variables relative to local (x, y)-coordinate axes centred on

(a, b) with new coordinates $(\delta x', \delta y')$ measured along axes parallel to the eigenvectors ($\mathbf{T}$ represents a *rotation* of the local axes, or a *rotation plus reflection* in (a, b) according as $\det \mathbf{T} = \pm 1$). In terms of the new coordinates (7.2.3) becomes

$$f(x, y) - f(a, b) \approx \delta^2 f = \frac{1}{2}\left(\lambda_1 \delta x'^2 + \lambda_2 \delta y'^2\right). \tag{7.2.7}$$

Evidently the point (a, b) is a *maximum* or *minimum* of $f(x, y)$ if the eigenvalues are either *both negative* or *both positive*.

On the other hand if, for example, $\lambda_1 > 0$ and $\lambda_2 < 0$, then the value of $f(x, y)$ will *increase* as we move away from (a, b) along the $\delta x'$-axis, but it will *decrease* as we move away along the $\delta y'$-axis. Thus, although the surface $z = f(x, y)$ becomes 'horizontal' at (a, b) (where $\mathbf{n} = (0, 0, 1)$), this point is neither a maximum nor a minimum; it is called a *saddle point*.

Now, because $\det \mathbf{A} = f_{xx}f_{yy} - f_{xy}^2 = \lambda_1 \lambda_2$ (§6.8), a sufficient condition for *either* a maximum or a minimum at (a, b) is that $\det \mathbf{A} = f_{xx}f_{yy} - f_{xy}^2 > 0$. When this is satisfied, the actual behaviour of $f(x, y)$ near (a, b) can be deduced by considering its behaviour as a function of x alone, by applying rules (1) and (2) of (7.2.2) to $f_{xx}(a, b)$: $f(a, b)$ is a minimum or a maximum according as $f_{xx}(a, b) \gtrless 0$. When $f_{xx}f_{yy} - f_{xy}^2 \equiv \lambda_1 \lambda_2 < 0$, (a, b) is a *saddle point*. If, however, either or both of the eigenvalues vanishes, so that $f_{xx}f_{yy} - f_{xy}^2 = 0$, then $\delta^2 f = 0$ and the second-order expansion (7.2.3) is not accurate enough to define properly the behaviour of $f(x, y)$ near the stationary point. The issue must then be decided by the consideration of higher-order terms of the Taylor series (7.1.2). In summary, we can say that for a stationary point (a, b) of a smooth function:

1. $f_x = f_y = 0$, $f_{xx} < 0$, $f_{xx}f_{yy} - f_{xy}^2 > 0$ at a maximum
2. $f_x = f_y = 0$, $f_{xx} > 0$, $f_{xx}f_{yy} - f_{xy}^2 > 0$ at a minimum
3. $f_x = f_y = 0$, $f_{xx}f_{yy} - f_{xy}^2 < 0$ at a saddle point
4. $f_x = f_y = 0$, $f_{xx}f_{yy} - f_{xy}^2 = 0$ further investigation required

$$\tag{7.2.8}$$

Example Find the stationary points of $z = x^3 + y^3 - 3xy + 1$.

We have

$$f_x = 3x^2 - 3y, \quad f_y = 3y^2 - 3x, \quad f_{xx} = 6x, \quad f_{xy} = -3, \quad f_{yy} = 6y.$$

The stationary points are the solutions of

$$x^2 - y = 0, \quad y^2 - x = 0, \quad \text{i.e.} \quad (x, y) = (0, 0) \text{ or } (1, 1).$$

At $(0, 0)$ $\qquad f_{xx}f_{yy} - f_{xy}^2 = -9 < 0$ $\qquad \therefore \quad (0, 0)$ is a saddle point

At $(1, 1)$ $f_{xx} = 6 > 0$, $f_{xx}f_{yy} - f_{xy}^2 = 36 - 9 > 0$ $\quad \therefore \quad (1, 1)$ is a minimum

Problems 7A

Find the stationary points of the following functions and determine their natures.

1. $z = (x^2 + y^2)^2 - 2(x^2 - y^2)$. $\quad [(0, 0)$, saddle point; $(\pm 1, 0)$, minima]

2. $f(x, y) = x^4 + y^4 - 4x^2$. $\quad [(0, 0)$, saddle point; $(1, 1)$, $(-1, -1)$ minima]

3. $f(x, y) = x^4 + y^3 - 3x^2y$. $\quad [(0, 0)$, saddle point; $(\pm\frac{3}{2}, \frac{3}{2})$ minima]

4. $z = x^3 - x^2y - x^2 + y^2$. $\quad [(0, 0)$, saddle point; $(1, \frac{1}{2})$, minimum; $(2, 2)$, saddle point]

5. $z = x\sqrt{(x-1)^2 + (y-1)^2}$. $\quad [(\frac{1}{2}, 1)$, saddle point; $(1, 1)$, non-analytic minimum]

6. $f(x, y) = 2x^3 - 2y^3 + xy - 1$. $\quad [(0, 0)$, saddle point; $(\frac{1}{6}, -\frac{1}{6})$, minimum]

7. $f(x, y) = 3x^2 - 6xy + 2y^3$. $\quad [(0, 0)$, saddle point; $(1, 1)$, minimum]

8. $z = 3x^2 - 12y^2 + 2x^3 + 3x^2y^2$. $\quad [(-2, 1)$, $(0, 0)$, saddle points; $(-1, 0)$, maximum]

9. The function $f(\mathbf{x})$ of the n independent variables $x_1, x_2, \ldots, x_n$ is stationary at $\mathbf{x} = \mathbf{a}$ if

$$f_i \equiv \left(\frac{\partial f}{\partial x_i}\right)_{\mathbf{x}=\mathbf{a}} = 0, \quad 1 \leq i \leq n.$$

Suppose that $f(\mathbf{x})$ has a minimum at $\mathbf{a}$. Then, according to (7.1.4), near $\mathbf{x} = \mathbf{a}$

$$f(\mathbf{x}) \approx f(\mathbf{a}) + \frac{1}{2} \sum_{i,j=1}^{n} f_{ij}(x_i - a_i)(x_j - a_j),$$

where the quadratic form

$$Q_n = \sum_{i,j=1}^{n} f_{ij}(x_i - a_i)(x_j - a_j), \tag{7.2.9}$$

is *positive definite*.

By observing that the quadratic form Q_m $(1 < m < n)$, obtained by deleting from Q_n the terms involving $x_{m+1} - a_{m+1}, x_{m+2} - a_{m+2}, \ldots, x_n - a_n$, must also be positive definite (and must therefore have *positive* eigenvalues, see §6.9), deduce that

$$f_{11} > 0, \quad \begin{vmatrix} f_{11} & f_{12} \\ f_{21} & f_{22} \end{vmatrix} > 0, \quad \begin{vmatrix} f_{11} & f_{12} & f_{13} \\ f_{21} & f_{22} & f_{23} \\ f_{31} & f_{32} & f_{33} \end{vmatrix} > 0, \ldots, \quad \begin{vmatrix} f_{11} & f_{12} & \cdots & f_{1n} \\ f_{21} & f_{22} & \cdots & f_{2n} \\ \cdot & \cdot & \cdots & \cdot \\ f_{n1} & f_{n2} & \cdots & f_{nn} \end{vmatrix} > 0.$$

$$(7.2.10)$$

10. The function $f(\mathbf{x}), \mathbf{x} = (x_1, x_2, \ldots, x_n)$ is stationary at $\mathbf{x} = \mathbf{a}$. Show that the conditions (7.2.10) are sufficient to ensure that $f(\mathbf{x})$ has a minimum at $\mathbf{a}$. [Consider in turn the quadratic forms Q_m, $1 \leq m \leq n$. The first of conditions (7.2.10) implies that $f(\mathbf{x})$ has a minimum at $\mathbf{a}$ with respect to variations in x_1 alone. This and the second of (7.2.10) then imply that the eigenvalues of Q_2 are both positive, and therefore that Q_2 is also positive definite. Hence, the third of (7.2.10) necessarily implies that the three eigenvalues of Q_3 are positive, etc.].

11. If $f(\mathbf{x}), \mathbf{x} = (x_1, x_2, \ldots, x_n)$ is stationary at $\mathbf{x} = \mathbf{a}$, show that $f(\mathbf{a})$ is a maximum if

$$f_{11} < 0, \quad \begin{vmatrix} f_{11} & f_{12} \\ f_{21} & f_{22} \end{vmatrix} > 0, \quad \begin{vmatrix} f_{11} & f_{12} & f_{13} \\ f_{21} & f_{22} & f_{23} \\ f_{31} & f_{32} & f_{33} \end{vmatrix} < 0, \quad \begin{vmatrix} f_{11} & f_{12} & f_{13} & f_{14} \\ f_{21} & f_{22} & f_{23} & f_{24} \\ f_{31} & f_{32} & f_{33} & f_{34} \\ f_{41} & f_{42} & f_{43} & f_{44} \end{vmatrix} > 0, \ldots, \text{ etc.}$$

12. Show that $f(x, y, z) \doteq x^4 + y^4 + z^4 - 4xyz$ has a minimum at $(1, 1, 1)$.

13. Show that $f(x, y, z) = xyz(1 - x - y - z)$ has a maximum at $(\frac{1}{4}, \frac{1}{4}, \frac{1}{4})$.

14. Calculate the eigenvalues of the quadratic form $f(x, y, z) = x^2 + 3xy + z^2$. Hence deduce that $(0, 0, 0)$ is a saddle point of $f(x, y, z)$. [$\lambda = 2, 1 \pm \sqrt{10}$]

15. Show that $f(x, y, z) = \ln(x^2 + y^2 + z^2 + 1)$ has a minimum at $(0, 0, 0)$.

16. **Maximum principle** Deduce from (7.2.8) that the solution $\varphi(x, y)$ of Laplace's equation $\varphi_{xx} + \varphi_{yy} = 0$ within a region $\mathcal{D}$ cannot attain maximum or minimum values in $\mathcal{D}$. Extend this conclusion to the n-dimensional equation $\varphi_{x_1 x_2} + \varphi_{x_2 x_2} + \cdots + \varphi_{x_n x_n} = 0$.

7.3 Constrained Maxima and Minima: Lagrange Multipliers

Consider the problem of locating the extremals of $f(\mathbf{x})$ when the position vector $\mathbf{x}$ is constrained to lie within some fixed region.

Example 1 Find the maximum value of

$$f(x, y) = 1 - (x^2 + 2y^2)$$

when (x, y) lies on the straight line

$$x + y + 3 - 0.$$

Obviously, $f(x, y)$ attains its maximum value at $(0, 0)$ when (x, y) can vary freely over the whole of the xy-plane. It assumes the constant value $f(x, y) = 1 - C$ when (x, y) lies on the *ellipse*

$$x^2 + 2y^2 = C = \text{constant} > 0.$$

The ellipse grows in size as C increases, causing the value of $f(x, y)$ on the ellipse to become progressively smaller. The required maximum will therefore correspond to the smallest value of C for which the ellipse (called the *maximal ellipse*) just touches the line $x + y + 3 = 0$.

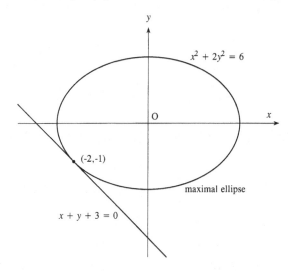

To determine the position (a, b) of the maximum in a systematic manner, let $\delta x = x - a$, $\delta y = y - b$, and consider the vanishing of the first variation $\delta f = f_x \delta x + f_y \delta y$ at (a, b):

$$f_x \delta x + f_y \delta y = 0. \tag{7.3.1}$$

In the absence of any constraints on the admissible orientation of the displacement $(\delta x, \delta y)$, the stationary values of f would be determined by setting $f_x = f_y = 0$. But this is only permissible when δx and δy are independent, whereas the constraint $x + y + 3 = 0$ implies that

$$\delta x + \delta y = 0. \tag{7.3.2}$$

However, we can use this equation to eliminate δy, say, from (7.3.1). To do this multiply (7.3.2) by a quantity λ and add the result to (7.3.1) to obtain

$$(f_x + \lambda)\delta x + (f_y + \lambda)\delta y = 0. \tag{7.3.3}$$

The term in δy is now removed by choosing λ to make $f_y + \lambda = 0$. We are then left with $(f_x + \lambda)\delta x$ and, because δx can vary freely along the line $x + y + 3 = 0$, the condition for a stationary point must be $f_x + \lambda = 0$.

Collecting together these results, we see that the value of λ and the stationary point $(a, b) = (x, y)$ are determined by solution of the three equations

$$f_x + \lambda = 0, \quad f_y + \lambda = 0, \quad x + y + 3 = 0. \tag{7.3.4}$$

Now $f_x = -2x$, $f_y = -4y$, and therefore

$$-2x + \lambda = 0, \quad -4y + \lambda = 0, \quad x + y + 3 = 0,$$

$$\therefore \quad x = -2, \quad y = -1, \quad \lambda = -4.$$

Hence, $(a, b) = (-2, -1)$ and $f(a, b) = -5$, so that the maximal ellipse corresponds to $C = 6$.

Equations (7.3.4) possess the following very simple geometrical interpretation. Admissible values of (x, y) for which $f(x, y)$ is a maximum are required to lie on the line $x + y + 3 = 0$, and $\delta\mathbf{x} = (\delta x, \delta y)$ must therefore represent a displacement along this line. Equation (7.3.1) states that $\nabla f \cdot \delta\mathbf{x} = 0$ at the stationary point, i.e. at (a, b) the normal to the maximal ellipse $f(x, y) = f(a, b)$ (i.e. $x^2 + 2y^2 = 6$) is *parallel* to the normal $(1, 1)$ of the straight line $x + y + 3 = 0$. Thus, $\nabla f = -\lambda(1, 1)$ on $x + y + 3 = 0$ for a suitable value of the constant of proportionality $-\lambda$, which corresponds precisely to (7.3.4).

The factor λ in (7.3.4) is called a *Lagrange multiplier*. The geometrical argument makes it clear why the apparently non-symmetrical formal procedure, whereby λ is introduced to eliminate δy rather than δx, leads, nonetheless, to a system of equations (7.3.4) for the stationary point that is *symmetric* in f_x, f_y.

The following remarkable conclusion should also be noted. Equations (7.3.4) are precisely those that would be obtained if, instead of considering the stationarity condition for $f(x, y)$ subject to the constraint $x + y + 3 = 0$, we had considered the symmetrical *unconstrained* condition for the function

$$F(x, y, \lambda) = f(x, y) + \lambda(x + y + 3),$$

where x, y *and* λ are regarded as *independent variables*. The vanishing of the first variation of F

$$\delta F = F_x \delta x + F_y \delta y + F_\lambda \delta \lambda = 0$$

when $\delta x, \delta y$ and $\delta \lambda$ vary independently then implies that

$$F_x \equiv f_x + \lambda = 0, \quad F_y \equiv f_y + \lambda = 0, \quad F_\lambda \equiv x + y + 3 = 0,$$

i.e. Equation (7.3.4).

Maxima and minima subject to one constraint Let us now repeat the above argument for the more general problem of determining the extremals of $f(x, y, z)$ when $\mathbf{x} = (x, y, z)$ is constrained to lie on the surface

$$g(x, y, z) = 0. \tag{7.3.5}$$

Let $\mathbf{a} = (a, b, c)$ be a stationary point, at which the first variation must vanish:

$$f_x \delta x + f_y \delta y + f_z \delta z = 0. \tag{7.3.6}$$

Only two components of the infinitesimal displacement $\delta x = x - a$, $\delta y = y - b, \delta z = z - c$ may be regarded as independent, because both $\mathbf{x}$

and **a** lie on the *two-dimensional* surface (7.3.5). We therefore eliminate δz by the λ-method of Example 1.

Because **x** lies on $g(x, y, z) = 0$

$$g_x\delta x + g_y\delta y + g_z\delta z = 0, \qquad (7.3.7)$$

where the derivatives $g_x = \partial g/\partial x$, etc, are evaluated at the stationary point. Multiply (7.3.7) by the Lagrange multiplier λ and add to equation (7.3.6):

$$(f_x + \lambda g_x)\delta x + (f_y + \lambda g_y)\delta y + (f_z + \lambda g_z)\delta z = 0.$$

The coefficient of δz is made to vanish by requiring λ to satisfy $f_z + \lambda g_z = 0$ at **a**. The remaining displacements δx and δy on $g(x, y, z) = 0$ are independent, and therefore $f_x + \lambda g_x = 0$, $f_y + \lambda g_y = 0$. Collecting results, the stationary point $\mathbf{a} = \mathbf{x}$ and the multiplier λ are seen to be determined by the *symmetric* system of equations

$$f_x + \lambda g_x = 0, \quad f_y + \lambda g_y = 0, \quad f_z + \lambda g_z = 0, \quad g(x, y, z) = 0.$$
$$(7.3.8)$$

Once again, the symmetry of these equations becomes obvious from a simple geometrical argument. When **x** and therefore $\delta\mathbf{x}$ both lie on $g(\mathbf{x}) = 0$, equation (7.3.6) states that $\nabla f(\mathbf{x})$ is parallel to the normal $\nabla g(\mathbf{x})$ of the constraint surface at the stationary point, and therefore, for some suitable constant of proportionality $-\lambda$,

$$\nabla f + \lambda\nabla g = 0 \quad \text{at } \mathbf{x} = \mathbf{a} \quad \text{on} \quad g(\mathbf{x}) = 0. \qquad (7.3.9)$$

Evidently the maximal surface $f(\mathbf{x}) = f(\mathbf{a})$ and the surface of constraint $g(\mathbf{x}) = 0$ *touch* one another at the stationary point.

Furthermore, equations (7.3.8) or (7.3.9) are equivalent to the *unconstrained* stationary condition for

$$F(\mathbf{x}, \lambda) = f(\mathbf{x}) + \lambda g(\mathbf{x}), \qquad (7.3.10)$$

where **x** and λ can vary independently. At the outset, therefore, the original constrained problem for $f(\mathbf{x})$ can be replaced by the unconstrained problem for $F = f + \lambda g$.

Example 2 Use the method of the Lagrange multiplier to find the maximum value of

$$f(\mathbf{x}) = 1 - (x^2 + 2y^2 + 4z^2) \quad \text{on the plane} \quad x + y + z + 7 = 0.$$

The stationarity conditions for $F(\mathbf{x}, \lambda) = f(\mathbf{x}) + \lambda(x + y + z + 7)$, where $\mathbf{x}$ and λ vary independently, are

$$-2x + \lambda = 0, \quad -4y + \lambda = 0, \quad -8z + \lambda = 0, \quad x + y + z + 7 = 0.$$

Hence the stationary point is at $\mathbf{a} = (-4, -2, -1)$ where f assumes the maximum value $f(\mathbf{a}) = -27$. The *maximal ellipsoid* $x^2 + 2y^2 + 4z^2 = 28$ (i.e. $f(\mathbf{x}) = f(\mathbf{a})$) touches the plane $x + y + z + 7 = 0$ at $\mathbf{a}$.

The extremal problem for $f(\mathbf{x}) \equiv f(x, y, z)$ can be solved in a similar fashion when $\mathbf{x}$ is required to satisfy the *two* conditions of constraint:

$$g^\alpha(\mathbf{x}) = 0, \quad g^\beta(\mathbf{x}) = 0. \tag{7.3.11}$$

The stationary point $\mathbf{a} = (a, b, c)$ lies on the curve Γ, say, defined by the intersection of these surfaces, and the stationarity condition at $\mathbf{a}$ requires that the first variation $\delta f = \nabla f \cdot \delta \mathbf{x} = 0$ for an infinitesimal displacement $\delta \mathbf{x} = \mathbf{x} - \mathbf{a}$ along Γ. But, $g^\alpha = g^\beta = 0$ on Γ, so that we must actually satisfy the following conditions:

$$\nabla f \cdot \delta \mathbf{x} = 0, \quad \nabla g^\alpha \cdot \delta \mathbf{x} = 0, \quad \nabla g^\beta \cdot \delta \mathbf{x} = 0, \quad \text{on } \Gamma.$$

Thus, at the stationary point each of the vectors $\nabla f, \nabla g^\alpha, \nabla g^\beta$ is normal to the direction of Γ, and there must therefore exist constants $\lambda_\alpha, \lambda_\beta$ such that

$$\nabla f + \lambda_\alpha \nabla g^\alpha + \lambda_\beta \nabla g^\beta = 0,$$

$$g^\alpha(\mathbf{x}) = 0, \quad g^\beta(\mathbf{x}) = 0 \quad \text{at the stationary point.} \tag{7.3.12}$$

The constants $\lambda_\alpha, \lambda_\beta$ are the Lagrange multipliers of the problem, and equations (7.3.12) are precisely the stationarity conditions that would be obtained by application of the Lagrange elimination procedure. They can also be derived, just as before, by formally considering the unconstrained extremal problem for

$$F(\mathbf{x}, \lambda_\alpha, \lambda_\beta) = f(\mathbf{x}) + \lambda_\alpha g^\alpha(\mathbf{x}) + \lambda_\beta g^\beta(\mathbf{x}), \tag{7.3.13}$$

where $\mathbf{x}, \lambda_\alpha, \lambda_\beta$ are regarded as independently varying quantities.

Example 3 Find the maximum and minimum distances from the origin to the curve defined by the intersection of

$$x^2 + 2y^2 + z^2 = 5, \quad z = 1.$$

The extremals of the distance $|\mathbf{x}|$ from the origin correspond to the extremals of $f(\mathbf{x}) = x^2 + y^2 + z^2$. We can therefore consider the unconstrained extremal problem for

$$F = x^2 + y^2 + z^2 + \lambda_1(x^2 + 2y^2 + z^2 - 5) + \lambda_2(z - 1).$$

The Lagrange equations (7.3.12) are

$$2x + 2\lambda_1 x = 0, \quad 2y + 4\lambda_1 y = 0, \quad 2z + 2\lambda_1 z + \lambda_2 = 0, \quad x^2 + 2y^2 + z^2 = 5, \quad z = 1,$$

with the two sets of solutions

$$x = \pm 2, \quad y = 0, \quad z = 1, \quad \lambda_1 = -1, \quad \lambda_2 = 0, \quad \therefore \quad |\mathbf{x}| = \sqrt{5},$$

$$x = 0, \quad y = \pm\sqrt{2}, \quad z = 1, \quad \lambda_1 = -\frac{1}{2}, \quad \lambda_2 = -1, \quad \therefore \quad |\mathbf{x}| = \sqrt{3}.$$

Problems 7B

Solve by the method of Lagrange multipliers:

1. Find the maximum and minimum distances from the origin to the curve $3x^2 + 3y^2 + 4xy = 2$. $[\sqrt{2}; \sqrt{2/5}]$

2. Find the stationary points of $f(x,y) = x^2 + y^2 + z$ when $x^2 - z^2 = 1$. $[(\pm\frac{\sqrt{5}}{2}, 0, -\frac{1}{2})]$

3. Find the minimum distance in $x > 0$, $y > 0$ from the origin to the curve $xy = 1$. $[\sqrt{2}]$

4. Find the minimum value of $f(x,y,z) = x^2 + 4y^2 + 16z^2$ on the surface $xyz = 1$. $[12]$

5. Find the maximum and minimum distances from the origin to the curve defined by the equations $x + y - 1 = 0$, $x^2 + 2y^2 + z^2 - 1 = 0$. $[\frac{\sqrt{5}}{3}; 1]$

6. Find the coordinates of the point on the surface $x^2yz = 1$ in $x, y, z > 0$ closest to the origin. $[(2^{\frac{1}{4}}, 2^{-\frac{1}{4}}, 2^{-\frac{1}{4}})]$

7. Find the minimum value of $f(\mathbf{x}) = x^3 + y^3 + z^3$ in $x, y, z > 0$ on the surface $1/x + 1/y + 1/z = 1$. $[81$ at $(3,3,3)]$

8. Show that when $\mathbf{x}$ lies on the surface $1/x^2 + 1/y^2 + 2/z^2 = 4$ in x, y, $z > 0$, the function $f(\mathbf{x}) = z(x + y)$ has a minimum at $(1,1,1)$.

9. Find the stationary point of the function $f(\mathbf{x}) = x^2 - y^2 + z^2 - 2x$ when $x + y - z = 0$, $x + 2y = 1$. $[(\frac{1}{2}, \frac{1}{4}, \frac{3}{4})]$

10. Find the point on the circle $(x-1)^2 + (y-1)^2 = 1$ that is closest to the origin. $[(1 - \frac{1}{\sqrt{2}}, 1 - \frac{1}{\sqrt{2}})]$

11. The hypotenuse of a right-angled triangle has unit length. If the lengths of the other sides are denoted by x and y, for what values of these variables is the area of the triangle a maximum? $[x = y = \frac{1}{\sqrt{2}}]$.

7.4 Stationary Definite Integrals

Let the curve $y = f(x)$ join the two points A and B which have the coordinates (a, y_a), (b, y_b) in the xy-plane, and suppose

$$\mathcal{L} = \mathcal{L}\left(x, y, \frac{dy}{dx}\right) \equiv \mathcal{L}(x, y, y')$$

is a *given* function of x, $y = f(x)$ and $y' = f'(x)$. Then the value of the integral

$$I = \int_a^b \mathcal{L}(x, y, y')dx, \qquad (7.4.1)$$

generally changes as the shape of the curve $y = f(x)$ between A and B is changed. The fundamental problem of the *Calculus of Variations* is to find the curve between A and B that makes the value of the integral *stationary*. This curve will correspond to a definite functional form of $f(x)$, that may make the value of the integral a maximum or a minimum relative to the values computed for neighbouring curves, or merely stationary.

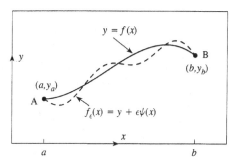

Probably the simplest such problem occurs when $\mathcal{L} = \sqrt{1 + y'^2}$, for which

$$I = \int_a^b \sqrt{1 + y'^2}\,dx = \text{distance along the curve between A and B.}$$

$$(7.4.2)$$

The stationary value of the integral is then a minimum, corresponding to the shortest distance between A and B, and the obvious solution $y = f(x)$ of the minimisation problem is the straight line

$$y = y_a + \left(\frac{y_b - y_a}{b - a}\right)(x - a). \tag{7.4.3}$$

This result can be derived by considering the *variation* of the integral, in a manner similar to that described at the beginning of §7.2 for locating the stationary point $x = a$ of a function $f(x)$. The quantity I defined by the integral (7.4.1) is called a *functional*, whose value depends on the choice of the *function* $y = f(x)$ over the complete range $a < x < b$, rather than at just a single point.

Suppose that $y = f(x)$ gives the required stationary value of the integral (7.4.1). We calculate the value of I for a 'neighbouring' function (illustrated by the broken line curve in the figure) by considering the value of the integral for

$$f_\epsilon(x) = y + \epsilon\psi(x),$$

where ϵ is a small positive quantity that can be made to tend to zero, and the function $\psi(x)$ is arbitrary, except that it must vanish at $x = a$ and b, where the values of y are fixed (respectively equal to y_a and y_b). The difference between $f_\epsilon(x)$ and $y = f(x)$ defines the *variation* δy of y: it is the variation in y at a *fixed value* of x, i.e.

$$\delta y = f_\epsilon(x) - y \equiv f_\epsilon(x) - f(x) = \epsilon\psi(x).$$

This is an infinitesimal quantity of order ϵ. The corresponding difference in the values of I is

$$\int_a^b \Big(\mathcal{L}(x, y + \epsilon\psi(x), y' + \epsilon\psi'(x)) - \mathcal{L}(x, y, y')\Big)dx,$$

which can be expanded in powers of ϵ by first using Taylor's theorem to write:

$$\mathcal{L}(x, y + \epsilon\psi(x), y' + \epsilon\psi'(x)) = \mathcal{L}(x, y, y') + \epsilon\psi(x)\frac{\partial\mathcal{L}}{\partial y} + \epsilon\psi'(x)\frac{\partial\mathcal{L}}{\partial y'} + \cdots ,$$

where $\partial\mathcal{L}/\partial y$ and $\partial\mathcal{L}/\partial y'$ are evaluated for the stationary function (x, y, y').

We therefore obtain the variation δI of I

$$\delta I = \int_a^b \left(\mathcal{L}(x, y + \epsilon\psi(x), y' + \epsilon\psi'(x)) - \mathcal{L}(x, y, y') \right) dx$$

$$= \epsilon \int_a^b \left(\psi(x)\frac{\partial\mathcal{L}}{\partial y} + \psi'(x)\frac{\partial\mathcal{L}}{\partial y'} \right) dx + \cdots \qquad (7.4.4)$$

The 'rate of change' of the integral is now obtained by dividing by ϵ and letting $\epsilon \to 0$

$$\frac{\delta I}{\epsilon} = \int_a^b \left(\psi(x)\frac{\partial\mathcal{L}}{\partial y} + \psi'(x)\frac{\partial\mathcal{L}}{\partial y'} \right) dx.$$

This must vanish if $y = f(x)$ makes the integral stationary for arbitrary functions $\psi(x)$. But to apply this condition it is first necessary to eliminate the derivative $\psi'(x)$ from under the integral sign by integration by parts:

$$\int_a^b \psi'(x)\frac{\partial\mathcal{L}}{\partial y'} dx = \left[\psi(x)\frac{\partial\mathcal{L}}{\partial y'} \right]_a^b - \int_a^b \psi(x)\frac{d}{dx}\left(\frac{\partial\mathcal{L}}{\partial y'} \right) dx.$$

The first term on the right makes no contribution because $\psi(x) = 0$ at the end-points a and b, and therefore the rate of change of the integral becomes

$$\frac{\delta I}{\epsilon} = \int_a^b \psi(x)\left(\frac{\partial\mathcal{L}}{\partial y} - \frac{d}{dx}\frac{\partial\mathcal{L}}{\partial y'} \right) dx, \qquad (7.4.5)$$

which is zero for arbitrary $\psi(x)$ only if

$$\frac{\partial \mathcal{L}}{\partial y} - \frac{d}{dx}\frac{\partial \mathcal{L}}{\partial y'} = 0 \quad \text{for all } x \text{ in } a < x < b. \tag{7.4.6}$$

This is because it is always possible to choose a differentiable function $\psi(x)$ which is non-zero only within an interval of length Δ enclosing any given value of $x = x_\Delta$ in $a < x < b$. By multiplying $\psi(x)$ by a suitable scale factor we can make its mean value in the interval $\bar{\psi} = 1$. When Δ is small the integral (7.4.5) can then be approximated by

$$\Delta \bar{\psi} \left(\frac{\partial \mathcal{L}}{\partial y} - \frac{d}{dx}\frac{\partial \mathcal{L}}{\partial y'} \right)_{x=x_\Delta} = 0, \quad \text{so that} \quad \left(\frac{\partial \mathcal{L}}{\partial y} - \frac{d}{dx}\frac{\partial \mathcal{L}}{\partial y'} \right)_{x=x_\Delta} = 0$$

for any value of x_Δ in $a < x_\Delta < b$.

Equation (7.4.6) is the *Euler–Lagrange* differential equation, whose solution determines the function $y = f(x)$ that makes the integral (7.4.1) *stationary*.

Example We may now verify the intuitive straight line solution (7.4.3) for the curve of minimum length joining two points A and B. The distance is given by the integral (7.4.2), so that

$$\mathcal{L}(x, y, y') = \sqrt{1 + y'^2},$$

which does not depend explicitly on x or y. The Euler–Lagrange equation (7.4.6) is therefore

$$\frac{\partial \mathcal{L}}{\partial y} - \frac{d}{dx}\frac{\partial \mathcal{L}}{\partial y'} \equiv -\frac{d}{dx} \left(\frac{y'}{\sqrt{1 + y'^2}} \right) = 0.$$

Hence, the stationary curve $y = f(x)$ is given by

$$\frac{dy}{dx} = A = \text{constant}$$
$$\therefore \quad y = Ax + B, \quad B = \text{constant}.$$

This straight line reduces to (7.4.3) when the values of the arbitrary constants are chosen to make the line pass through A and B. (Strictly speaking an additional calculation should be performed to determine the *second variation* $\delta^2 I$ of the integral I in order to verify that the stationary value of the integral furnished by the straight line actually represents a minimum. This is often very difficult to do in practice, and we must then appeal, as here, to geometrical intuition to confirm that the derived stationary value does correspond to an extremal.)

Example Find the function $y = f(x)$ that satisfies $f(0) = 0$, $f(\frac{\pi}{2}) = 1$ for which $I = \int_0^{\frac{\pi}{2}} (y^2 - y'^2 + 4ye^x)dx$ is stationary.

Take $\mathcal{L} = y^2 - y'^2 + 4y e^x$ in the Euler–Lagrange equation. Then

$$\frac{\partial \mathcal{L}}{\partial y} - 2y + 4e^x, \qquad \frac{\partial \mathcal{L}}{\partial y'} = -2y'$$

$\therefore$ the Euler–Lagrange equation becomes

$$y'' + y = 2e^x$$

$\therefore \quad y = A \cos x + B \sin x + e^x, \quad$ where A, B are constant

Apply the boundary conditions,

$$0 = A + 1, \quad 1 = B + e^{\frac{\pi}{2}}$$

$$\therefore \quad y = f(x) = -\cos x + \left(1 - e^{\frac{\pi}{2}}\right) \sin x + e^x.$$

Special cases

1. $\mathcal{L} = \mathcal{L}(x, y')$ (no dependence on y). The Euler–Lagrange equation becomes

$$\frac{d}{dx}\frac{\partial \mathcal{L}}{\partial y'} = 0, \quad \therefore \quad \frac{\partial \mathcal{L}}{\partial y'} = A = \text{constant}.$$

This equation can be solved for y' in terms of x and A, y is then found by integrating

$$\frac{dy}{dx} = F(x, A)$$

where $F(x, A)$ is known.

2. $\mathcal{L} = \mathcal{L}(y, y')$ (no dependence on x). The Euler–Lagrange equation is

$$\frac{\partial \mathcal{L}}{\partial y} - \frac{d}{dx}\frac{\partial \mathcal{L}}{\partial y'} \equiv \frac{\partial \mathcal{L}}{\partial y} - y'\frac{\partial^2 \mathcal{L}}{\partial y' \partial y} - y''\frac{\partial^2 \mathcal{L}}{\partial y'^2} = 0.$$

Multiply by y':

$$y'\frac{\partial \mathcal{L}}{\partial y} - y'^2\frac{\partial^2 \mathcal{L}}{\partial y' \partial y} - y'y''\frac{\partial^2 \mathcal{L}}{\partial y'^2} \equiv \frac{d}{dx}\left(\mathcal{L} - y'\frac{\partial \mathcal{L}}{\partial y'}\right) = 0.$$

Hence, a first integral of the Euler–Lagrange equation is

$$\mathcal{L} - y'\frac{\partial \mathcal{L}}{\partial y'} = \text{constant}, \quad \text{when } \mathcal{L} = \mathcal{L}(y, y'). \tag{7.4.7}$$

3. $\mathcal{L} = \mathcal{L}(x, y)$ (no dependence on y'). The Euler–Lagrange equation is

$$\frac{\partial \mathcal{L}}{\partial y} = 0.$$

This is *not a differential equation*, but may be solved immediately to give $y = f(x)$.

Example   Find the curve $y = f(x)$ through the points $(0, 1)$, $(1, 0)$ for which

$$I = \int\limits_0^1 \frac{\sqrt{1 + y'^2}}{x}\, dx$$

is stationary.

This is a Case 1 problem, for which $\partial \mathcal{L}/\partial y' = $ constant:

$$\therefore \quad \frac{y'}{x\sqrt{1 + y'^2}} = A$$

$$\therefore \quad \frac{dy}{dx} = \frac{Ax}{\sqrt{1 - A^2 x^2}}$$

$$\therefore \quad y - B = -\frac{1}{A}\sqrt{1 - A^2 x^2}, \quad B = \text{constant}$$

$$\therefore \quad x^2 + (y - B)^2 = \frac{1}{A^2}.$$

The end conditions give $\quad (1 - B)^2 = \frac{1}{A^2}, \quad 1 + B^2 = \frac{1}{A^2}$

$$\therefore \quad A = 1 \quad \text{and} \quad B = 0,$$

and the stationary curve is the circle $x^2 + y^2 = 1$.

Example   Find the curve through the points (a, y_a), (b, y_b) that generates the surface of minimum area when rotated about the x-axis.

The surface area is the integral $I = 2\pi \int\limits_a^b y\sqrt{1 + y'^2}\, dx.$

Thus, we have an example of Case 2, because $\mathcal{L} = y\sqrt{1 + y'^2}$ does not depend on x. The first integral (7.4.7) of the Euler–Lagrange equation gives

$$y = A\sqrt{1 + y'^2}, \quad A = \text{constant},$$

$$\text{so that} \quad y' = \sqrt{\frac{y^2 - A^2}{A^2}},$$

$$\therefore \quad dx = \frac{A\,dy}{\sqrt{y^2 - A^2}},$$

$$\therefore \quad y = A\cosh\left(\frac{x+B}{A}\right), \quad B = \text{constant}$$

The surface of rotation for this curve is called a *catenoid*. The values of A and B are calculated from the condition that the curve must pass through the end points (a, y_a), (b, y_b). For some pairs of points, however, it is not possible to find A and B to satisfy this condition. In that case there is no *smooth* surface that minimises the surface area.

Problems 7C

1. Show that $I = \int_{-1}^{1} \sqrt{y(1+y'^2)}\,dx$ where $y = 1$ at $x = \pm 1$ is stationary when $y = \frac{1}{2}(1+x^2)$.

2. Find the function $y = f(x)$ which satisfies $f(0) = 0$, $f(1) = 1$ for which $I = \int_0^1 (y^2 + y'^2 - 4ye^{-x})\,dx$ is stationary. $[f(x) = \frac{2\sinh x}{(1+e)} + xe^{-x}]$

3. Show that $I = \int_0^1 \frac{1+y^2}{y'^2}\,dx$ where $y(0) = 0$, $y(1) = 1$ is stationary when $y = \sinh\left(x\ln[1+\sqrt{2}]\right)$.

4. Show that $y = A + Bx^4$, where A, B are constant, makes $I = \int_a^b \frac{y'^2}{x^3}\,dx$ stationary.

5. Verify that $y = \frac{1}{2}\cosh x$ is a stationary curve for the functional $I = \int_0^{\frac{\pi}{2}} (y^2 - y'^2 - 2y\cosh x)\,dx$, where $y(0) = 0$, $y(\frac{\pi}{2}) = \frac{1}{2}\cosh\frac{\pi}{2}$.

6. Find the curve in the (x, y)-plane for which $I = \int_0^{\frac{\pi}{2}} \sqrt{E - y^2}\sqrt{1+y'^2}\,dx$ is stationary, where $E > 1$, given that $y(0) = 0$, $y(\pi) = 0$. $[y = \sqrt{E-1}\sin x]$.

7. Show that the Euler–Lagrange equation for $I = \int_a^b \mathcal{L}(x, y, y', y'')\,dx$ is

$$\frac{\partial \mathcal{L}}{\partial y} - \frac{d}{dx}\left(\frac{\partial \mathcal{L}}{\partial y'}\right) + \frac{d^2}{dx^2}\left(\frac{\partial \mathcal{L}}{\partial y''}\right) = 0.$$

8. By using polar coordinates to write $I = \int \sqrt{x^2 + y^2}\sqrt{1+y'^2}\,dx = \int r\sqrt{1+r^2p^2}\,dr$, $p = d\theta/dr$, show that the integral is stationary for the family of curves

$$x^2\sin\alpha - 2xy\cos\alpha - y^2\sin\alpha = \beta, \quad \alpha, \ \beta = \text{constant}.$$

9. *Principle of least action.* The position $x = x(t)$ at time t of a particle of mass m in one-dimensional motion along the x-axis subject to a conservative force

field with potential $V(x)$ satisfies the variational condition

$$\delta \int_{t_1}^{t_2} \mathcal{L}(t, x, \dot{x})dt = 0, \quad \dot{x} = \frac{dx}{dt},$$

where $\mathcal{L} = \frac{1}{2}m\dot{x}^2 - V(x)$ is called the *Lagrangian* of the system.
Deduce the equation of motion $m\ddot{x} = -dV/dx$.

10. Calculate the stationary value of $I = \int_0^{\frac{\pi}{2}} (y'^2 + 2xy - y^2)dx$ where $y(0) = 0$, $y(\frac{\pi}{2}) = \frac{\pi}{2}$. $[y = x, \ I = \frac{\pi}{2}\left(1 + \frac{\pi^2}{12}\right)]$

11. Find the curve in the (x, y)-plane for which $I = \int_0^1 y'^2(1 + y'^2)dx$ is stationary, given that $y(0) = 0$, $y(1) = 2$. $[y = 2x]$.

12. Show that $\int_a^b \frac{y'^2}{1+y^2} dx$ is stationary when $y = \sinh(\alpha x + \beta)$, where α, β are constants.

13. Show that the Euler–Lagrange equation

$$\text{for} \quad I = \int_0^1 (y'^2 + 2xy - xy^2)dx \quad \text{is} \quad \frac{d^2y}{dx^2} + xy = x.$$

A solution of the differential equation is required that satisfies $y(0) = y(1) = 0$. By making the 'guess' $y = \alpha x(1 - x)$, calculate the constant α by evaluating I and applying the stationarity condition in the form $dI/d\alpha = 0$. $[\alpha = -\frac{5}{19}]$.

7.5   Isoperimetric Problems

A stationary problem in which the function $y = f(x)$ is required to satisfy

$$\delta \int_a^b \mathcal{L}(x, y, y')dx = 0,$$

together with the integral *constraint*

$$\int_a^b \mathcal{G}(x, y, y')dx = \mathcal{G}_o = \text{constant}, \tag{7.5.1}$$

is called an *isoperimetric problem*. The original problem of this name (the problem of Dido of Carthage, 814 BC) was to find the shape of the closed curve of given length that encloses the maximum area.

Suppose $y = f(x)$ is required to pass through the endpoints (a, y_a), (b, y_b), then according to (7.4.5) y must satisfy

$$\int_a^b \psi(x)\left(\frac{\partial \mathcal{L}}{\partial y} - \frac{d}{dx}\frac{\partial \mathcal{L}}{\partial y'}\right) dx = 0, \qquad (7.5.2)$$

for an arbitrary function $\psi(x)$. Furthermore, because the value of the integral (7.5.1) must be unchanged by this variation, we must also have

$$\int_a^b \psi(x)\left(\frac{\partial \mathcal{G}}{\partial y} - \frac{d}{dx}\frac{\partial \mathcal{G}}{\partial y'}\right) dx = 0. \qquad (7.5.3)$$

Now divide the integration range into n equal intervals of width Δ_n, and let x_i be the midpoint of the ith interval. Equations (7.5.2) and (7.5.3) are then the limiting forms (as $n \to \infty$) of

$$\sum_{i=1}^n \psi_i \mathcal{E}_i \, \Delta_n = 0, \qquad \sum_{i=1}^n \psi_i \mathcal{F}_i \, \Delta_n = 0,$$

where

$$\mathcal{E}_i = \left(\frac{\partial \mathcal{L}}{\partial y} - \frac{d}{dx}\frac{\partial \mathcal{L}}{\partial y'}\right)_i, \qquad \mathcal{F}_i = \left(\frac{\partial \mathcal{G}}{\partial y} - \frac{d}{dx}\frac{\partial \mathcal{G}}{\partial y'}\right)_i,$$

and the subscript i denotes evaluation at $x = x_i$. Thus, the arbitrary n-dimensional vector $\boldsymbol{\psi} = (\psi_1, \psi_2, \ldots, \psi_n)$ is simultaneously orthogonal to $\mathbf{E} = (\mathcal{E}_1, \mathcal{E}_2, \ldots, \mathcal{E}_n)$ and $\mathbf{F} = (\mathcal{F}_1, \mathcal{F}_2, \ldots, \mathcal{F}_n)$. The components of $\boldsymbol{\psi}$ are not independent and there must therefore exist a constant of proportionality λ such that $\mathbf{E} = -\lambda\mathbf{F}$, provided $\mathbf{F} \neq \mathbf{0}$. This condition leads to the following modified Euler–Lagrange equation

$$\frac{\partial \mathcal{L}}{\partial y} - \frac{d}{dx}\frac{\partial \mathcal{L}}{\partial y'} + \lambda\left(\frac{\partial \mathcal{G}}{\partial y} - \frac{d}{dx}\frac{\partial \mathcal{G}}{\partial y'}\right) = 0 \quad \text{for all } x \text{ in } a < x < b.$$
$$(7.5.4)$$

The constant λ is just the Lagrange multiplier of §7.3. If we set

$$L = \mathcal{L} + \lambda(\mathcal{G} - \hat{\mathcal{G}}_o), \qquad I_L = \int_a^b \Big(\mathcal{L}(x, y, y') + \lambda[\mathcal{G}(x, y, y') - \hat{\mathcal{G}}_o]\Big) dx,$$
$$(7.5.5)$$

where $\hat{\mathcal{G}}_o = \mathcal{G}_o/(b-a)$, and consider the variational problem in which we vary the function y and λ *independently*, then corresponding to (7.4.4) we have

$$\delta I_{\mathrm{L}} \approx \epsilon \int_a^b \left(\psi(x)\frac{\partial L}{\partial y} + \psi'(x)\frac{\partial L}{\partial y'} \right) dx \;+\; \delta\lambda \left(\int_a^b \mathcal{G}(x,y,y')dx - \mathcal{G}_o \right) = 0.$$

The vanishing of the first integral leads to (7.5.4) and the vanishing of the coefficient of $\delta\lambda$ yields the constraint equation (7.5.1).

Example   Find the curve of length ℓ in $y > 0$ with endpoints $(\pm 1, 0)$ that encloses the maximum area between itself and the x-axis.

We have to minimise $I = \int_{-1}^1 y \, dx$ subject to $\int_{-1}^1 \sqrt{1+y'^2}\, dx = \ell$. We consider

$$I_{\mathrm{L}} = \int_{-1}^1 \left(y + \lambda\left[\sqrt{1+y'^2} - \frac{\ell}{2} \right] \right) dx.$$

Then (cf. Case 2 of §7.4) the Euler–Lagrange equations are

$$y - \frac{\lambda}{\sqrt{1+y'^2}} = A \quad (A = \text{constant}), \qquad \int_{-1}^1 \sqrt{1+y'^2}\, dx = \ell.$$

The first of these is easily integrated, and gives the portion of the circle

$$(x - B)^2 + (y - A)^2 = \lambda^2 \quad \text{of radius } \lambda \text{ in } y > 0.$$

The condition that the circle passes through $(\pm 1, 0)$ then yields

$$B = 0, \quad 1 + A^2 = \lambda^2.$$

Finally, the constraint $\int_{-1}^1 \sqrt{1+y'^2}dx = \int_{-1}^1 \frac{\lambda}{\sqrt{\lambda^2-x^2}}dx = \ell$, gives the equation

$$\frac{1}{\lambda} = \sin\left(\frac{\ell}{2\lambda} \right)$$

for the radius λ. When $\ell = \pi$ the radius $\lambda = 1$ and $A = 0$ and the curve is the semi-circle $x^2 + y^2 = 1$.

Problems 7D

1. Find the curve $y = f(x)$ between $(0,0)$ and $(1,0)$ such that $\int_0^1 y^2 dx = \ell$ and $\int_0^1 y'^2 \, dx$ is stationary.

$$\left[\lambda = -n^2\pi^2, n = 1, 2, \ldots, y = \sum_{n=1}^{\infty} A_n \sin(n\pi x), \text{ where } \sum_{n=1}^{\infty} A_n^2 = 2\ell \right]$$

2. Find the curve $y = f(x)$ between $(0,0)$ and $(1,0)$ such that $\int_0^1 y^2 dx = \ell$ and $\int_0^1 (y'^2 + x^2) dx$ is a minimum.

$$\left[\lambda = -n^2\pi^2, \ n = 1, 2, \ldots . \text{ Minimum value} = \frac{1}{3} + \pi^2\ell \quad \text{for } n = 1, \right.$$

$$\left. y = \sqrt{2\ell} \sin(\pi x) \right]$$

3. Show that, if $y(0) = 0$, $y(1) = 1$ and $\int_0^1 y dx = \frac{1}{3}$, then the minimum value of $\int_0^1 y'^2 dx$ is $\frac{4}{3}$, and that it occurs when $y = x^2$. $[\lambda = 4]$.

4. The potential energy of a chain hanging between two fixed points is represented by $I = \int_a^b y\sqrt{1+y'^2} dx$, where $y(x)$ is the height of the chain at x. The shape assumed by the chain makes I a minimum subject to the condition that its total length $\ell = \int_a^b \sqrt{1+y'^2} dx$ is fixed. Show that $y = -\lambda + \alpha \cosh\left(\frac{x-\beta}{\alpha}\right)$, where the values of the constants α, β, λ depend on ℓ and the positions of the ends of the chain.

5. Show that the function $y = \frac{1}{\sqrt{2\pi}}e^{-x^2/2}$ maximizes the integral $I = -\int_{-\infty}^{\infty} y \ln y dx, y \geq 0$, provided $\int_{-\infty}^{\infty} y dx = \int_{-\infty}^{\infty} x^2 y dx = 1$. [Use two Lagrange multipliers.]

6. Show that the method of Lagrange multipliers applied to the variational problem

$$\delta \int_0^1 y'^2 dx = 0, \quad y(0) = 0, \quad y(1) = 1, \quad \int_0^1 \sqrt{1+y'^2} dx = 5$$

yields the solution $y = x$, but that this does *not* satisfy the integral constraint. $\left[\text{For } y = x, \ \int_0^1 \sqrt{1+y'^2} dx = \sqrt{2} \neq 5. \text{ The curve } y = x \text{ also makes } I_G = \int_0^1 \sqrt{1+y'^2} dx \text{ stationary, i.e. the 'vector' } \mathbf{F} \text{ used in the proof of the multiplier method is null, so that it cannot be asserted that } \mathbf{E} = -\lambda\mathbf{F}. \right]$

7. If $\pi \int_{-a}^a y^2 dx = \frac{4}{3}\pi R^3$, where $y(\pm a) = 0$ and the surface area of revolution about the x-axis $S = 2\pi \int_{-a}^a y\sqrt{1+y'^2} dx$ is a minimum, show that $a = R$ and that the surface is the sphere obtained by rotating $x^2 + y^2 = R^2$ about the x-axis. $[\lambda = \frac{2}{R}]$.

8. Find the curve in the (x, y)-plane for which $I = \int_0^{\frac{\pi}{2}} y'(y' + 2xy) dx$ is stationary, given that $y(0) = 0$, $y(\frac{\pi}{2}) = 1$, and $\int_0^{\frac{\pi}{2}} y dx = \frac{\pi}{2} - 1$. $[y = 1 - \cos x]$.

USEFUL FORMULAE

Trigonometric

$$\sin(x \pm y) = \sin x \cos y \pm \cos x \sin y$$

$$\cos(x \pm y) = \cos x \cos y \mp \sin x \sin y$$

$$2 \sin x \cos y = \sin(x + y) + \sin(x - y)$$

$$2 \cos x \cos y = \cos(x + y) + \cos(x - y)$$

$$2 \sin x \sin y = \cos(x - y) - \cos(x + y)$$

$$\sin 2x = 2 \sin x \cos x$$

$$\cos 2x = \cos^2 x - \sin^2 x = 2 \cos^2 x - 1 = 1 - 2 \sin^2 x$$

$$\sin 3x = 3 \sin x - 4 \sin^3 x$$

$$\cos 3x = 4 \cos^3 x - 3 \cos x$$

$$\sin x = \frac{e^{ix} - e^{-ix}}{2i} = x - \frac{x^3}{3!} + \frac{x^5}{5!} - \frac{x^7}{7!} + \cdots$$

$$\cos x = \frac{e^{ix} + e^{-ix}}{2} = 1 - \frac{x^2}{2!} + \frac{x^4}{4!} - \frac{x^6}{6!} + \cdots$$

Hyperbolic

$$\sinh(x \pm y) = \sinh x \cosh y \pm \cosh x \sinh y$$

$$\cosh(x \pm y) = \cosh x \cosh y \pm \sinh x \sinh y$$

$$2 \sinh x \cosh y = \sinh(x + y) + \sinh(x - y)$$

$$2 \cosh x \cosh y = \cosh(x + y) + \cosh(x - y)$$

$$2 \sinh x \sinh y = \cosh(x - y) - \cosh(x + y)$$

$$\sinh 2x = 2\sinh x \cosh x$$

$$\cosh 2x = \cosh^2 x + \sinh^2 x = 2\cosh^2 x - 1 = 1 + 2\sinh^2 x$$

$$\sinh 3x = 3\sinh x + 4\sinh^3 x$$

$$\cosh 3x = 4\cosh^3 x - 3\cosh x$$

$$\sinh(ix) = i\sin x$$

$$\cosh(ix) = \cos x$$

$$\sinh x = \frac{e^x - e^{-x}}{2} = x + \frac{x^3}{3!} + \frac{x^5}{5!} + \frac{x^7}{7!} + \cdots$$

$$\cosh x = \frac{e^x + e^{-x}}{2} = 1 + \frac{x^2}{2!} + \frac{x^4}{4!} + \frac{x^6}{6!} + \cdots$$

Definite integrals

$$\int_0^\infty e^{-\alpha x^2}\,dx = \frac{1}{2}\sqrt{\frac{\pi}{\alpha}}. \quad \alpha > 0$$

By differentiation with respect to α:

$$\int_0^\infty x^2 e^{-\alpha x^2}\,dx = \frac{1}{4}\sqrt{\frac{\pi}{\alpha^3}}. \quad \alpha > 0$$

$$\int_0^\infty e^{-\alpha x}\sin Ax\,dx = \frac{A}{\alpha^2 + A^2}, \quad \alpha > 0$$

$$\int_0^\infty e^{-\alpha x}\cos Ax\,dx = \frac{\alpha}{\alpha^2 + A^2}, \quad \alpha > 0$$

$$\int_0^\infty \frac{e^{-\alpha x}}{x}\sin Ax\,dx = \tan^{-1}\frac{A}{\alpha}, \quad \alpha > 0$$

$$\int_0^\infty \frac{e^{-\alpha x}}{x}\sin Ax \sin Bx\,dx = \frac{1}{4}\ln\left(\frac{\alpha^2 + (A+B)^2}{\alpha^2 + (A-B)^2}\right), \quad \alpha > 0$$

$$\int_0^\infty e^{-\alpha x^2}\cos Ax\,dx = \frac{1}{2}\sqrt{\frac{\pi}{\alpha}}e^{-A^2/4\alpha}, \quad \alpha > 0$$

$$\int_0^\infty x e^{-\alpha x^2}\sin Ax\,dx = \frac{A}{4}\sqrt{\frac{\pi}{\alpha^3}}e^{-A^2/4\alpha}, \quad \alpha > 0$$

$$\int_0^\infty \frac{e^{-\alpha x^2}}{x} \sin Ax \, dx = \frac{\pi}{2} \operatorname{erf}\left(\frac{A}{2\sqrt{\alpha}}\right), \quad \alpha > 0$$

$$\int_0^\infty e^{-\alpha x} \sin^2 Ax \, dx = \frac{2A^2}{\alpha(\alpha^2 + 4A^2)}, \quad \alpha > 0$$

$$\int_0^\infty e^{-\alpha x} \cos^2 Ax \, dx = \frac{\alpha^2 + 2A^2}{\alpha(\alpha^2 + 4A^2)}, \quad \alpha > 0$$

$$\int_0^\infty \cos \alpha x^2 \, dx = \frac{1}{2}\sqrt{\frac{\pi}{2\alpha}}, \quad \alpha > 0$$

$$\int_0^\infty \sin \alpha x^2 \, dx = \frac{1}{2}\sqrt{\frac{\pi}{2\alpha}}, \quad \alpha > 0$$

$$\int_0^\infty \cos \alpha x^2 \cos Ax \, dx = \frac{1}{2}\sqrt{\frac{\pi}{\alpha}} \cos\left(\frac{\pi}{4} - \frac{A^2}{4\alpha}\right), \quad \alpha > 0$$

$$\int_0^\infty \sin \alpha x^2 \cos Ax \, dx = \frac{1}{2}\sqrt{\frac{\pi}{\alpha}} \sin\left(\frac{\pi}{4} - \frac{A^2}{4\alpha}\right), \quad \alpha > 0$$

Example: Differentiation under the integral sign

To evaluate:
$$I = \int_0^\infty \frac{e^{-\alpha x}}{x} \sin Ax \, dx$$

Observe that $I = 0$ when $A = 0$ and differentiate with respect to A:

$$\frac{dI}{dA} = \int_0^\infty e^{-\alpha x} \cos Ax \, dx$$

$$= \frac{\alpha}{\alpha^2 + A^2}$$

$$\therefore \quad I = \int \frac{\alpha}{\alpha^2 + A^2} dA + C \quad (C = \text{constant})$$

$$= \tan^{-1}\left(\frac{A}{\alpha}\right) + C.$$

But, $I = 0$ when $A = 0$, $\therefore$ $C = 0$

$$\text{i.e. } I = \tan^{-1}\left(\frac{A}{\alpha}\right).$$

Series

$$\frac{1 - x^n}{1 - x} = 1 + x + x^2 + x^3 + \cdots + x^n, \quad (n + 1 \text{ terms})$$

$$\frac{1}{1 - x} = 1 + x + x^2 + x^3 + \cdots + x^n + \cdots, \quad |x| < 1$$

(infinite geometric series)

$$(1 + x)^\alpha = 1 + \frac{\alpha}{1!}x + \frac{\alpha(\alpha - 1)}{2!}x^2 + \frac{\alpha(\alpha - 1)(\alpha - 2)}{3!}x^3 + \cdots$$

$$+ \frac{\alpha(\alpha - 1)(\alpha - 2)\cdots(\alpha - n + 1)}{n!}x^n + \cdots,$$

$$|x| < 1, \ \alpha \neq 0, 1, 2, \ldots \text{ (infinite binomial series)}$$

$$(1 + x)^n = 1 + \frac{n}{1!}x + \frac{n(n - 1)}{2!}x^2 + \frac{n(n - 1)(n - 2)}{3!}x^3 + \cdots + x^n,$$

$$n = \text{(positive integer)}$$

$$\ln(1 + x) = x - \frac{x^2}{2} + \frac{x^3}{3} - \frac{x^4}{4} - \cdots \quad |x| < 1$$

$$e^x = 1 + x + \frac{x^2}{2!} + \frac{x^3}{3!} + \frac{x^4}{4!} + \cdots + \frac{x^n}{n!} + \cdots$$

$$\frac{n(n + 1)}{2} = 1 + 2 + 3 + 4 + \cdots + n$$

$$\frac{n(n + 1)(2n + 1)}{6} = 1^2 + 2^2 + 3^2 + 4^2 + \cdots + n^2$$

Vector analysis

$$\mathbf{a} \times (\mathbf{b} \times \mathbf{c}) = (\mathbf{a} \cdot \mathbf{c})\mathbf{b} - (\mathbf{a} \cdot \mathbf{b})\mathbf{c} \quad \text{(triple vector product)}$$

$$(\mathbf{a} \times \mathbf{b}) \times (\mathbf{c} \times \mathbf{d}) = (\mathbf{a} \cdot \mathbf{b} \times \mathbf{d})\mathbf{c} - (\mathbf{a} \cdot \mathbf{b} \times \mathbf{c})\mathbf{d}$$

Taylor's theorem in three dimensions

$$f(\mathbf{x} + \mathbf{h}) = f(\mathbf{x}) + (\mathbf{h} \cdot \nabla) f(\mathbf{x}) + \frac{1}{2!} (\mathbf{h} \cdot \nabla)^2 f(\mathbf{x}) + \cdots + \frac{1}{n!} (\mathbf{h} \cdot \nabla)^n f(\mathbf{x}) + \cdots$$

The divergence theorem

$$\text{div } \mathbf{F} = \frac{\partial F_1}{\partial x} + \frac{\partial F_2}{\partial y} + \frac{\partial F_3}{\partial z} \equiv \nabla \cdot \mathbf{F}$$

If $\mathbf{F}(\mathbf{x})$ is defined on and within the interior V of a closed surface S:

$$\int_V \operatorname{div} \mathbf{F} \, dV = \oint_S \mathbf{n} \cdot \mathbf{F} dS \equiv \oint_S \mathbf{F} \cdot d\mathbf{S}.$$

Stokes' theorem

$$\mathbf{curl\,F} \equiv \nabla \times \mathbf{F} = \begin{vmatrix} \mathbf{i} & \mathbf{j} & \mathbf{k} \\ \dfrac{\partial}{\partial x} & \dfrac{\partial}{\partial y} & \dfrac{\partial}{\partial z} \\ F_1 & F_2 & F_3 \end{vmatrix}$$

$$= \left(\frac{\partial F_3}{\partial y} - \frac{\partial F_2}{\partial z} \right) \mathbf{i} + \left(\frac{\partial F_1}{\partial z} - \frac{\partial F_3}{\partial x} \right) \mathbf{j} + \left(\frac{\partial F_2}{\partial x} - \frac{\partial F_1}{\partial y} \right) \mathbf{k}.$$

C is a closed contour and S an open, two-sided surface bounded by C; $\mathbf{n}$ is orientated in the positive sense with respect to C:

$$\oint_C \mathbf{F} \cdot d\mathbf{r} = \int_S \mathbf{curl\,F} \cdot d\mathbf{S} \equiv \int_S \mathbf{n} \cdot \mathbf{curl\,F} dS.$$

Vector identities

$$
\begin{aligned}
\mathbf{curl}(\varphi \mathbf{F}) &= \nabla \times (\varphi \mathbf{F}) &= \varphi \mathbf{curl\,F} + \nabla \varphi \times \mathbf{F}; \\
\mathbf{curl}(\mathbf{curl\,F}) &= \nabla \times (\nabla \times \mathbf{F}) = \nabla(\operatorname{div} \mathbf{F}) - \nabla^2 \mathbf{F}; \\
\operatorname{div}(\mathbf{F} \times \mathbf{G}) &= \nabla \cdot (\mathbf{F} \times \mathbf{G}) &= \mathbf{curl\,F} \cdot \mathbf{G} - \mathbf{F} \cdot \mathbf{curl\,G}; \\
\mathbf{curl}(\mathbf{F} \times \mathbf{G}) &= \nabla \times (\mathbf{F} \times \mathbf{G}) = (\mathbf{G} \cdot \nabla)\mathbf{F} - (\mathbf{F} \cdot \nabla)\mathbf{G} + \mathbf{F} \operatorname{div} \mathbf{G} - \mathbf{G} \operatorname{div} \mathbf{F}; \\
\operatorname{grad}(\mathbf{F} \cdot \mathbf{G}) &= \nabla(\mathbf{F} \cdot \mathbf{G}) &= (\mathbf{G} \cdot \nabla)\mathbf{F} + (\mathbf{F} \cdot \nabla)\mathbf{G} + \mathbf{G} \times \mathbf{curl\,F} + \mathbf{F} \times \mathbf{curl\,G}.
\end{aligned}
$$

Integral transformations

$$\int_V \nabla \varphi \, dV = \oint_S \mathbf{n} \, \varphi \, dS,$$

$$\int_V \nabla \cdot \mathbf{F} \, dV = \oint_S \mathbf{n} \cdot \mathbf{F} dS,$$

$$\int_V \nabla \times \mathbf{F} \, dV = \oint_S \mathbf{n} \times \mathbf{F} dS.$$

Surface integrals

$$\int_S \mathbf{F} \cdot d\mathbf{S} = \int_S \mathbf{F} \cdot \mathbf{n} \, dS = \int_S (F_1 n_1 + F_2 n_2 + F_3 n_3) \, dS$$

$$\mathbf{r} = x(u, v)\mathbf{i} + y(u, v)\mathbf{j} + z(u, v)\mathbf{k}$$

$$\mathbf{n} = \pm \frac{(\mathbf{r}_u \times \mathbf{r}_v) du dv}{|(\mathbf{r}_u \times \mathbf{r}_v) du dv|} = \pm \frac{\mathbf{r}_u \times \mathbf{r}_v}{|\mathbf{r}_u \times \mathbf{r}_v|}$$

$$dS = ndS = \pm(\mathbf{r}_u \times \mathbf{r}_v)dudv$$

$$\int_S \mathbf{F} \cdot \mathbf{n}\, dS = \pm \int_S \mathbf{F} \cdot \mathbf{r}_u \times \mathbf{r}_v\, dudv.$$

Cauchy–Riemann equations

If $f(z) = u(x,y) + iv(x,y)$ is regular:

$$\frac{\partial u}{\partial x} = \frac{\partial v}{\partial y}, \quad \frac{\partial u}{\partial y} = -\frac{\partial v}{\partial x}.$$

Upper bound for a contour integral

$|f(z)| \le M$ on C, L = length of contour:

$$\left| \int_C f(z)dz \right| \le ML.$$

Cauchy's theorem

If $f(z)$ is continuous on a simple closed contour C and regular within C:

$$\oint_C f(z)dz = 0.$$

Poles

$$f(z) = \sum_{n=0}^{\infty} a_n(z - z_0)^n + \sum_{n=1}^{m} \frac{b_n}{(z - z_0)^n}$$

has a pole of order m at z_0 with *residue* b_1.

Residue theorem

If $f(z)$ is continuous on a simple closed contour C and regular within C except for isolated singularities at $z_1, z_2, \ldots, z_n$:

$$\oint_C f(z)dz = 2\pi i \sum_{m=1}^{n} \mathcal{R}_m$$

$$\mathcal{R}_m = \text{residue at } z = z_m.$$

Residue at a simple pole

If $\quad f(z) = \dfrac{P(z)}{Q(z)}\quad$ has a simple pole at $z = z_0$ then residue $= \dfrac{P(z_0)}{Q'(z_0)}$.

Residue at a double pole

If $\quad f(z) = \dfrac{P(z)}{(z - z_0)^2}\quad$ then at z_0 residue $= \left(\dfrac{\partial}{\partial z}\left\{(z - z_0)^2 f(z)\right\} \right)_{z=z_0} = P'(z_0).$

D'Alembert's solution of the wave equation

If $\quad \dfrac{\partial^2 u}{\partial x^2} - \dfrac{1}{c^2}\dfrac{\partial^2 u}{\partial t^2} = 0, \quad$ with $u = f(x)$ and $\dfrac{\partial u}{\partial t} = g(x)$ at $t = 0$:

then: $\quad u(x,t) = \dfrac{1}{2}\Big[f(x-ct) + f(x+ct)\Big] + \dfrac{1}{2c}\displaystyle\int\limits_{x-ct}^{x+ct} g(\eta)\,d\eta.$

One-dimensional delta function

$$\int\limits_a^b \delta(x-y)\,dy = \begin{cases} 1, & \text{when } a < x < b, \\ 0, & \text{otherwise,} \end{cases}$$

$$\int\limits_a^b f(y)\delta(x-y)\,dy = f(x), \quad \text{when } a < x < b,$$

$$\delta(x-y) = \lim_{\epsilon \to +0} \frac{\epsilon}{\pi[(x-y)^2 + \epsilon^2]}.$$

Fourier transform

$$\hat{f}(k) = \frac{1}{\sqrt{2\pi}}\int\limits_{-\infty}^{\infty} f(x)\mathrm{e}^{-ikx}\,dx, \quad f(x) = \frac{1}{\sqrt{2\pi}}\int\limits_{-\infty}^{\infty} \hat{f}(k)\mathrm{e}^{ikx}\,dk,$$

Fourier transform of a derivative

Fourier transform of $\ f^{(n)}(x) = (ik)^n \hat{f}(k).$

Fourier transform of unity

$$\frac{1}{\sqrt{2\pi}}\int\limits_{-\infty}^{\infty} \mathrm{e}^{-ikx}\,dx = \sqrt{2\pi}\,\delta(k).$$

Fourier sine transform

$$\hat{f}_\mathrm{s}(k) = \sqrt{\frac{2}{\pi}}\int\limits_0^{\infty} f(x)\sin kx\,dx, \quad f(x) = \sqrt{\frac{2}{\pi}}\int\limits_0^{\infty} \hat{f}_\mathrm{s}(k)\sin kx\,dk.$$

Fourier cosine transform

$$\hat{f}_\mathrm{c}(k) = \sqrt{\frac{2}{\pi}}\int\limits_0^{\infty} f(x)\cos kx\,dx, \quad f(x) = \sqrt{\frac{2}{\pi}}\int\limits_0^{\infty} \hat{f}_\mathrm{c}(k)\cos kx\,dk.$$

BIBLIOGRAPHY

Abramowitz, M. and Stegun, I. A. (editors) *Handbook of Mathematical Functions*, 10th corrected printing, US Department of Commerce, National Bureau of Standards, Applied Mathematics Series No. 55, Washington, DC, 1972.

Aitken, A. C. *Determinants and Matrices*, Greenwood Press, Westport CN, 1983.

Andrews, L. C. *Special Functions of Mathematics for Engineers*, 2nd edition, Oxford University Press, Oxford, 1998.

Arthurs, A. M. *Calculus of Variations*, Routledge & Kegan Paul, London, 1975.

Bateman, H. *Partial Differential Equations of Mathematical Physics*, Dover Publications, New York, 1944.

Birkhoff, G. and Mac Lane, S. *A Survey of Modern Algebra*, 2nd edition, The Macmillan Company, New York, 1953.

Bolza, O. *Lectures on the Calculus of Variations*, Chelsea Publishing Co., New York, 1973.

Burkill, J. C. *The Theory of Ordinary Differential Equations*, 2nd edition, Oliver and Boyd, London, 1962.

Carathéodory, C. *Calculus of Variations and Partial Differential Equations of the First Order*, 3rd edition, Chelsea Publishing Co., New York, 1989.

Carrier, G. F. and Pearson, C. E. *Partial Differential Equations*, Academic Press, New York, 1976.

Carrier, G. F., Krook, M. and Pearson, C. E. *Functions of a Complex Variable*, McGraw-Hill, New York, 1966.

Copson, E. T. *Theory of Functions of a Complex Variable*, Oxford University Press, Oxford, 1935.

Copson, E. T. *Partial Differential Equations*, Cambridge University Press, Cambridge, 1975.

Coulson, C. A. *Electricity*, 5th edition, Oliver and Boyd, London, 1961.

Courant, R. and Hilbert, D. *Methods of Mathematical Physics*, Volume I, Interscience Publishers, New York, 1953.

Cramér, H. *Mathematical Methods of Statistics*, Princeton University Press, Princeton NJ, 1946 (reprinted 1999).

Dirac, P. A. M. *The Principles of Quantum Mechanics*, 4th edition, Oxford University Press, Oxford, 1958.

Forsyth, A. R. *Theory of Functions of a Complex Variable*, Volumes I and II, Dover Publications, New York, 1965.

Forsyth, A. R. *A Treatise on Differential Equations*, 6th edition, Dover Publications, New York, 1996.

Franklin, P. *Functions of Complex Variables*, Pitman and Sons, London, 1959.

Frazer, R. A., Duncan, W. J. and Collar, A. R. *Elementary Matrices*, Cambridge University Press, Cambridge, 1965.

Gel'fand, I. M. and Fomin, G. E. *Calculus of Variations*, Prentice-Hall, Englewood Cliffs, NJ, 1963.

Gel'fand, I. M. and Shilov, G. E. *Generalized Functions*, Volume 1, Academic Press, New York, 1964.

Gerrish, F. *Pure Mathematics Volume 1: Calculus*, Cambridge University Press, Cambridge, 1970.

Gillespie, R. P. *Partial Differentiation*, 2nd edition, Oliver and Boyd, London, 1960.

Goldstein, H. *Classical Mechanics*, Addison Wesley, Reading MA, 1950.

Goldstein, M. E. and Braun, W. H. *Advanced Methods for the Solution of Differential Equations*. NASA Special Publication SP-316, Scientific and Technical Information Office, Washington, DC, 1973.

Gourset, E. *Functions of a Complex Variable*, Dover Publications, New York, 1959.

Hadamard, J. *Lectures on Cauchy's Problem in Linear Partial Differential Equations*, Dover Publications, New York, NJ, 1952.

Hildebrand, F. B. *Advanced Calculus for Applications*, 2nd edition, Prentice-Hall, Englewood Cliffs, NJ, 1976.

Ince, E. L. *Ordinary Differential Equations*, Dover Publications, New York, 1956.

Ince, E. L. *Integration of Ordinary Differential Equations*, 7th edition, Oliver and Boyd, London, 1956.

Jeffreys, H. *Cartesian Tensors*, Cambridge University Press, Cambridge, 1931 (reprinted 1969).

Jeffreys, H. and Jeffreys, B. S. *Methods of Mathematical Physics*, 3rd edition, Cambridge University Press, Cambridge, 1972.

Jones, D. S. *Theory of Generalised Functions*, Cambridge University Press, Cambridge, 1982.

Kellogg, O. D. *Foundations of Potential Theory*, Dover Publications, New York, 1953.

Kreyszig, E. *Advanced Engineering Mathematics*, 7th edition, John Wiley and Sons, New York, 1993.

Lanczos, C. *The Variational Principles of Mechanics*, 4th edition, Dover Publications, New York, 1986.

Lanczos, C. *Applied Analysis*, Dover Publications, New York, 1988.

Ledermann, W. *Multiple Integrals*, Routledge & Kegan Paul, London, 1966.

Lighthill, M. J. *Fourier Analysis and Generalised Functions*, Cambridge University Press, Cambridge, 1958.

Marcus, M. and Minc, H. *A Survey of Matrix Theory and Matrix Inequalities*, Dover Publications, New York, 1992.

McLachlan, N. W. *Bessel Functions for Engineers*, corrected edition, Oxford University Press, Oxford, 1941.

Milne-Thomson, L. M. *Theoretical Hydrodynamics*, 5th edition, Macmillan, London, 1976.

Nering, E. D. *Linear Algebra and Matrix Theory*, 2nd edition, John Wiley and Sons, New York, 1970.

Perlis, S. *Theory of Matrices*, Dover Publications, New York, 1991.

Phillips, E. G. *A Course of Analysis*, 2nd edition, Cambridge University Press, Cambridge, 1939.

Phillips, E. G. *Functions of a Complex Variable*, 8th edition, Oliver and Boyd, London, 1961.

Piaggio, H. T. H. *Elementary Treatise on Differential Equations*, revised edition, Bell and Sons, London, 1962.

Poole, E. G. C. *Introduction to the Theory of Linear Differential Equations*, Oxford University Press, Oxford, 1936.

Rutherford, D. E. *Vector Methods*, 9th edition, Oliver and Boyd, London, 1951.

Schwartz, L. *Mathematics for the Physical Sciences*, Dover, New York, 2008.

Smithies, Frank. *Cauchy and the Creation of Complex Function Theory*, Cambridge University Press, Cambridge, 1997.

Sneddon, I. N. *Special Functions of Mathematical Physics and Chemistry*, 2nd edition, Oliver and Boyd, London, 1961.

Sneddon, I. N. *The Use of Integral Transforms*, Tata McGraw-Hill, New York, 1974.

Sommerfeld, A. *Partial Differential Equations in Physics*, Academic Press, New York, 1964.

Stephenson, G. *Partial Differential Equations for Scientists and Engineers*, 3rd edition, Longman, London and New York, 1974.

Stephenson, G. *Worked Examples in Mathematics for Scientists and Engineers*, Longman, London and New York, 1985.

Stephenson, G. and Radmore, P. M. *Advanced Mathematical Methods for Engineering and Science Students*, Cambridge University Press, Cambridge, 1990.

Titchmarsh, E. C. *The Theory of Functions*, 2nd corrected edition, Oxford University Press, Oxford, 1952.

Tranter, C. J. *Techniques of Mathematical Analysis*, English Universities Press, London, 1957.

Tranter, C. J. *Integral Transforms in Mathematical Physics*, 3rd edition, Chapman and Hall, London, 1971.

Watson, G. N. *Complex Integration and Cauchy's Theorem*, Cambridge University Press, Cambridge, 1914.

Weatherburn, C. E. *Advanced Vector Analysis*, Bell and Sons, London, 1924.

Webster, A. G. *Partial Differential Equations of Mathematical Physics*, 2nd corrected edition, Dover Publications, New York, 1955.

Whittaker, E. T. and Watson, G. N. *A Course of Modern Analysis*, 4th edition, Cambridge University Press, Cambridge, 1927.

Zwillinger, D. (editor) *Standard Mathematical Tables and Formulae*, 30th edition, CRC Press, Boca Raton FL, 1996.

INDEX

Printed in the United States
By Bookmasters